Praise for *The Eighth Continent*

"Whether meeting a leaf-tailed gecko eye-to-eye, searching for the extinct giant lemur, discussing the biogeography of animals, or describing conservation issues and local customs, Peter Tyson provides a delightful narrative, the most informative, fascinating, and enjoyable book on the natural and cultural history of Madagascar I have read."

> —George B. Schaller, director for science,
> Wildlife Conservation Society

"Madagascar always seems somehow separate—so oddly beautiful, so completely strange. But Peter Tyson, in this remarkable book, bridges that distance. He eloquently makes real and familiar both the island and the island-dwellers, and he makes us realize how much we would all lose if they were not part of our far-flung extended family."

> —Deborah Blum, Pulitzer Prize–winning author of
> *The Monkey Wars* and *Sex on the Brain*

"This is indeed a lovely book. It is simultaneously authoritative and quirky. It teases the reader with surprises to discover and ideas to savor—whether the leaf-tailed gecko plastered to a tree trunk or the oratory of Malagasy poetry. In a country counted among the world's poorest, Tyson unveils riches."

> —Alison Jolly, author of
> *Lucy's Legacy: Sex and Intelligence in Human Evolution*

"This compendium of ecological information and personal anecdote illuminates a land that, even in the jet age, stands aside from the modern world."

> —*Booklist*

"For readers who have not experienced Madagascar, *The Eighth Continent* will provide a quantum leap in their understanding of the Malagasy culture and the island's diverse landscape and wildlife."

> —*Wildlife Conservation*

"In this impressive volume, [Tyson] writes about what he learned on [his] visits, successfully conveying both the flavor of field research and the biological mysteries of the island nation."

> —*Publishers Weekly*

"By turns thoughtful and vivacious, science writer and *NOVA* producer Tyson draws an anecdotally rich portrait of the biological wonderland known as Madagascar. . . . A fine portrait."

—*Kirkus Reviews*

"This often droll travel book has an old-fashioned appeal."

—*Outside* magazine

"Tyson's book is an engrossing testament to one of the planet's most astonishing places."

—*Discover*

"A sharp picture of an intriguing place."

—*Scientific American*

"A thorough researcher, Tyson packs this book with solid information. . . . And he covers an amazing amount of terrain with a scholar's love of detail."

—*Boston Globe*

"Tyson's book makes a strong case that our planet would be a poorer place if [Madagascar] were to continue to wash into the sea."

—*The Sciences*

"Science writer Tyson gives us a feel for the breadth and complexity of the world's fourth-largest island and tells us why it is worth saving."

—*Natural History*

"Tyson accompanied four scientists . . . on separate field expeditions to the island. . . . In *The Eighth Continent*, readers relive these marvelous journeys."

—*Science News*

"[Tyson's] book is a worthy companion to David Quammen's *The Song of the Dodo* and valuable reading for anyone concerned with the world environment."

—Amazon.com

The Eighth Continent

The Eighth Continent

LIFE, DEATH, AND DISCOVERY IN THE LOST WORLD OF MADAGASCAR

Peter Tyson

Perennial
An Imprint of HarperCollinsPublishers

Portions of Part II originally appeared in "The Lizard King" (*The Sciences*, March/April 1996). Reprinted by permission of *The Sciences*.

Poems on pp. 265–67 from *Hainteny: The Traditional Poetry of Madagascar*, edited by Leonard Fox (1990). Reprinted by permission of Associated University Presses.

Poems on pp. 268–70 from *Translations from the Night: Selected Poems of Jean-Joseph Rabéarivelo* by Jean-Joseph Rabéarivelo (1975). Courtesy of Heinemann Educational Publishers.

Plan of Malagasy house on p. 234 from *Taboo: A Study of Malagasy Customs and Beliefs* by Jørgen Ruud (1960). Courtesy of Scandinavian University Press.

All photographs courtesy of the author unless otherwise noted.

Map on p. xxviii by Anne Marcotty

A hardcover edition of this book was published in 2000 by William Morrow, an imprint of HarperCollins Publishers.

HarperCollins books may be purchased for educational, business, or sales promotional use. For information please write: Special Markets Department, HarperCollins Publishers Inc., 10 East 53rd Street, New York, NY 10022.

FIRST PERENNIAL EDITION PUBLISHED 2001.

Designed by Kim Llewellyn

Library of Congress Cataloging-in-Publication Data is available.

ISBN 0-380-79465-9

01 02 03 04 05 RRD 10 9 8 7 6 5 4 3 2 1

For Melissa

Contents

Author's Note

After introducing Malagasy whom I met with their full names, I subsequently use their first names only. I would be mortified if anyone took this as a sign of paternalism or disrespect, for I have only the deepest regard for the people of Madagascar. I've done this because most surnames of Malagasy who appear in this book begin with Ra-, the equivalent of Mr. or Mrs. in the Malagasy language, and contain a mouthful of letters (see, e.g., the Acknowledgments). Western readers may find it difficult to pronounce, much less distinguish between these lengthy names, so I use their simpler and more accessible first names.

Also, during field research for this book, I proved unable to find one or more Malagasy scientists to whom I could devote one or more Parts, and so I have focused solely on the work of four Western scientists. This occurred as the result of various logistical problems, with which anyone who has lived or worked in Madagascar is all too familiar, and of the unfortunate dearth of Malagasy researchers who have the means to conduct their own fieldwork. It is my fervent hope that the singular Malagasy voice comes through nevertheless in these pages.

Acknowledgments

FIRST AND FOREMOST, I'd like to thank the four scientists who generously invited me along on their field expeditions, attended to my every need and request, and reviewed their respective parts of the book for accuracy. Without the openness and conscientiousness of Chris Raxworthy, Dave Burney, Bob Dewar, and Pat Wright, this book would not be.

Other Madagascar researchers and experts who patiently put up with everything from lengthy interviews to nagging questions include Josephine Andrews, Richard Carroll, Garfield Dean, Lisa Gaylord, Sabrina Hardenbergh, Georges Heurtebise, Helen James, Susan Kus, Torie Lines, Ross MacPhee, Sheila O'Connor, Ron Nussbaum, Michael O'Dea, Joe Peters, Alison Richard, Paul Siegel, Elwyn Simons, Eleanor Sterling, Bob Sussman, Linda Sussman, Norman Uphoff, and Anne Yoder. I'm grateful to Russ Mittermeier for writing the Foreword, and I'd like to express special thanks to Alison Jolly, who graciously reviewed the manuscript and, over many years, has freely given of herself and her decades of experience on the island.

This book got its start in magazine articles. I'd like to thank Barbara Wallraff at the *Atlantic Monthly*, Burkhard Bilger at *The Sciences*, and Mark Cherrington at *Earthwatch* for sensitive editing of articles on Madagascar, bits and pieces of which have wound up, in modified form, in this book.

I'd like to express my deep gratitude for the many Malagasy who suffered through interviews, guided me through forests, answered questions, or otherwise helped me during my travels through their country. These include Thera Bruno Andriamaro, Ellis Fandrorota, Julien Mohamed Jules, Jean-Gervais Rafamantanantsoa, Emile Rajeriarison,

Toussaint Rakotondrazafy, Leva Rakotovao, Jean-Baptiste Ramana-manjata, Jocelyn Ramanitrandrasana, Parson Ramiandrisoa, Ramilison-ina, Roland Ranaivo-Ratsitohaina, Gabriel Randriambahoaka, Gervais Randrianasolo, Loret Rasabo, Achille Raselimanana, Darsot Léon Rasolofomampianina, Joelisoa Ratsirarson, Françoise Ravaoarimalala, and the ever-smiling twins Angelin and Angeluc Razafimanantsoa. A special thanks to M. and Mme. André Peyrieras of Maroantsetra, Mada-gascar, who kindly put me up for four nights in their home overlooking Nosy Mangabe.

For unusual source material, I will forever be indebted to James Sibree and the other missionary-editors of the *Antananarivo Annual and Madagascar Magazine,* a scientific and literary journal published in the Malagasy capital between 1875 and 1900. I dug up ragged-edged copies of this obscure and fascinating periodical, not to mention a bevy of other venerable and rarely seen works on Madagascar, in the inspiringly cav-ernous recesses of Harvard University's Widener Library, which I'd like to acknowledge for opening my eyes to a much wider range of material on the island than I'd previously thought existed. I'm obliged to Pierre Verin for his thorough *History of Civilization in North Madagascar,* from which I drew liberally for Part Four.

I owe an enormous debt of gratitude to Monique Rodriguez of Cortez Travel (Solana Beach, California), who gave her unstinting sup-port throughout my travels to Madagascar, and to her entire staff, par-ticularly Christophe de Comarmond in Tana. Air Madagascar magnani-mously provided transportation to and within the island on all four trips. At Earthwatch, I'd like to thank Betty Parfenuk for helping to arrange my trips; Nini Bloch, Mark Cherrington, and Anne Marcotty for their friendship and support through several articles on Madagascar (and Anne for designing the beautiful map that appears in this book); and also Bear Burnes, David Lowe, and Blue Magruder. Thanks also to Lauren Aguirre at *NOVA* for granting me a short leave to finish the man-uscript.

Theresa Park, my agent, called me one day in March 1996 and asked if I'd ever considered writing a book on Madagascar. You're holding the result of that phone conversation. I'm immensely grateful for Theresa's unbridled support during this project. My editor, Stephen Power, did a thoughtful edit of the manuscript, vastly improving the book and saving me from not a few embarrassing blunders. I'd also like to thank editor Bret Witter for seeing the book through to publication.

Thanks to my mother, Patricia McCurdy, for taking time out from her own book to read a draft of mine, and for always being there. And hugs to my three kids, Christopher, Olivia, and Nick, for uncomplainingly putting up with my long absences.

Finally, Melissa Banta, my wife, is the only one who can possibly know the degree to which I owe her and love her for her enthusiasm, good humor, early readings and recommendations, and general unqualified support during this project. She was, and continues to be, like fresh air. Thanks, M.

Foreword

MADAGASCAR IS A LIVING laboratory unlike anyplace else on Earth. The world's fourth-largest island and the largest oceanic island, it is about the size of France or the state of Texas, covering roughly 225,000 square miles or 0.4 percent of the land surface of the planet. Although it is located only about 250 miles off the eastern coast of Africa, it has been isolated from other landmasses for more than 80 million years, meaning that most of the plant and animal species found there evolved apart from the rest of the world and are unique to the island, as is described in more depth in Part One of this book. Biologically, it is in many ways a minicontinent, with very high diversity in certain groups of plants and animals and amazingly high levels of endemism (species that occur nowhere else) that few if any other countries can match. (How it got that way is another matter, which is investigated in Part Two.)

All of these factors contribute to elevate Madagascar to very high status within the world of biodiversity conversation. Madagascar is one of the top eight *megadiversity countries* on Earth, sharing this title with several much larger countries, including Brazil, Indonesia, and China. In addition, it ranks as one of the world's top four or five threatened *biodiversity hotspots*, those ecoregions that harbor by far the highest concentration of Earth's plants and animals and especially those at greatest risk of extinction. These hotspots, twenty-five of them in all, occupy only 1.4 percent of the land surface of the planet, and yet have within them at least 60 percent of the world's terrestrial biodiversity and an even higher percentage of the diversity at greatest risk—and Madagascar occupies a very special position even within this very elite group.

Without a doubt, the most striking feature of Madagascar's biodiversity is its degree of endemism, both at the species level and at the level of genera and families, and this is what places it so high on international priority lists. Of the estimated 10,000 to 12,000 species of vascular plants, more than 80 percent are thought to be endemic. Furthermore, of the 160 to 181 plant families and 1,289 genera of plants in Madagascar, fully 10 families and 260 genera are endemic. Compare this to Brazil, which is the world's leader in total plant species diversity (with 50,000 to 56,000 species) but has no endemic families. Only Australia (which is more than thirteen times larger than Madagascar) with thirteen endemic families, South Africa (which is more than twice the size of Madagascar) with six families, and tiny New Caledonia, with five endemic plant families, can compare. Even within Madagascar, the levels of regional endemism are striking. For example, fully 95 percent of the plant species and 48 percent of the plant genera of the southern spiny desert are endemic not just to Madagascar but to that small region of the country! Furthermore, Madagascar's overall plant diversity is also quite impressive. Indeed, it is home to roughly 25 to 30 percent of all plants in the African region, even though it occupies less than 2 percent of the land area of that part of the world.

Looking at the animals, we find comparably impressive figures. Among the birds, which are low in total diversity but very high in endemism, fully 115 of 250 species (46 percent) and 35 of 162 genera (21.6 percent) are endemic, and there are also an amazing 5 endemic bird families as well. The reptiles and amphibians are both highly diverse and rich in endemics. Fully 274 of 300 species (91 percent) and 36 of 64 genera (56 percent) of reptiles are endemic, a degree of higher level endemism that few countries can match. With the frogs, which are the only amphibians represented on the island, endemism is almost complete. The 178 species of frogs include only 1 or 2 introduced species, so endemism is 99 percent; the number of endemic genera stands at 16 or 17 out of a total 18.

The most striking example of both diversity and endemism is without a doubt that of the nonhuman primates, represented on Madagascar by a fantastic array of lemurs. Fully 33 species and 51 taxa of lemurs are found on the island, and they are 100 percent endemic (with only two of them found on the neighboring Comoros, where they were almost certainly introduced by our own species). Furthermore, the 33 species are divided into 14 endemic genera and 5 endemic families, a level of primate endemism at higher taxonomic levels that no other country can approach. Even Brazil, the world's leader in total primate species diversity, with 77, has only two endemic genera and no endemic families. Looking at Madagascar's great primate diversity in yet another way, we see that this small country has great global responsibility. Although it is only one of ninety-two countries with wild primate populations, it alone is responsible for 21 percent (14 of 65) of all primate genera and 36 percent (5 of 14) of all primate families, making it the single highest priority on Earth for the conservation of these, our closest living relatives. Madagascar is so important for primates that primatologists divide the world into four major regions, the whole of South and Central America, all of southern and Southeast Asia, mainland Africa, and Madagascar, which ranks as a full-fledged region all by itself.

As if these figures were not impressive enough, new species are still being described. Indeed, recent investigations indicate that the number of primate species in Madagascar may increase by ten to twenty in the next decade. The wonderful ongoing discovery of frogs and reptiles, with new species coming to light every few years, is discussed further in Parts One and Two of this book and is worthy of particular attention from the reader.

Madagascar's endemism is so striking that it has the highest ratio of endemic to nonendemic species, both for plants and vertebrates, of any of the twenty-five hotspots—ecoregions that were chosen in the first place in large part because of their high endemism. Madagascar has a striking 1:0.24 endemics to nonendemics for plants and 1:0.28 endemics to nonendemics for vertebrates (excluding fish), figures that no other hotspot exceeds and only a handful of other top priority hotspots such as the Philippines, the Cape Floristic region of South Africa, and New Caledonia can even approach.

In simple terms, all this means is that most of Madagascar's wonderfully unique species are found nowhere else and that the ultimate responsibility for maintaining them rests with the people of Madagascar themselves. The international community can and should do as much as humanly possible to help, but the final decision will be in the hands of the Malagasy people.

Unfortunately, Madagascar is also a global leader in environmental degradation. Given its isolation, it was one of the world's last large landmasses to be colonized, with humans arriving there no more than 2,000 years ago. As a result of this late arrival by people, the animals and plants living on the island did not coevolve with humans. They were quite naïve and had virtually no escape or defense behaviors against human predation, making them easy prey for early hunters. Indeed, even to this day, one is impressed by how easily one can approach most species of wild lemurs and other animals, in striking comparison to their counterparts on the African mainland and other continental areas.

The saddest and most striking examples of recent extinctions in Madagascar are those of the giant lemurs and the elephant birds. As remarkable as Madagascar's living lemur fauna is in global terms, it pales in comparison to what existed as recently as a thousand years ago. Since the arrival of humans on Madagascar, fully 8 genera and at least 15 to 17 species of lemur have become extinct, representing almost one-third of all known species (living and extinct) and fully 36 percent (8 of 22) of known genera. Furthermore, virtually all the extinct species were diurnal and larger than any of those surviving today. They include some spectacular creatures like *Megaladapis,* which resembled a large Australian koala and grew to the size of a calf, *Babakotia,* which looked like a large primate version of the South American tree sloth, and the truly amazing *Archaeoindris,* which was larger than an adult male gorilla and probably occupies a niche similar to that of the now extinct North American ground sloths. Along with them went the elephant birds, including the giant elephant bird (*Aepyornis maximus*), at approximately 990 to 1,100 pounds the heaviest bird that ever lived and exceeded in height only by the large moa of New Zealand (now also extinct). The elephant bird grew to be nearly ten feet tall, had eggs that weighed twenty pounds, and represented the largest single cell known.

The loss of the giant lemurs, the elephant birds, the pygmy hippo, and the other creatures that went with them on Madagascar (so nicely described in Part Three) represents one of the three or four most striking extinction spasms in recent times. (The others are the losses of much of Hawaii and New Zealand's native flora and fauna following first the arrival of the Polynesians and then the Europeans, and the disappearance of many of Australia's endemic mammals since the arrival of Europeans.) These episodes have occurred within the period of written history and are not events of the distant past like the extinction of the dinosaurs. They all demonstrate that the much discussed possibility of major extinction spasms in our lifetime is not a figment of the conservationist's fertile imagination; it is very real and looming on the horizon, if not already under way. The risk of this recent extinction spasm continuing in Madagascar and eliminating much of what still survives there is very real indeed.

Arresting the spasm in Madagascar is especially challenging considering the nation's extreme poverty. The Malagasy love their "island of the ancestors" as deeply as any people could their native land, as is made abundantly clear in Part Four of this book, which is devoted to the Malagasy and their unique culture. If they had the means to safeguard their island's treasures, both natural and cultural, they would surely do so. But for all too many Malagasy, life is a matter of bare survival, and conservationist thinking, much less action, is difficult under such circumstances. Nonetheless, development of such an ethic is essential at all levels of Malagasy society, since without it, the situation of Madagascar's poor will only worsen as watersheds are destroyed and forests that could provide a host of products on a sustainable basis are cut for slash-and-burn agriculture.

Tragically, most of Madagascar's natural vegetation has already been destroyed or seriously modified over the past two millennia, and especially in the past century. This is due both to a long-standing need for farmland and fuelwood among Malagasy, and to poorly thought-out international development programs, which have often seriously exacerbated degradation problems. Rough estimates indicate that about 90 percent of the original natural vegetation is now gone, with only about 10 percent remaining in more or less intact condition. This level of destruction is one of the most severe on Earth. It is especially striking in the central plateau region of the country, which in the dry season looks more like a moonscape than a healthy, functioning part of planet Earth. The barren red slopes of denuded, eroded lateritic soil and the heavily silted rivers present a very depressing sight and have led to one of Madagascar's tragic nicknames, *"Le Grande Île Rouge"* (the Great Red Island).

The combination of high biodiversity, very high endemism, great ecosystem diversity, and high level of threat all combine to make every little piece of what remains in Madagascar extremely important in global terms and in many ways "the most valuable biodiversity real estate on our planet." Indeed, if one looks at that remaining intact 10 percent, where all of Madagascar's biodiversity is now concentrated, we see that it has 4 percent of all the planet's plants and animals in just 0.04 percent of its land surface, a degree of concentration of life two orders of magnitude greater than what one would expect.

Fortunately, Madagascar does have a protected-area network aimed at maintaining its great biodiversity. This system was established in 1927, and its more important categories are national parks, strict nature reserves, and special reserves. Together they cover a total of 2,770,060 acres, or 1.9 percent of the country. Unfortunately, as important as these parks and reserves are, they do not cover a fully representative portion of Madagascar's biological heritage. As is stressed in Part Five, which delves deeply into the conservation question, many more protected areas need to be created, with as much as possible of that remaining 10 percent of natural vegetation being added to the protected-area network as soon as possible.

Furthermore, only a handful of the existing parks and reserves are adequately protected, mainly those that have received major infusions of international aid over the past fifteen years, such as Ranomafana, the national park profiled in Part Five. Most of the others, including the Lokobe strict nature reserve featured in Part One, are greatly under-staffed, simply because Madagascar is a very poor country and has lacked the resources to fully implement its own conservation legislation. Much more needs to be done, and it must happen very soon.

What then can we in the developed world do for Madagascar? First and foremost, we have to learn more about it and recognize its impor-tance to the planet. Since it has been of little geopolitical significance in this century, and since there are very few Malagasy living here in the United States, Madagascar has rarely come to the attention of the out-side world. Indeed, it is really only the biologists who have focused heavily on it, and we have been reasonably successful at highlighting its importance over the past decade. Nonetheless, much more needs to be done to make this megadiversity country more visible in the international arena and to make it the household word that it so richly deserves to be.

Closely associated with this recognition is the importance of visiting Madagascar. I believe that ecotourism is one of the real conservation solutions for Madagascar, because it generates income for the country and demonstrates strongly and clearly to Malagasy decision makers and local people alike that biodiversity conservation pays, that it is good business, and that it offers a real competitive advantage for Madagas-car—perhaps the greatest competitive advantage that the country pos-sesses. Indeed, I believe that there is no reason why Madagascar could not become one of the world's premier ecotourism destinations over the next decade, and that ecotourism could and should become the country's number-one foreign-exchange earner.

This pressing need for greater awareness of Madagascar and interest in its natural and human wonders is why this book by Peter Tyson is really so important. His vivid and engaging accounts of his own experi-ences in Madagascar, his marvelous adventures, the intriguingly obscure historical material he has dug up on the island's natural and cultural riches, and his respect for both the people and the wildlife help bring this dis-tant island to life and convey, through his eyes, why it is so special and so different from anyplace else. His experiences with individuals like Chris Raxworthy, Patricia Wright, and the twins Angelin and Angeluc Razafimanantsoa, who have become part of the biodiversity folklore in

Madagascar, are especially appealing. And his factual accuracy and balanced coverage of some of the leading scientific issues now being addressed in Madagascar make this book useful as a reference as well.

Indeed, even though I have been to Madagascar more than fifteen times already, this book makes me want to call my travel agent and arrange yet another trip. I hope that it has the same impact on you, and that after many pleasant hours of reading Peter's accounts, you will also have that same urge, make the trip, and become part of a growing cadre of Madagascar fans, advocates, and supporters, who in many ways, along with the people of Madagascar, hold the future of this earthly paradise in their hands. Good reading, and I hope to see you there some day.

RUSSELL A. MITTERMEIER
President, Conservation International
Chairman, IUCN/SSC Primate Specialist Group

Preface

If one incident during the four years in which I traveled to Madagascar could symbolize why I wanted to write this book, it would be the time I made a big mistake in Isalo National Park.

I had just come out of the Canyon of the Monkeys, one of the many steep-sided canyons that make Isalo, a spectacularly eroded limestone canyonland in southern Madagascar, the island's Grand Canyon. It was a beautiful morning, not a cloud in the sky. The southern edge of the Isalo massif reared up behind me like a muddy tidal wave caught in midcurl. I followed in the bare footprints of my guide, a tall, proud youth of the Bara tribe, reputedly the fiercest warriors on Madagascar. Shouldering a stout stick with a basket slung at the end (our lunch), Claude wore a flower-print shirt held together by one button, a gray panama, and a huge knife stuck in a sheath at his belt.

As we made our way along the raised edges of rice fields, their harvestable plants drooping from the weight of the grain, Claude suddenly stopped and spoke to a farmer standing in the middle of the paddy, then turned and asked if I wanted to see something. Our exchange was really more a series of gestures, since neither of us spoke the other's language. But I caught one unmistakable word in Malagasy, as both the language and the people of Madagascar are known: *manditra*, snake. We stepped into the paddy, careful not to crush any rice plants. Following the farmer's motions, Claude delicately pushed aside some rice stalks with his stick. There, coiled in a small depression, lay a boa constrictor. I immediately bent down and picked it up.

That was my big mistake.

Claude, the farmer, and an older woman and child standing nearby suddenly jumped as if jabbed with electric cattle prods. With a shriek, the woman broke into a sprint, dragging the wide-eyed child along by the wrist. She didn't stop until she reached the high grass at the edge of the paddy. The farmer, too, dashed off to stand uncertainly some distance away, hands straight down at his sides. Claude, perhaps conscious of preserving his status as a fearless Bara warrior, only fell back out of arm's reach, but his big brown eyes bore the look of a startled deer. Then, led by the near-hysterical woman, the three adults began cackling among themselves about what I had just done. That brought a cascade of Bara children pouring toward us out of nearby fields, their curious parents not far behind.

I should have known better. The Malagasy people have a morbid fear of snakes. They call them *kakalava,* or "long enemy," and believe that whoever encounters one will hear about death. "The snake and the man: they are both afraid" goes a Malagasy proverb. Alison Jolly, a primatologist who has worked on Madagascar since the early 1960s, holds that this national dread is either "a legacy from ancestors who came from lands of cobras and puff adders and mambas, or else Freud is right that the mere shape of a snake makes it supernatural." Missionaries may be partly to blame, having made a native hognose snake known as the *menarana* the snake that tempted Eve in the Malagasy Bible. But the Malagasy have their own beliefs: One snake with a red tail is thought to drop tail-first out of trees and pierce the hearts of cattle.

I calmly asked Claude if he'd like me to put the manditra in the forest. He nodded, and I set off for a nearby stand of trees with the snake coiled around my daypack. A knot of children frolicked at my heels, squealing with delight.

That broke the ice. The old woman began laughing in relief, slapping her hands on her skirts. The farmer followed suit, and even stalwart Claude managed a smile. A major faux pas had fortunately become a festive break from routine. With the kids dancing at my back, I felt like the Pied Piper.

The incident possesses a little of everything I find fascinating about Madagascar: The island's singular people and culture. The striking beauty of the landscape. And the wonder of the wildlife, largely unique to Madagascar and of wholly enigmatic origins. (How is it that boas, for one, are found only there and half a world away in the Americas?)

This book is based on four trips I took to Madagascar between October 1993 and October 1997. These journeys took me right around the country, from the rainforest island of Nosy Be in the northwest to the mountainous peninsula at Fort Dauphin in the southeast; from the heart of vanilla country on the northeast coast, to the scorching spiny desert of the southwest. Along the way I joined four scientists—a herpetologist, a paleoecologist, an archeologist, and a primatologist—as they sought, with the help of Malagasy and Western colleagues, to solve some of the foremost mysteries on an island rife with them:

🏵 Where did the island's singular animals and plants, from the spiny tenrec to the octopus tree, come from (Part One)?

🏵 What caused them to speciate so wildly (Part Two)?

🏵 Who or what forced every native animal weighing more than twenty-five pounds—among them elephant birds, a pygmy hippo, and a host of giant lemurs—into extinction sometime in the past 2,000 years (Part Three)?

🏵 Why was Madagascar apparently uninhabited until the time of Christ? What accounts for the predominantly Indonesian aspects of the culture and language in this island abutting Africa (Part Four)?

🏵 Finally, with Madagascar ranked among the world's poorest nations and faced with an environmental crisis of truly staggering proportions, can the island's natural and human treasures be saved, and if so, how (Part Five)?

In this book, I will address each of these questions in depth. At the same time, I hope to give you a taste of what it's like, among other things, to watch a group of Bara tribespeople go wild over a snake.

PETER TYSON
Arlington, Massachusetts
September 1999

The Eighth Continent

Vila de Mocimboa
da Praia

Grand
Comore
Moroni ● Foumbouni

Anjouan

COMOROS

Moheli

● Pemba

MOZAMBIQUE

Nacala

Lumbo
Moçambique

*Île
Glorieuses
(FRANCE)*

*Cap
d'Ambre*

Dzaoudzi
Mayotte

Nosy Mitsio

Nosy Bé
Helliville ● Lokobe
Reserve

Ambilobe

Antsiranana (Diégo-Suarez)
Amber Mountain
National Park

Vohimarina
(Vohemar)

ANTSIRANANA

Benemevika
Sambava
Benavony

Antsohihy

Andapa

Antalaha

Befandriana

Maroantsetra

Cap Est

Mahajanga
(Majunga)
Antsohihy
Lava

Marovoay

MAHAJANGA

*Cap
St. André*

*Nosy
Chesterfield*

*Juan
de Nova
(FRANCE)*

Maevatanana

Andilamena

*Masoala
National
Park*

*Nosy
Boraha*

Fenerive

L. Alaotra

Ambatondrazaka

Toamasina (Tamatave)

Maintirano

*Nosy
Barren*

Ankavandra
Tsiroanomandidy
ANTANANARIVO

ANTANANARIVO

Miandrivazo

Ambatolampy

Moramanga

Brickaville

TOAMASINA

**INDIAN
OCEAN**

Morondava

Antsirabe

Mania

Mahanoro

Belo-sur-Mer

Ambositra

Nosy Varika

Morombe

TOLIARA

Ambomahasoa

Fianarantsoa

FIANARANTSOA

Ranomafana
National
Park

Mananjary

Ambalavao

Manakara

Mangoky

*Isalo
National
Park*

Ihosy

Farafangana

*Île Europa
(FRANCE)*

▲ PK-32

Onilahy

Toliara
(Tuléar)

▲ Beza
Mahafaly
Reserve

Betenty
Private
Reserve

Ampanihy

TROPIC OF CAPRICORN

Ambovombe

Tolagnaro (Fort Dauphin)

Cap Ste. Marie

Mozambique Channel

Madagascar

- – · – Province boundary
- ★ National capital
- ◉ Provincial capital
- —— Road

0   50   100 Kilometers
0   50   100 Miles

Introduction

MANY HAVE TRIED TO get at the heart of what makes Madagascar so different from every other place in the world. "May I announce to you that Madagascar is the naturalist's promised land?" the Frenchman Philibert de Commerson wrote in 1771. "Nature seems to have retreated there into a private sanctuary, where she could work on different models from any she has used elsewhere." David Attenborough, who brought some of the first images of the island to the outside world in the early 1960s, described "a place where antique outmoded forms of life that have long since disappeared from the rest of the world still survive in isolation." To me, the most evocative metaphor comes from Alison Jolly, the doyenne of lemur studies, who once wrote that on Madagascar it is as if "time has broken its banks and flowed to the present down a different channel."

Down that channel have come a cornucopia of curious beasts. When the first people arrived on the island around the time of Christ, they found a real-world *Jurassic Park*. Flightless elephant birds, standing ten feet tall and weighing half a ton, thundered through the island's wooded savannas on legs like tree trunks. Lemurs the size of apes nibbled leaves high in rainforest trees, while Galápagos-sized tortoises and dwarf hippopotamuses grazed below. Sadly, these creatures and the rest of Madagascar's so-called megafauna are now gone, rendered extinct sometime in those 2,000 years.

Yet countless other types of animals and plants remain to astonish the visitor. Madagascar is a place where lizards scream, giant cockroaches hiss, and a handsome beast called the *indri* sings a song of inexpressible beauty. Fully fifty kinds of lemurs, early models of primates that were the smartest living things 40 million years ago, live there—and only there. One is so small you could cup it in your palm. The fossa ("foo-sa"), Madagascar's stab at a mountain lion, slinks around the jungle at night, seeking to plant its retractile claws into unwary prey. In the trees, two-foot-long chameleons zap bugs with tongues as long as their bodies, while cryptic snakes with noses shaped like sharpened pencils slide through the branches, veritable twigs on the move. The vangas, just one of five families of birds found nowhere else, rival Darwin's finches for diversity, and many members of the island's phantasmagoria of inverte-brates, epitomized in my mind by the truly freakish giraffe-necked wee-vil, might have come straight out of the bar scene in *Star Wars.*

Plants are just as singular. Eight out of ten of them grow naturally only on Madagascar. Like the animals, they are the dinosaurs of the plant world, relics from a time long past. The traveler's tree, a fan-shaped banana relative that serves as a kind of national symbol, counts its closest relatives in South America. The comet orchid, one of more than a thousand varieties of orchid that decorate the island, has a four-teen-inch spur. When Darwin saw this species in 1862, he predicted that a giant hawkmoth must exist with a proboscis long enough to take advantage of the nectar at the spur's base; entomologists found it forty-one years later. Madagascar even has an entire floral ecosystem all its own, the spiny desert. A searing, otherworldly landscape in which virtu-ally all the plants exist nowhere else on Earth, the spiny desert might have sprung from the imagination of Henri Rousseau. Here, the finger-like stalks of "octopus trees" wiggle at the sky below massive baobabs rearing overhead like vegetable elephants.

The human world is as exceptional as the natural. Though the island lies just over 250 miles off Africa, its people, customs, and language orig-inally hail from Indonesia. "This strikes me as the single most astonishing fact of human geography for the entire world," the physiologist and bio-geographer Jared Diamond has written. "It's as if Columbus, on reaching Cuba, had found it occupied by blue-eyed, blond-haired Scandinavians speaking a language close to Swedish, even though the nearby North American continent was inhabited by Native Americans speaking Amerindian languages." Signs that the forebears of the Malagasy came from Southeast Asia are pleasantly rife throughout the island: terraced rice paddies, outrigger canoes, and ancestor worship, to name just a few. While passing Malagasy in the streets of Antananarivo, the capital, you might swear you were in Jakarta. Over the two millennia since the first Malagasy settled the island, a significant African component has blended in, so that the culture and people can seem at times a perfect mix of both.

Africans brought the island-wide obsession with cattle, for instance, and the widespread regard for spirit possession. Malagasy society is also spiced with strong Arabian, South-Indian, and European influences.

All told, there are eighteen officially recognized tribes. (The term "tribe" is not considered pejorative in Madagascar.) Living in the center of the island in and around Tana, as everyone abbreviates the name of the capital, the Merina are the most Asian-looking of the tribes and the country's political and economic leaders. Just to the south of them in the mountainous central Highlands are the Betsileo ("the invincibles"), experts in rice terracing and wood carving. The largest tribe is the Sakalava, a cattle-herding people who range up and down the west coast. They owe their affinities primarily to Africa, as do the tribes that scratch out a living in the scorched south, including the Bara, Mahafaly, and Antandroy. The latter two build the island's most elaborate stone tombs, huge painted monuments that are among the country's most out-standing architectural treasures. Along the northeast coast lie the Bet-simisaraka, Madagascar's second most numerous tribe, who tend fields of sugarcane, cloves, and vanilla they have carved out of the rainforest. Farther south you'll find the Antaimoro, who guard sacred theological texts known as *Sorabe* ("Great Writings") and claim an Arab ancestry.

Many customs boggle the Western mind. In wild celebrations known as *famadihana* ("turning of the bones"), the Merina and several other tribes regularly disinter and reenshroud the bodies of their dead relatives (whom all tribes consult on important events in their lives). The Sakala-va specialize in *tromba*, during which the ancestors speak to the living through entranced mediums. Malagasy in remote areas believe that white-skinned Westerners are *mpakafo*, or "heart-takers," who have come to the island to kill Malagasy, especially women and children, and eat their vital organs. For most foreigners, such beliefs contrast incongru-ously with other aspects of Malagasy culture that, by Western standards, seem highly refined. Malagasy music, played on home-grown instru-ments such as the zitherlike *valiha*, took the international scene by storm a decade ago through the work of classical masters like Rakotofra and popular groups such as Tarika Sammy. Malagasy verbal arts are espe-cially cultivated, with proverbs, oratory, and the traditional poetry known as *hainteny* rivaling the finest examples of similar forms elsewhere.

Madagascar's natural and cultural history owes its uniqueness to the island's long isolation in the Indian Ocean. Once sandwiched between Africa and Australia in the supercontinent known as Gondwana, the island broke free 150 to 165 million years ago and began drifting to its present spot off the southeast coast of Africa. At least 40 million years ago, and perhaps earlier, the Mozambique Channel became too wide for African animals to accidentally float over on storm-fashioned rafts of vegetation. Thus the island has none of the nearby continent's lions, elephants, gazelles, monkeys, or poisonous snakes; their kind evolved too late to make it aboard this natural ark. Isolated ever since, the island today provides a glimpse of the world as it looked during the age of the dinosaurs.

The island is not only old but big. The world's fourth-largest after Greenland, New Guinea, and Borneo, it is fully 1,000 miles long and 350 miles across at its widest point. It would stretch from New York City to Tampa, Florida, with its western edge pushing well into Ohio. Madagascar's enormous size is only the most obvious of several reasons, including its exceptional diversity of habitats and species, why some observers feel the island could justly be called the world's eighth continent. Oriented to the northeast—a sacred direction for the Malagasy, as it turns out—its shape has been likened to a brick, a left foot, even a badly made omelette. To me, the island has always looked like an upended side of beef, which seems an apt symbol of both the island's rich resources and the wanton abuse of such that continues unabated.

Madagascar is very much a Southern Hemisphere island. If you were to fly due west around the globe from Tana, which lies in the center of the island at about 19° South latitude, you would pass over the Mozambique Channel and hit Africa at the mouth of the Zambezi River in Mozambique. You'd cross over Zimbabwe, Botswana, and Namibia before leaving the continent. Crossing the Atlantic, you'd enter Brazil about halfway between Rio and Brasilia, then continue across Bolivia and the northern tip of Chile. In the Pacific, you'd fly over the Society Islands just south of Tahiti and cross the International Dateline at Tonga. You'd hit Australia at Townsville, Queensland, then push on over the Great Sandy Desert of Western Australia. Once in the Indian Ocean, you'd find nothing but ocean until Rodrigues Island, then another 800 landless miles until you reached Madagascar.

Though it lies so far south that its southern end dips into the temperate zone, Madagascar is very much a tropical island. It is a land of humid rain forests and hellishly hot deserts. There are no glacier-clad peaks, indeed no evidence that the island was glaciated anytime in the past million years. While temperatures in the austral winter can drop to near freezing in the central Highlands, Malagasy who have never left the island have never seen snow. Within that essentially tropical climate exists a range of subclimates, however. Beyond sheer size, this is due to the island's long north-south orientation—Madagascar stretches from

12° south latitude to well south of 25° south latitude—and to its geographic layout. A rugged mountain range running down the island's spine east of center neatly divides Madagascar in two. East of the range lie the island's rain forests, west of it savannah, dry forest, and in the deep south, the spiny desert.* As befits a minicontinent, the island features a host of other habitats as well, from mangrove forests to alpine heath to coral reefs.

By the same token, Madagascar has a range of distinguishing geologic features. Most of these features consist of ancient African crystalline basement, which makes up the eastern two-thirds of the island. (The western third holds dinosaur-bearing sedimentary rock.) If you were to walk south down the island from its northern end, you'd hit a variety of distinct mountain ranges, for example. On the northernmost tip, you'd find the rain-forest-clad Montagne d'Ambre, a range born of onetime volcanic activity on Madagascar. Just south of it, you'd come upon the Ankarana, a limestone massif with more than sixty miles of caverns and underground rivers. Just south of that, you'd hit the island's tallest peak, 9,490-foot-tall Maromokotro, which is part of the Tsaratanana Mountains. Farther along the spine, south of Tana, lie the ancient granitic ranges of the Ankaratra and Andringitra, while off by itself in the south-central part of the island stands the Isalo Massif. Equally isolated in the remote west lie the *tsingy* of Bemaraha, which arguably put the eroded spires of Bryce Canyon National Park in the U.S. to shame. Tsingy comes from the Malagasy word for tiptoeing, as the karst pinnacles of this craggy plateau lie so close together that the Malagasy say you can't put your foot down.

To a large extent, Madagascar's diversity of landscapes has helped shape its history. From the settlement of the first Malagasy along the coasts to the rise of the Merina monarchy in the central Highlands over a millennium and a half later, the lay of the land has dictated who succeeded where, and why.

The earliest signs of humans on Madagascar have turned up on the southwest and northeast coasts. The admittedly scanty archeological evidence suggests that no one lived on the island prior to the Christian era—an astonishing fact considering Madagascar's enormous size and its

---

*The one exception to this east-west split is the northwest coast in and around the island of Nosy Be, which a climatic anomaly lends a humid climate and tropical forest.

proximity to the cradle of humankind in nearby Africa. Very little is known about the first settlers. Were they pure Indonesians who came directly from Southeast Asia in one or more migrations across the Indian Ocean? Or were they Indonesian-Africans whose Asian ancestors lived and traded along the north coast of the Indian Ocean before arriving on the northeast coast of Africa, where they intermarried with black Africans? No one knows.

What is known is that an indigenous Malagasy fishing and cattle-herding culture developed on and near the coasts through the first millennium A.D. Arab merchants began trading along the northern coastlines about the ninth or tenth centuries, though they apparently never penetrated into the interior. It wasn't until the thirteenth century that the first Malagasy even reached the high, mountainous center of the island around Tana, which today represents the heart of the Malagasy nation.

The first written accounts of the island appeared around the tenth century, when Arab geographers began recording information they had picked up from returning Arab traders. Based on hearsay, these accounts were far from accurate, but they planted the notion of a huge island off Mozambique, which the Arabs called the "Island of the Moon." Marco Polo, while traveling through Arabia on his return from China in the thirteenth century, heard these tales as well and penned his own description of the island. His account was equally devoid of fact—the island bore lions and camels, he said, as well as elephants that were lifted into the sky and devoured by giant birds—but it includes the first use of the name Madagascar.

Scholars believe Polo's misnomer is a corruption of Mogadishu, a town on the Somalian coast that Polo would likely have heard of. Certainly it is not a Malagasy word. The Malagasy language does not have a letter *c*, for instance, and all words end in a vowel. When the first European missionaries arrived on the island in the nineteenth century, they found that the Malagasy had several names for their island. Some called it *Nosin-dambo* or "Isle of the Wild Boars." Others used the more lyrical *Ny aninvon' ny riaka*, literally "The (land) in the midst of the moving waters." Most Malagasy, however, called it simply *Izao rehetra izao*, "This All," or *Izao tontolo izao*, "This Whole." Nevertheless, the name that stuck was one bestowed by a European who never laid eyes on the island.

The first European to do so was a Portuguese sea captain named Diogo Dias. While sailing round the Cape of Good Hope on his way to India in the year 1500, Dias was blown off course in a storm. On August 10, he landed on a coast he assumed to be that of Mozambique. When the land ended in the north, he realized he'd discovered a large island. He promptly named it St. Lawrence, after the saint on whose feast day he had sighted the island, but the name eventually melted away with all the others.

Within a few years, the Portuguese, followed by the English and French, set about trying to establish trading settlements on what they saw as rich virgin turf. In a gushing 1647 pamphlet, an English merchant named Richard Boothby declared Madagascar "the chiefest paradice [*sic*] this day upon Earth" and urged his countrymen to build plantations on the island without delay. Many would heed his call, yet all such early attempts at colonization ended in disaster, mostly from disease, starvation, and constant battles with local Malagasy. Just two years after Boothby's pamphlet appeared, fellow Englishman Powle Waldegrave wrote a scathing response: "I could not but endeavour to dissuade others from undergoing the miseries that will follow the persons of such as adventure themselves for Madagascar . . . from which place God divert the residence and adventures of all good men."

Good men, yes, but for bad men it was Boothby's "paradice." Indeed, European and American pirates were the first outsiders to gain a successful foothold in Madagascar. The vast island contained an abundance of food, fresh water, and building materials as well as countless bays in which to hide from British men-of-war. As Lord Bellamont, then governor of New York, declared in 1699, "The vast riches of the Red Sea and Madagascar are such a lure to seamen that there's almost no withholding them from turning pirates." Buccaneer activity in Madagascar peaked in the late seventeenth and early eighteenth centuries, as chronicled in *A General History of the Robberies and Murders of the Most Notorious Pyrates*, by Capt. Charles Johnson, a pseudonymous author once thought to be Daniel Defoe. On the island's northern tip, two pirates even formed a utopian community they dubbed the Republic of Libertalia. Based on democratic principles of government, equal profit-sharing, and the abolition of slavery—not to mention the plunder of innocent merchant ships—the republic, which was little more than a settlement, was eventually attacked and destroyed by local Malagasy.

Though the pirates were "but scurvy School-masters to teach [the Malagasy] Morals," as one early chronicler phrased it, they did have a lasting effect on Malagasy history. It was European pirates, for instance, who inspired the only attacks the Malagasy people are known to have ever made on another land. In the late eighteenth and early nineteenth centuries, the Betsimisaraka and Sakalava tribes made annual slave-raiding assaults on the neighboring Comoros Islands and even the African coast. The largest expeditions boasted up to 500 outrigger

canoes and perhaps 18,000 men. The most significant pirate contribution, however, came in the form of Ratsimilaho, the son of a European
pirate and a Malagasy princess. Born in 1712 and educated in London,
Ratsimilaho succeeded in unifying the many disparate tribes of the
northeast coast, who subsequently took the name Betsimisaraka ("the
many inseparables").

While the Betsimisaraka never managed to form an actual kingdom
or state, the Sakalava of the west coast did. Just as the Betsimisaraka
owe their rise to pirates, so the Sakalava owe theirs to trade with foreigners, particularly Arabs, along the west coast. "The independent
Sakalavas . . . seem to have no needs," wrote a European in the mid-
nineteenth century. "So long as the Americans or the Europeans bring
them, from time to time, rifles (English models), powder, a little cloth . . .
as a luxury, a few pieces of glass for necklaces, and alcohol, they have no
other desires." Yet it was just these supplies, particularly the weapons,
that helped the Sakalava become the most powerful kingdom Madagascar had yet seen. In the early eighteenth century, the Sakalava empire
stretched from Tuléar on the southwest coast to the northernmost tip of
the island. But the kingdom's far-flung nature and constant infighting
among its various subtribes eventually led to its dissolution in the mid-
1900s.

The kingdom that had the most lasting impact on Malagasy history
and culture was that of the Merina. The most light-skinned of the eighteen tribes, the Merina are thought to have arrived in the Highlands in
the early part of the sixteenth century. Over the next two centuries, they
consolidated their power around Antananarivo. The "city of a thousand"
became the Merina capital in the late eighteenth century during the reign
of Andrianampoinimerina, the greatest Malagasy ruler of all time. Andrianampoinimerina, whose name means "the prince in the heart of Imerina," built the extensive rice paddies that make Imerina, the region in and
around the capital, the breadbasket of Madagascar. (He once famously
said, "Rice and I are one.") Andrianampoinimerina established the
fokonolona, the village committees that still run community affairs to this
day. Through skillful diplomacy and enlightened leadership, he succeeded in greatly expanding the Merina kingdom, including subduing the
Betsileo tribe to the south. But it wasn't enough for him. On his
deathbed, the king told his son and successor, "Imerina has been gathered into one, but behold the sea is the border of my rice-fields, O
Radama."

Radama I, who assumed the throne in 1810, took up his father's challenge. With the help of the British, whose emissaries began furthering

their nation's trade interests in Madagascar by 1820, Radama managed to unify most of the island. The only regions he failed to conquer before his untimely death at age thirty-six was the far west under the Sakalava and the far south, inhabited by the hotly independent Bara, Mahafaly, and Antandroy tribes. In return for the military assistance he received from the British, Radama opened the country to English missionaries, who built churches, reduced the Malagasy language to a written form, and spread the Christian faith throughout the island. Through the person of James Cameron, a brilliant Scottish engineer, he also launched a miniature Industrial Revolution that featured the construction of an aqueduct and reservoir, the manufacture of bricks and soap, and the advent of the first printing press in Madagascar.

Radama's widow and successor, however, was to reverse much of her late husband's work. Ranavalona I has been likened to a female Caligula, who "waded to the throne through streams of blood." Since her claim to the crown was weak, her first act as queen was to have all rivals to the throne put to death, including her mother-in-law. Her killing didn't stop there. In 1835, she abolished Christianity—its practices, she argued not unreasonably, "are not the customs of our ancestors"—and ordered the slaughter of all who would not renounce their Christian beliefs. The first martyr, a young woman named Ralama, was speared to death in 1837 as she knelt in prayer. Hundreds more met a gruesome end in the coming years, a time Malagasy Christians would long remember as *ny tany maizina*, "the time when the land was dark." Ranavalona's most ruthless attempt to eradicate the foreign faith came in 1849, when she had 2,000 adherents arrested. Most were fined, but more than 100 were flogged and sentenced to life in chains and hard labor, and eighteen were killed. Fourteen of them were thrown off "the place of hurling," a sheer cliff below the Queen's Palace in Tana, while four of noble birth, including a pregnant woman, were given the "honor" of burning at the stake.

Later in her thirty-three-year reign, Ranavalona cut off all foreign trade and made forced labor for the government a requirement even of soldiers. She revived the abhorrent customs of killing babies born on unlucky days and of forcing people suspected of crimes to drink a poison from the *tangena* tree. If they survived, they were innocent; most didn't. In the end, few Malagasy were sorry when she finally died in 1861.

Some Malagasy have since sought to paint Ranavalona I in a better light, "a heroine of Malagasy nationalism" as the historian Mervyn Brown has put it. After all, she was only advocating a return to customary beliefs and practices. These efforts included a push to collect the traditional poetry of Madagascar. In addition, she greatly furthered the industrial work of Cameron and Radama, with the help of another foreigner, a remarkable Frenchman named Jean Laborde. Shipwrecked on Madagascar in 1831, Laborde went on to build a factory that produced weapons, pottery, candles, sealing wax, bricks, rum, sugar—basically all the products a small country would need to be largely self-sufficient. He also constructed the Queen's Palace, a distinctive wooden structure with a peaked roof held up by a 130-foot tree trunk, which took 5,000 laborers to transport from the eastern forests.*

When Radama II succeeded his mother, he was true to the memory of his namesake and nullified all her edicts. He reopened contacts with the West, including welcoming back the Christian missionaries and their religion. In fact, Radama's short-lived reign—he was assassinated a year after taking office—marked the beginning of a decades-long rivalry between the English and French churches in Madagascar. Neither gained the upper hand, and today most Malagasy towns feature two churches, one Protestant and one Catholic, staring each other down across the main street.

Madagascar had three more queens—Radama II's widow Rasoherina assumed the throne in 1863, Ranavalona II in 1868, and Ranavalona III in 1883. But power during their reigns lay in the hands of the prime minister, a calculating man named Rainilaiarivony, who, following tradition, married each of the queens in succession. Rainilaiarivony did great things for his country, including overseeing the construction of buildings and roads, improving education, and increasing medical services and the training of Malagasy doctors. He even managed to negotiate treaties with both England (1865) and France (1868).

But Rainilaiarivony could not hold back the juggernaut that was France, with its desire to add Madagascar to its collection of African colonies. Britain had never evinced a colonial interest in the island, and in 1890 it even signed a treaty with France recognizing the latter's protectorate over Madagascar in exchange for French recognition of the British protectorate over Zanzibar. With its only possible rival out of the way, France made its move.

French troops began their assault in December 1894 with an attack on Tamatave, a port on the east coast, followed two months later by

---

*Alas, both structures are now gone. The factory was destroyed in the 1850s during an uprising by the 1,200 laborers Ranavalona I forced to work there without pay, and the Queen's Palace burned to the ground in November 1995. By that time part of a vast royal complex built up around Laborde's original structure, the palace's destruction, thought to be the work of an arsonist, was a national tragedy. One commentator likened it to the simultaneous loss of Westminster Abbey, Buckingham Palace, and the Tower of London.

another on Majunga on the west coast. The latter expedition marched all the way to Tana, surprising Malagasy military commanders, who had expected an assault from the much nearer east coast. On October 1, 1895, the French took the capital with little resistance. Astonishingly, they lost but twenty men to fighting during the entire campaign (though nearly 6,000 perished from disease). The French commander forced the queen's representatives to sign away the country as a protectorate, and the following year the French Parliament voted to annex the island as a colony.

To France's credit, it helped usher Madagascar into the modern world during its six and a half decades of official control. The French unified the whole island for the first time, thereby finishing the work Andrianampoinimerina had begun in the eighteenth century. They vastly improved the country's basic infrastructure by building roads and airports and installing a communications system. In 1927, they formed the first reserve system in the African region in Madagascar. Most important, they established the administrative, military, and economic framework of a modern nation-state.

As colonizers, however, the French were no better than most. They instituted a dual judicial system—French laws for colonists and a mercilessly unjust code known as the *indigénat* for Malagasy—and a forced-labor mandate that had all Malagasy men between the ages of sixteen and sixty work for the French for fifty days a year at minimal wages. Not surprisingly, the Malagasy took exception to such racist policies, which essentially left them second-class citizens in their own country. During World War I, while 46,000 Malagasy fought beside the French in Europe (and more than 2,000 died), some of their compatriots launched the first of several rebellions against French rule. It failed, but in the early 1920s a Malagasy schoolteacher named Jean Ralaimongo, who had shared a room in Paris with Ho Chi Minh, began stirring things up again.

World War II, in turn, interrupted Ralaimongo's efforts and those of other nationalists. Attention was drawn both to Europe, where 34,000 Malagasy soldiers were serving in France by June 1940, and even to the coasts of Madagascar itself. In May 1942, the British navy attacked and took control of Diégo-Suarez in the far north; four months later another British force overwhelmed Majunga on the west coast. The British wanted to keep control of the Mozambique Channel, a strategic passageway for Allied ships that was frequented by German U-boats.*

*The Nazis had their own designs on Madagascar: a secret plan to resettle 5 million European Jews on the island.

The end of the war brought renewed nationalist activities. These culminated in a fierce rebellion in 1947, which began simultaneously in several parts of the country. The French succeeded in quashing the insurrection, and estimates of the number of Malagasy lives lost range up to 80,000. But the rebellion made a return to independence only a matter of time. During a visit to Tana in 1958, Gen. Charles de Gaulle pointed to the Queen's Palace and proclaimed metaphorically, "Tomorrow you will once more be a State, as you were when that palace was inhabited." Madagascar finally regained its independence on June 26, 1960.

Initially, independence proved to be in name only. The first president, Philibert Tsiranana, was pro-Western, and France continued to oversee trade and financial institutions, and maintain several military bases. But after a series of coups between 1972 and 1975—one president was shot dead six days after entering office—Didier Ratsiraka took control in 1975 and established a paradoxical Christian-Marxist state. (He even wrote his own Mao-inspired *Boky Mena* or Red Book.) He nationalized banks, insurance companies, and other institutions without compensation, and launched a bevy of social and economic reforms based on Communist models. Even as he turned his back on the West, he began currying favor with Communist nations, particularly the Soviet Union and North Korea. He booted out the last of the French administrators, and most other French citizens left as well, taking with them much-needed technical know-how. Rumors circulated of up to 20,000 "disappeared" Malagasy. The economy began a long decline.

When Communist states began to fall in the late 1980s, a savvy Ratsiraka continued an about-face he had begun in 1979, when he'd introduced an "all-out investment" policy of industrialization, and tried to court the West. But it was too late. The economy had collapsed, and his desperate people were up in arms. In 1991, they rose up in largely peaceful demonstrations, calling for Ratsiraka's resignation. It was a long time in coming—and many Malagasy died in the process—but he was finally ousted in a general election in 1993, ending seventeen years of despotic rule. Several weak democratic governments followed in the mid-1990s before, astonishingly, Ratsiraka was voted back into office in early 1997.

Today, the Republic of Madagascar faces an uncertain future. The population is over 14 million strong, with more than half of Malagasy under the age of fifteen. With a growth rate in 1998 of 2.8 percent, the population will double by 2025. This is not good news for a country that is already one of the ten or twelve poorest nations on Earth, with a per-capita income of less than $250 and an external debt exceeding $4 billion. With eight out of ten people on the island farmers who grow principally rice and specialty export items like vanilla, cloves, and coffee, there is little chance that the Malagasy will ever earn enough to pay off such a crushing debt load.

Throughout Madagascar, the farming and herding way of life is spelling doom for the environment, a crisis that Alison Jolly deems "a tragedy without villains." The island shows how perilously close human beings can come to eating themselves out of house and home. Rampant deforestation over the centuries, coupled perhaps with certain natural forces, have left most of the island a sterile grassland, which Malagasy herders burn every year to bring forth the "green bite" for their livestock. So denuded is the Great Red Island, the sobriquet born of Madagascar's iron-rich soil, that in wintertime it looks in many places like the surface of Mars. With few trees to hold the red soil in place, huge erosion gullies gouge the hillsides like suppurating wounds; when the rains come, the soil runs like blood into the seas. (Astronauts flying overhead have said that Madagascar appears to be bleeding to death.)

Despite a bleak prognosis for any quick cures of the island's ills, a surprising amount of hope exists in Madagascar. The Malagasy people cherish their "land of the ancestors" so much that, whenever they leave it, many thrust a handful of soil in their pockets to ensure their safe return. If one of their kind dies overseas, they will do anything to retrieve the body for burial in the family tomb. A host of Western scientists and conservationists, and happily a rising number of Malagasy counterparts, have staked their careers and their dreams on saving this exceptional natural and cultural ark.

That hope is what this book is about.

Deep into a Primordial Land

THE PERFUMED ISLE

Behave like the chameleon:

Look forward and observe behind.

—*Malagasy proverb*

1.

IT IS MY FIRST NIGHT EVER in Madagascar, and just like that, my prayers are being answered. I arrived this afternoon on the island of Nosy Be off the northwest coast, and already I'm hiking on a rainforest trail with Christopher Raxworthy, a herpetologist from the University of Michigan, searching for "herps"—reptiles and amphibians. Moments ago, through the darkness farther up the trail, someone yelled, *"Uroplatus!"* I felt a surge of adrenaline, for *Uroplatus* is the generic name of the leaf-tailed gecko, one of Madagascar's most otherworldly creatures.

I'm in a somnambulant state, having just flown halfway around the world. From Boston, I flew across the Atlantic to Paris, through Zurich to Nairobi, then over the Mozambique Channel to Tana, where I boarded a 727 for the final leg here to Nosy Be ("Big Island"). From the island's sleepy airport, I had a forty-five-minute drive through plantations of vanilla, pepper, and *ylang ylang*, a fragrant tree that lends Nosy Be its sobriquet "the Perfumed Isle." Thirty-two hours after leaving Boston, I finally reached my destination, the strict nature reserve known as Lokobe. I was beyond exhausted, but how could I resist an inaugural foray into Madagascar's storied jungle?

I've come to help Raxworthy conduct the first-ever comprehensive survey of herps in Lokobe ("*loo*-koo-bay"). The team also includes two Malagasy graduate students, two Malagasy guides and cooks, and a crew of Earthwatch volunteers.* Our survey will help Malagasy wildlife managers better protect this last surviving patch of rain forest on Nosy Be as well as its menagerie of unique plants and animals. It is also the critical first step in answering questions of evolution that nag at biologists, including Raxworthy.

I hurry up the red dirt path. The two Malagasy guides and several of the volunteers already have their headlamps trained on an indistinct form clinging to a vine rising into the canopy like the Beanstalk. I crane my neck to see it, a whitish blob perhaps fifty feet up. One of the Malagasy fights his way downhill through thick vegetation to the base of the liana and begins shaking it.

"It's off!" someone yells, and just then I hear a plunk. One of Madagascar's most bizarre and wonderful experiments in evolution has just landed at my feet.

As Raxworthy gently lifts the leaf-tailed gecko off the ground, I whisper its scientific name to myself: *Uroplatus henkeli.* I've wanted to handle this creature ever since I saw a photo of one back in the States. With its gray, splotched skin, suction-grip feet, and crenellated, body-length fringe, the lizard looks like a lichen come alive. This reptile has taken the art of camouflage to extraordinary lengths. During the day, when it naps head-down on tree trunks, its color matches that of bark so well that, unless you happen to catch the bulge of its body from an angle, you won't see it even from steps away. The fringe, which even encircles the bulbous, night-vision eyeballs, keeps telltale shadows from forming.

"I think it's a male," Raxworthy says in his native Hertfordshire accent. "Big, bulged eye. Who would like to handle him?"

Before I can even open my mouth, one of the volunteers, a Briton named Garfield Dean, reaches over and carefully pries the gecko from Raxworthy's arm. As if affronted, the animal cranes its alien-looking head toward him, opens its mouth, sticks out a crimson,

---

*Based in Maynard, Massachusetts, the Earthwatch Institute funds scientific research expeditions around the world by signing up "volunteers" who share the costs and labor of the fieldwork.

grublike tongue, and screeches in his face: *Aaaaaaaaahhhhhhhhhh!
Aaaaaaaaahhhhhhhh!*

The sound sends chills down my spine, but Raxworthy reacts
without missing a beat. He starts screaming back at the lizard, trying to imitate its cat-loses-battle-with-a-screen-door shriek. The volunteers look at Raxworthy with a mixture of amusement and apprehension.

To the uninitiated, it might seem incongruous to find the foremost
herpetologist working in Madagascar today screaming at a lizard in
the dark of night. But Chris Raxworthy is a man who probably feels
most fulfilled when he's screaming at lizards. With his lean, six-foot-
four frame, his elongated hands and feet, and the roving gaze of his
walleye, Raxworthy could almost pass for an overgrown chameleon.
If he shaved off the shag of black curls on his head, the effect would
be complete. Watching him move through the forest, his large feet in
beat-up Keds stepping soundlessly through the leaf litter, and seeing
how lovingly he handles his herps, I have a fleeting thought that perhaps in a previous life he *was* a lizard.

"It's Malagasy tradition," Raxworthy says, the light from a cluster
of headlamps finding his smile. "It's bad luck to leave the animal's call
unanswered."

Thirty-one years old, Raxworthy has spent almost ten years docu-
menting the herpetological treasures of Madagascar, discovering and
collecting more species than any biologist before him. He has focused
on the island's "necklace of pearls," as German biologist Bernhard
Meier has dubbed Madagascar's string of nature reserves, as well as
other areas bearing an unusually high biodiversity, in order to pinpoint forests deserving of immediate protection. Though only twice
the size of New York's Central Park, Lokobe Reserve is just such a
gem, a patch of rain forest teeming with rare plants and animals, most
of them endemic to Madagascar and several endemic to the reserve
itself. When it comes to reptiles and amphibians, Madagascar, including protected areas like Lokobe, remains largely terra incognita, and
Raxworthy has become its Captain Cook. John Behler, a herpetologist with the Wildlife Conservation Society, says that a few field scientists working on Madagascar have "indefatigable drive, a seemingly
unquenchable thirst for discovery and the ability to survive—no,
flourish—under the most dreadful field conditions . . . Chris Raxworthy is one of them. He is without question the dean of herpetology on
Madagascar."

To become so, he has had to roll with the *fady*, or taboos, that dictate virtually every act in rural Madagascar. While working in the Tsaratanana Mountains in northern Madagascar, he agreed not to wear solid-black clothes, not to do anything on Tuesdays, and not to climb the island's highest peak without taking along a white chicken as an offering to the mountain gods. He has worked in areas where cattle thieves are hunted down and shot, and where tribes have killed foreigners whom they believe are "heart-takers." And he screams at lizards.

Raxworthy's nonchalance in the face of what many would consider quite serious impediments can be disconcerting. After returning from one of his epic collecting journeys in Madagascar, for instance, he wrote casually to a mutual friend, "Just returned to Michigan with a malaria fever that reached 106°F and lots of hookworms!" Or take the time a leech crawled into his eye. "It went round to the back and started feeding," he told me. "Oddly enough, I couldn't tell if it was feeding on the eyeball or on the eye socket." It was after dark and he was in the rain forest, an hour's walk from the nearest village, with a guide who spoke no English. "I thought, how the hell am I going to get this thing out? Normally we use DEET or a cigarette butt. Other people on the expedition had had them up their noses, and they just waited for the things to gorge with blood and drop out. But my eye already felt pretty uncomfortable." He and the guide ran through a downpour to the village, where an old woman, making sense of his smattering of Malagasy and French, poured water mixed with salt crystals into his eye, forcing the leech out.

"That," he said, with the ghost of a smile on his lips, "would make most people a little bit twitchy, don't you think?"

A lot of what Raxworthy does would make most people twitchy. For six months of every year he roots under logs, slogs through marshes, climbs mountains, and treks across deserts in pursuit of blind snakes, legless skinks, and, yes, screaming geckos. The leaf-tailed gecko and everything else we hope to catch on this expedition —tonight we've already snagged a thumb-sized chameleon like a toy dinosaur, a small gecko resembling a pale E.T., and a snake with the species name every other herp around here seems to bear (*madagascariensis*) —will wind up in a five-gallon jar back at our oceanside camp, embalmed in preservative. Later, the specimens will end up in museum drawers around the world. Conservation biology, I'm quickly learning, can call for a good deal of killing.

For all the terror that Raxworthy inspires among Madagascar's herps, however, he is really their champion. "Either you collect a few

specimens and document what's here and make a strong case for conservation," he told me, "or you leave all the animals here, don't touch
any of them, and watch the forest burn." In a land already stripped of
perhaps 90 percent of its original forests by Malagasy seeking fuelwood and farmland, in a land where national parks and nature
reserves like Lokobe protect less than 2 percent of what remains, conservation groups urgently need reliable information on species and
habitats in order to decide where to focus their limited resources.
Since 1985, when he cobbled together his first expedition to Madagascar as an undergraduate at the University of London, Raxworthy
has gone to unheard-of lengths to gather that information.

Reliable data on Malagasy herps are hard to come by. For one
thing, many species remain unknown to scientists. Raxworthy and his
University of Michigan colleague Ronald Nussbaum are currently
describing close to 150 new species they have turned up. That will
add a third again to the roughly 500 reptiles and amphibians biologists have already documented on the island. And particulars on those
500 known animals are sparse at best. Even simple lists of what lives
in the country's protected areas, in which Raxworthy has concentrated his efforts over the past decade, have proven woefully threadbare.
For example, even after a century of scientific collecting at Montagne
d'Ambre, a national park that Raxworthy surveyed with an Earthwatch crew in 1991–92, 56 percent of the herp species they captured
were new records for the park.

Indeed, by Raxworthy's reckoning, between 25 and 40 percent of
the published literature on Malagasy reptiles and amphibians is inac-
curate. For one thing, early collectors were not as thorough as modern biologists like Raxworthy and Nussbaum. Moreover, in the 19th
century, when many herp species were first described, many weeks or
months might have passed between the time a specimen was caught
and pickled in rum and when it arrived, often by way of a third party,
in the hands of museum curators who would do the actual describing
and cataloguing. "You can imagine a lot of species getting muddled in
jars," Raxworthy says.

One such species is *Zonosaurus boettgeri*. Even as I marvel at the
leaf-tailed gecko, I know that the real prize to be found in Lokobe
Reserve is this plated lizard. For all its star quality, the leaf-tailed
gecko is as common in Lokobe as a garter snake in the Catskills. But
only two specimens of *Z. boettgeri* have ever been found, both in the
1890s, and in the years since, both specimens have inexplicably vanished in Western museums. All that biologists like Raxworthy have

had left to go on in a search for this long-lost species is a mere mention of where each specimen was originally collected.

And that's where the problem lies. According to original notes housed in the natural history museum in Paris, the two specimens turned up on opposite sides of Madagascar, one on Nosy Be and the other near Vohemar on the northeast coast. Raxworthy has combed the eastern forests in vain for "butchery," as the species name is pronounced. "It may have occurred there and gone extinct, or it may have never occurred there, or it may still be there," he told me with the kind of apologetic inexactitude that crops up in most discussions of Madagascar's wildlife. Moreover, scant notes on the one found at Nosy Be in 1893 do not specify *where* on the island it was discovered or in what kind of habitat. Today, Nosy Be's rain forests outside of Lokobe are a thing of the past. When Raxworthy first arrived here a few weeks ago, he wondered, does *Z. boettgeri* still survive on Earth?

The answer came a few days ago. Achille Raselimanana, one of Raxworthy's graduate students, was fighting his way through a ship's cordage of vines near camp when he caught sight of a long-tailed green lizard scuttling up a tree. Before the reptile could vanish into the heights, Raselimanana fired one of the industrial-strength rubber bands Raxworthy uses to stun fleeing herps, bringing it down. Back at camp, Raxworthy confirmed what Raselimanana had bagged: the first *Z. boettgeri* seen by scientists in a century.

Now, as we head back down to our oceanside camp, the leaf-tailed gecko safely stowed in a cloth sack, I decide that in the week I will spend tramping around this reserve, I will be the one to find the *second* "butchery" seen in a century.

2.

CREATURES LIKE THE leaf-tailed gecko and *Z. boettgeri* exemplify the uniqueness of Madagascar's wildlife. The island exhibits an exceptional degree of both biodiversity (the sheer range of plants and animals) and endemicity (the state of existing nowhere else). Madagascar boasts more than 300 species each of reptiles and amphibians, for example, putting the island in the top five of the world's countries for herps. And a full 99 percent of them are endemic.

To varying degrees, the same holds true for other kinds of wildlife. The island harbors whole categories of mammals, to say nothing of individual species, that many people in the West may have never even heard of. Everyone is familiar with lemurs, the sprightly primates that are the poster creatures of the Malagasy environmental scene. But who outside of Madagascar has ever seen a tenrec or a fossa? Tenrecs resemble shrews and hedgehogs; fossa are elusive carnivores like the offspring of a mongoose and a mountain lion. Primates in particular tip the scales in terms of both biodiversity and endemicity. Madagascar has the third-highest rate of primate diversity in the world, after Brazil and Indonesia, which are 93 percent and 75 percent larger than it is, respectively. The island features over 50 kinds of lemurs, and every one of them—from the fat-tailed dwarf lemur to the golden-crowned sifaka—lives only there. Flora can match those numbers.

All told, biologists estimate that eight out of ten of all living things on Madagascar exist only there in all the world. How did this come to pass? Where did the island's distinctive menagerie come from? When? And how did its assorted members get there?

The matter might be solved with a good fossil record, which might enable paleontologists to trace the family history of, say, the Malagasy giant jumping rat right back through the ages to its most distant recognizable relative in perhaps the late Cretaceous period (146 to 65 million years ago). Yet the land-animal fossil record on Madagascar is all but a blank slate for the entire Cenozoic, our current era, which began about 65 million years ago when the dinosaurs breathed their last and mammals began to flourish. Cenozoic fossils that paleontologists have managed to dig up so far currently extend back no more than 26,000 years, the date of the earliest mammal remains, leaving the ancestors of the giant jumping rat and most other Malagasy beasts an enigma. The best fossil record on the island is that of the recently extinct megafauna—the elephant birds, giant lemurs, pygmy hippo, and so forth. But that catalogue gives few clues to evolutionary history. To complicate matters, Africa itself has virtually no mammalian fossil record for the first half of the Cenozoic, that is, the very period when lemurs were evolving from their primitive mammalian ancestors.

Even if a decent fossil record existed, there is the problem of absences. Is the absence of an animal from the fossil record an accident of fossilization or an indication that the animal never occurred there? This is an important point when seeking the biogeographic origins of species with such tangled histories as those on Madagascar.

In a recent paper, the paleontologist David Krause and two other researchers deemed the biogeographic origins of the island's modern land animals "one of the greatest unsolved mysteries of natural history."

That mystery is matched by another: how arriving species managed to speciate so wildly on the island. Today, Madagascar boasts fifty-one species and subspecies of lemur, and at least seventeen more once existed, all of them larger than any lemurs found today. At least sixty-seven species of primates coexisting on a single island—the number leaves biologists scratching their heads in amazement. The tenrecs are only slightly less prolific. Taxonomists have identified more than thirty species, from the bizarre aquatic tenrec to the streaked tenrec, which has been listed in the *Guinness Book of World Records* for the speed with which its kind reaches sexual maturity (thirty-five days). Chameleons exploded, radiating into sixty-two separate species, which represent over two-thirds of the world's complement. Madagascar's vangas, an endemic family of birds, rival Darwin's finches for diversity, with more pronounced morphological differences than beak size. There are 130-odd species of palm tree. Intense speciation can be relative. Take baobabs, for instance. Outside of Madagascar, these massive, bottle-shaped trees come in only one variety, a species native to Africa. Madagascar has eight species (including the African one). Such remarkable speciation is what one might expect of a continent, not an island.

On Madagascar, studies of speciation are even fewer than biogeographic studies. But the singular conditions attending on the island make it, as the herpetologist Charles Blanc wrote, "a vast natural laboratory particularly suited to research on speciation."

As the baobabs suggest, not all plants and animals on Madagascar speciated wildly. Many genera are known only from a few species. Scientists believe boas have been on the island for many millions of years, since they exist today only in Madagascar and the Americas. Yet the island has only three species of boa. Compare that to the island's colubrids, a cosmopolitan family of nonvenomous snakes, of which there are more than seventy species. There is a sound ecological reason for this: Only three niches were apparently available to boid-type snakes. For such species, Madagascar has served less as a continent in which to speciate than as an island refuge, "a sanctuary for species which have disappeared elsewhere tens of millions of years ago," as Blanc once wrote.

The wildlife's ancient pedigree shows up in the limited overall variety of creatures. Of primates, there are only lemurs. Of rodents, only cricetids (a family of small rodents that includes hamsters). Of carnivores, only viverrids (meat-eaters that preceded canids and felids, the dog and cat families). Madagascar has a glorious assortment of tenrecs, but they comprise the only kind of insectivore on the island. The world's artiodactyls, or even-toed ungulates, include swine, hippos, camels, giraffes, deer, antelopes, sheep, goats, and cattle. Yet of this broad group, Madagascar has but the bush pig, which scholars believe was introduced by the island's first colonizers, and the pygmy hippo, which is extinct. This same taxonomic poverty is found in all faunal groups.*

Then there are the unexplained presences. Like the boas, the iguanid family survives only in Madagascar and the Americas (along with representatives in Fiji and Tonga). Madagascar's national plant, the traveler's tree, counts its closest relative in Guiana. The *Myzopoda* bat resembles primitive bats of New Zealand and South America. The scholar Renauld Paulian might have been speaking for all these oddities when he wrote of *Myzopoda* that "the particular conditions and long isolation of the Malagasy microcontinent have provided this strange bat with a wholly exceptional evolutionary history."

Mysterious absences eclipse the mysterious presences. If you look at a map and see how close the island is to the continent that gave it birth—Cap Saint André on the west coast is but 266 miles from Mozambique, roughly the distance from Boston to Philadelphia—you'd think that Madagascar must share a similar assortment of creatures with Africa. Such is not the case; most of Africa's signature wildlife is altogether missing. Madagascar has no elephants. No rhinos, tapirs, horses, or other odd-toed ungulates. The continent's vast herds of antelopes and bovids, including duikers, dwarf antelopes, horse antelopes, gazelles, reedbucks, wildebeests, impala, and African buffalo—not there. Africa's great carnivores are nowhere to be found. You won't see lions, cheetahs, or other cats on Madagascar, nor foxes, jackals, or wild dogs. There are no hyenas. No genets and civets (except the introduced palm civet). No weasels.

The list goes on. Lemurs abound because the more advanced primates—the monkeys, baboons, and apes—never reached Madagascar. Nor did the rabbits, hares, and pikas. The squirrels, porcupines, springhares, and naked mole rats. The hyraxes. The pangolins. The aardvark. There is not a single toad, salamander, or caecilian (a bur-

---

*Madagascar's flora, by contrast, shows a corresponding richness that makes it one of the world's floral hotspots for its size.

rowing amphibian resembling a worm). Africa's widespread agamid and monitor lizards, and the small lizards of the family Lacertidae: missing. Adders and vipers: missing. Cobras, pythons, and mambas: missing. Whatever kept all these creatures out also kept out fish. Common African families, including those that icthyologists know as characids, cyprinids, mormyrids, and polypterids, cannot be found. Amazingly, there is no such thing as a native primary freshwater fish on Madagascar; that is, there are no fishes native to the country that are physiologically restricted to freshwater. Even some of Africa's most characteristic birds, which are the great dispersers of the animal kingdom, are absent. Madagascar, for instance, has no trogons or barbets, no hornbills or honeyguides, no oxpeckers or woodpeckers.

Madagascar's wildlife is characterized, then, by a dizzying assortment of creatures found nowhere else on the planet, including a host of large animals that vanished forever sometime in the past two millennia. A striking dissimilarity with the fauna of nearby Africa, and a zoo's treasure of floral and faunal curiosities whose nearest relatives lie on the other side of the planet. Finally, poverty at the family and genus level, which is more typical of islands, and extraordinary richness at the species level, which is more characteristic of continents. (Raxworthy, for one, thinks one should consider Madagascar a "continental island.")

How did all this happen? What caused such uniqueness? "If you think about it, you're trying to put together a scenario of what happened ten, twenty, maybe fifty million years ago," Raxworthy told me. "It's horrendously complicated." Nevertheless, he has been working to solve that mystery for ten years. His medium is the herp.

3.

"THE MALAGASY HAVE a lot of rumors about this snake," Raxworthy says now, cinching down the mouth of a large cloth sack with one hand while reaching in carefully with the other. It is the morning after my arrival at Lokobe Reserve, and Raxworthy is giving the Earthwatch team a live-animal briefing at our oceanside camp. The talk should give us a good feel, he says, for the spectacular radiation that has occurred amongst Madagascar's herps.

"When missionaries translated the Bible into Malagasy, they made this the snake that tempted Eve," he continues. I can see why.

The large brown-and-yellow snake he pulls out *looks* evil. It is the menarana, and it bears the pugnacious upturned snout that gives hognosed snakes their name, a long forked tongue like an animated strand of linguini, and a pair of shiny black eyes that, due to a fold of skin pressing down from the forehead, makes the serpent look downright pissed off.

"That's one mean-looking head," I say.

"I think he's got a sweet head," Raxworthy responds as he eases the snake back into the bag.

Glancing at our leader, I think, *He's blinded by love.* Raxworthy handles his "animals," as he calls them, with the tender touch of a veterinarian examining a kitten. Even while catching them with his outsized hands, he is remarkably gentle. Last night I spotted a bright-green gecko high on a palm leaf and began suggesting all manner of surely doomed ways to catch it. Almost in a whisper, Raxworthy said, "Let's just let him come along here," as he effortlessly ushered the fleet creature into his hands. He might as well have coaxed a squirrel into his lap.

As Raxworthy digs around for another live specimen, I look out over the Mozambique Channel. Aquamarine shading to dark blue, it stretches away from the beach we're on like the Indian Ocean, of which it is the westernmost arm. Our camp sits just back from the beach, facing south off Nosy Be. The mountainous west coast of Madagascar rears up about ten miles away. To the southwest, I can see Hellville, Nosy Be's main town, nestled in an estuary a short distance away. Somewhere over the horizon to the west lie the Comoros Islands and, beyond them, Africa. It was in the deep waters off the Comoros where, fittingly enough for a region with such an ancient pedigree of isolation, a fisherman in 1938 hauled up a coelacanth, a "fossil fish" previously thought extinct for 60 million years.

Behind us rises the redolent, tangled mess of rain forest, like a living fortress. A single trail, the one we took last night, climbs through the forest to a viewless summit about 1,500 feet up. We've pitched our camp on the seaward edge of the reserve, which covers the entire Lokobe Peninsula on the southeast corner of Nosy Be. I saw this emerald peninsula on the flight in on Air Mad, as everyone affectionately abbreviates the name of the national airline. On either side of Lokobe—that is, on the rest of Nosy Be and on the nearby island of Nosy Komba (Lemur Island)—I could see a land laid waste, rain forest replaced by drab secondary forest and savanna.

"If you look out, you'll see it's all a horrible, hideous mess," Raxworthy says, as if reading my thoughts. He points across a narrow strait to Nosy Komba, which looks brown and sere, a far cry from the deep green surrounding us at Lokobe. Beyond it I can see the mainland, gray with smoke from the annual springtime fires. We stand within a microcosm of Madagascar: a tiny island of original beauty in a rising sea of degradation.

Visitors have long perceived Nosy Be as just such a paradise, often to their peril. In 1649, a century and a half after Europeans first set eyes on Madagascar, the English set out to establish a colony on Nosy Be to increase trade and convert the heathen. It was their second attempt; their first, on the southwest coast at Saint Augustine's Bay a few years before, had ended in disaster. But Col. Robert Hunt, head of the new party, held out great hope for Nosy Be. "I do believe, by God's blessing," he wrote in a pamphlet published posthumously, "that not any part of the World is more advantagious [*sic*] for a Plantation, being every way as well for pleasure as for profit, in my estimation."

His estimation was off. Early in 1649, he set sail for Nosy Be with a group of settlers in the *Assada Merchant*. (At the time, Nosy Be was known as Assada.) The following year, in June, two additional ships left England to reinforce Hunt's party. But as they neared Nosy Be, the two ships encountered a small boat bearing a few desperate colonists, the only survivors of Hunt's expedition. For some reason, Hunt had decided to settle not on Nosy Be but on a nearby island he called Goat Island (probably Nosy Komba). There he and many other colonists apparently died from fever or attacks by hostile Malagasy. The rest sailed on the *Assada Merchant* for the mainland, where several were lured ashore with promises of trade and slaughtered. Perhaps they lost the ship to the Malagasy, or perhaps they decided they were unable to handle the *Assada Merchant* on their own. But the few remaining colonists, surely with little hope, fled in a small boat, and only by chance met up with their incoming compatriots.

Surprisingly, the would-be colonists decided to give it another try. The two ships dispatched a party onto Nosy Be, with supplies to last six months and under the command of one of Hunt's original crew, a Major Hartley. But history repeated itself. Anyone straying from the encampment wound up dead at the hands of warlike Malagasy, while others succumbed to "the contagion of the place." Hunt's dream of a plantation had become a nightmare. After seven weeks, his successors abandoned the settlement and sailed to India, thus ending English imperial plans in Madagascar for centuries to come.

Nosy Be remained a valuable place, however, for English and other foreigners to procure slaves. And that's how an English youth named Robert Everard, on his first journey to sea, wound up being stranded alone on the island, abandoned by his shipmates and left to the mercy of the none-too-friendly inhabitants. He chronicled his remarkable story in a delightfully ingenuous book he wrote upon his return, *A Relation of Three Years Suffering of Robert Everard, upon the coast of Assada near Madagascar, in a voyage to India, in the year 1686. And of his wonderful preservation and deliverance, and arrival at London, Anno 1693.*

Everard's troubles began almost as soon as the *Bauden,* the slaver he had joined, left London on August 5, 1686. Pirates boarded the ship on the high seas, and a fierce battle ensued on deck. Everard hid himself until the fighting ceased and all became quiet. When he emerged he found a well-armed pirate "almost dead upon our poop, with a fuzee, an axe, a cartouch-box, a stinkpot, a pistol, and a cut-lass." The captain, to whom Everard had been apprenticed, and four others lay dead nearby, and sixteen others were wounded, "whereof one was myself," Everard wrote. "So by the brave courage of our cap-tain and men the pirate was forc'd to leave us." The ship sailed onto Bombay, India, where its merchant owner decided to continue the voyage to Madagascar for slaves. "I told the merchant, I had a mind to go home to my own country, my master being kill'd," Everard wrote. "[B]ut he told me, I should not go home till the ship went home."

Alas, the *Bauden* would return long before Everard ever did. Arriving at Nosy Be, the new captain sent ashore a shore party, including Everard, to begin negotiations for trade. The local Mala-gasy, for their part, dispatched a delegation to the ship. "[T]he cap-tain having victuals brought him, asked if they would eat, but they refus'd," Everard later wrote, "and before he had done eating, they drew out their lances from underneath their clouts, and cut his throat from one ear to the other. . . . " The Malagasy also killed the mate and the purser before being driven off the ship. Meantime, Malagasy on the beach attacked the shore party, killing every man save Everard; perhaps he was spared because of his tender age. His captors thrust him into a small hut with the village slaves, but in the morning he was shoved out to fend for himself.

Later that day, after firing a series of rounds at the village, the

Bauden set sail. The crew likely thought that none of their compatriots had survived, but in any case they didn't stick around to find out. Later, Everard learned from one of the king's men, who spoke some English, what had transpired aboard the *Bauden* and why they had been betrayed. Apparently the Malagasy were taking revenge for the treachery of an English ship that had previously "play'd the rogue" with them.

Everard paid dearly for that treachery. The king had let him live, but what a life it became. He was free to go about as he pleased, but that meant fending entirely for himself in an alien environment, with neighbors whom he felt might do him harm at any moment. He subsisted on yams and potatoes he dug out of the ground with a stone, and on occasion crabs, fish, and sea turtles he managed to capture. He had no tools, no containers to hold food or water, no lodging, not even any clothes. With rain falling sometimes for "three months together," he slept on the bare ground beneath a tree "naked as ever I was born, lying under the water which came down upon me, for I could not help it, having no other covering but the branches and leaves of the tree." He suffered terribly from sores that broke out all over his body, and once, while forced to accompany the king on a slave raid to a neighboring village, "I was taken light-headed, so that sometimes I fell down, and all the skin of my back was burnt off as raw as a piece of beef, so that I could scarce travel nor stir myself, but with much difficulty, and extream [*sic*] pain."

One wonders at Everard's pluck. Couldn't he have built himself a shelter, fashion himself some clothes? The historian Mervyn Brown has gone so far as to question the veracity of his account, but his tale smacks of the truth. Everard survived on Nosy Be for two years and nine months before an Arabian slave ship purchased him from the king for twenty dollars. Though he was treated civilly and was eventually sold to an Englishman at Muscat, a town on the Arabian Peninsula, it took him several years and much journeying around the Indian Ocean before he finally arrived back home in Blackwall, England, "where I met with my father, to the great joy of us both." It had been seven years since he had left home.

In the ensuing centuries, Nosy Be continued to deny foreign attempts at long-term settlement. A boatload of Indians shipwrecked on the island in the seventeenth or eighteenth century built a small settlement out of coral and lime, but its jungle-covered ruins near Lokobe attest to its mortality. In 1839, as the Merina from the Highlands continued to gain control of Sakalava lands, a Sakalava queen fled to Nosy Be, where she asked the French for help. Seeing an opportunity, Adm. Anne-Chrêtien-Louis de Hell of Réunion—he after whom Hellville is named—signed a treaty of protectorate with the Sakalava royalty. Two years later, the Sakalava ceded both Nosy Be and Nosy Komba to France. The French maintained garrisons here throughout the nineteenth and twentieth centuries, until they left after granting Madagascar its independence in 1960. Today, a new kind of colonizer has taken over the island: international hoteliers, who run the string of beachside hotels on the island's western shore.

By contrast, where Raxworthy has set up camp on Nosy Be's south-eastern shore remains a sleepy holdout of Sakalava fishermen, per-haps descendants of the very Malagasy who did "mischief" to Hunt and Everard. These Sakalava hold strongly to tradition. For example, it was *fady* (taboo), Raxworthy learned after the fact, to build the camp's pit toilet in the forest. (Locals go in the ocean.) After we leave, local Sakalava plan to hold a ceremony to appease the affront to the ancestors. This morning on the beach, I saw personally how deeply age-old beliefs remain in force here. A fisherman had dragged a shark he had caught in his nets onto the beach. With a crowd of villagers silently watching, he cut off the dorsal fin and pushed the half-dead fish back into the sea. Through an English-speaking Malagasy, I asked why he took only the fin. Though "finning" is common the world over, I was surprised that in a country as poor as Madagascar the rest of the meat would be discarded. He said it was fady to eat the species, which legend held once rescued a drowning fisherman. The fin would grow back, he said. But surely the shark will die, I object-ed. He paused, then replied, "No, it will only get weak."

Jørgen Ruud, a Norwegian ethnographer whose 1960 book *Taboo: A Study of Malagasy Customs and Beliefs* remains the classic in the field, says that fady are "hammered in" from early childhood. It is impossible for *vazaha* (foreigners) traveling through Madagascar to stay on top of local fady, as they differ from village to village, even family to family. Some families even keep them secret. But their power is absolute, for they come from the ancestors, and anyone breaking a taboo makes the entire village *maloto* (unclean). Fady help hold together the moral fabric of society. As Ruud notes, "the taboos in all their many fields show a consistent tendency to point a damning finger at the contrast to the ideals which the individual and the group hold in all the different walks of life." So deeply ensconced are fady in the Malagasy world that the word for "excuse me" and "please" is *azafady,* which literally means "May it not be taboo to me."

Fady collected around the island over the past century and a half give an idea of how deeply superstitious the Malagasy are. One should not sing while eating, for it causes the teeth to grow long, while to skin a banana with the teeth results in extreme poverty. Never eat while lying down; your parents will be choked, just as kicking the walls of your house will kill your parents. A soldier should never eat tenrec, lest he assume the timid, shrinking disposition of an animal that typically rolls into a ball when attacked. Pregnant women must not consume eels for fear that the embryo will become as slippery as the eel and slide out of her body. Nor should she ever step over an ax: The child will become bowlegged. A newborn baby should be called ugly and likened to a pig or a dog so that it is not hated by ghosts, which will follow you at night if you whistle, shout, or carry mutton. Never point at a tomb or your fingers will fall off, and never build a tomb at the end of a valley, for it will draw the living toward itself. And so on.

Now, as Raxworthy returns yet another herp to its cloth sack, he tells us how he tried to explain fady to John Mittermeier, the eight-year-old son of our logistics coordinator.

"He wasn't getting it, so I said there are lots of fady in the States, only we don't call them fady," Raxworthy says. "He said, 'No there aren't.' I said, 'Yes there are. There are a lot of things you don't do because it's fady.' He said, 'Like what?' I said, 'Like if you wanted to take a pee in the classroom or the girl's bathroom, it would be fady.' And he sort of thought about it for a second and said, 'Yeah.'"

The volunteers' chuckles fill the air. We're standing in Raxworthy's field laboratory, where the only sounds come from lapping waves on the beach and the occasional *hoot-hoot* of black lemurs back in the forest. The lab consists of a crude wooden table under a blue plastic tarp tied to overarching beachside trees. Beneath the table lies the "library," a worn green knapsack stuffed with books on Madagascar's natural history, and jugs of formalin in which Raxworthy preserves his specimens. The formaldehyde's acrid smell mixes in the warm air with the rich organic aromas from the sea and forest. Along a thin rope strung between branches, a dozen or so small canvas bags dangle from their string ties like medieval moneybags. This is Raxworthy's detention center, where herps brought in from the forest wait their turn to be anesthetized and put into formalin for transport back to the universities of Michigan and Antananarivo, the two repositories for his specimens. Or they are released, depending on whether he has enough individuals of that species, usually five or six for each repository, to cover for age, sex, size, and other so-called intraspecies differences.

Raxworthy pulls out another snake. It is one of Madagascar's blind, burrowing snakes known by the evocative generic name of *Typhlops*. At first glance, the only physical feature distinguishing this wriggling, pinkish creature from a worm is the pair of tiny black dots at one end—the vestigial, unseeing eyes. Raxworthy thinks it might be a new species.

"The first *Typhlops* we found here was almost black with prominent eyes," he says. "This one is very pale with almost no eyes at all, and it has a very interesting white blotch here. The other one wasn't anything like as long as this one and was a bit more stocky as well."

Raxworthy's passion for herps goes way back. At an age when most children were playing with blocks in the nursery, he was busy catching newts behind his parents' house in Saint Albans, Hertfordshire. By the time he was fifteen he had assembled a private menagerie of frogs, lizards, and snakes. His talents as a taxonomist surfaced while he was earning a degree in zoology at the University of London. During summer vacations, he had been assisting archeologists in a Viking-era excavation at Repton, England. Handed a trove of tiny bones thought to be the remains of birds, he took them back to the university and identified them as the bones of frogs and toads. Not content to stop there, he went on to show, in a scientific paper eventually published in *Herpetological Journal*, how a change in the ecology of the area around the site since Viking times caused the fortunes of one of those species to decline and of another to rise.

He returned to his beloved newts for his doctorate, studying the courtship behavior of the smooth newt, *Triturus vulgaris*, at England's Open University. "You can't observe this in the field, because these places are dark and murky and everything's happening underwater," Raxworthy told me. So most of the work took place in the lab, but only after he had traveled to Italy, Yugoslavia, and Turkey to collect specimens—and get bitten by the fieldwork bug.

Now, as Raxworthy returns yet another victim to its pouch, I ask him whether he's disappointed that the island has no newts.

"No," he says. "Madagascar has plenty of other weird and wonderful things."

As if to illustrate his point, Raxworthy pulls the not-seen-in-a-hundred-years *Z. boettgeri* out of a sack. He's saved it for last.

"Now this is mind-blowing, cosmically mind-blowing, wouldn't you say?"

I realize with a start that Raxworthy is gazing at me, even though he's facing the others. With his walleye, he can look at you without turning his head, an act that leaves you feeling he is always a step ahead.

"He is a wise old lizard, isn't he?"

It does look wise, in the same way that Madagascar's signature chameleons, with their strangely human eyes, appear wise. I take mental note of the lizard's every feature for the search image I will employ over the coming week: dark green skin with yellow spots, long toes, a ten-inch body with a tail half again as long.

4.

THE EARLIEST INVESTIGATORS to look into the biogeographic origins of creatures such as *Zonosaurus boettgeri* knew that the singular nature of Malagasy wildlife owes much to Madagascar's long isolation far out in the Indian Ocean. Yet precisely how it all happened was the cause of protracted debate.

The most prevalent explanation of the mid-nineteenth century was the notion of a lost continent. The idea appears to have arisen initially as an attempt to explain the presence of lemurlike animals in Africa and Asia as well as on Madagascar. Lemurs are only found on the latter, but Africa has pottos and bushbabies and Asia lorises—all lemurlike prosimians, an early form of primate. Together, all these animals constitute the suborder Strepsirhini within the order Primates. How else to account for the presence of these related creatures in such distant places than to believe that those lands were once somehow attached to one another? In the April 1864 issue of the *Quarterly Journal of Science*, Philip Sclater, secretary of the Zoological Society of London, summed up the then-current thinking on this notion, even making so bold as to give it a name. Sclater declared that

the anomalies of the mammal fauna of Madagascar can be best explained by supposing that anterior to the existence of Africa in its present shape a large continent occupied parts of the Atlantic and Indian Oceans, stretching out towards (what is now) America to the west, and to India and its islands on the east; that this continent was broken up into islands, of which some have become amalgamated with the present continent of Africa, and some, possibly, with what is now Asia; and that in Madagascar and the Mascarene Islands [Mauritius, Réunion, and Rodrigues] we have existing relics of this great continent, for which, as the original focus of the *stirps lemurum,* I should propose the name Lemuria!

In the absence of a better idea, Lemuria took hold in the scientific imagination. Coming sixty years before Alfred Wegener proposed the theory of continental drift, and perhaps a century before the scientific community began to accept it, the idea of a conveniently placed continent that then conveniently vanished without a trace seemed a plausible if hardly defensible explanation. In 1878, the geologist Henry Blanford read a paper before England's Geological Society in which, based on similarities between fossil plants and reptiles in India and South Africa, he suggested that a continent formerly connected those two lands. Four years later, the ornithologist Gustav Hartlaub, basing his argument on the presence of a number of Malagasy birds of obvious Indo-Malayan affiliation, characterized Lemuria as "that sunken land, which, containing parts of Africa, must have extended far eastward over Southern India and Ceylon [Sri Lanka], and the highest points of which we recognize in the volcanic peaks of Bourbon [Réunion] and Mauritius, and in the central range of Madagascar itself—the last resorts of the mostly extinct Lemurine race which formerly peopled it."

Even the great Victorian naturalist Alfred Russel Wallace, the codiscoverer with Charles Darwin of the theory of evolution by natural selection, apparently bought initially into the idea of a lost continent. In his 1876 book *Geographical Distribution of Animals,* Wallace wrote, concerning the period before higher mammals had arisen, that "Madagascar was no doubt united with Africa and helped to form a great southern continent which must at one time have extended eastward as far as Southern India and Ceylon. . . ." (The similarity between his phrasing and Hartlaub's is striking.) But something soon changed his mind—possibly even an attack Hartlaub made on him in a paper published a year after Wallace's book appeared. In his next book, *Island Life,* published in 1880, Wallace strenuously backed off from any support of the notion of Lemuria:

I have gone into this question in some detail, because Dr. Hart-laub's criticism on my views has been reproduced in a scientific periodical, and the supposed Lemurian continent is constantly referred to by quasi-scientific writers, as well as by naturalists and geologists, as if its existence had been demonstrated by facts, or as if it were absolutely necessary to postulate such a land in order to account for the entire series of phenomena connected with the Madagascar fauna, and especially with the distribution of the Lemuridae.

Wallace, who conducted the first in-depth investigation into the origins of Madagascar's flora and fauna, was uncomfortable with the thought that an entire continent could simply vanish, Atlantis-like, beneath the Indian Ocean. For one thing, soundings showed that parts of that ocean were extremely deep, reaching 2,400 fathoms. Could a large landmass subside more than 14,000 feet down into the abyss? Wallace felt there was a far more reasonable explanation for how Hartlaub's Asian birds got to Madagascar: island-hopping.

A slew of islands lie scattered between India and Madagascar. Beginning with the Maldives southwest of India, these include the Chagos Islands, the Seychelles and Amirantes group, Aldabra, the Cargados, and finally the Mascarenes. Extensive shoals and coral reefs, features that in Wallace's day were thought to indicate subsidence, surround many of these islands. Wallace felt that in ancient times some of these isles may have been in size "not much inferior to Madagascar itself." While those deep soundings proved that these islands were never connected, they would have provided a line of communication for birds of "comparatively easy stages" of 400 or 500 miles.

While dispensing with Lemuria, Wallace's theory did not explain how lemurs got to be in both places. Wallace believed that whereas small animals such as insects and even lizards might accidentally hitch a ride from one land to another on floating rafts of vegetation washed out of rivers, mammals could not do so; they were either too big or not hardy enough to survive the long voyage. Obviously lemurs didn't get to Madagascar from India via the stepping-stone method. Then how did they?

Wallace worked from the same principle of subsiding land to suggest that Madagascar was once much larger—so large, in fact, that it was actually attached to Africa. The lemurs and other curious Malagasy mammals came to Madagascar that way, he concluded, though without much confidence. One can feel his struggle to make sense of a world in which the notion of continental drift was still decades away. In *Islands*, he wrote that "a land connection with *some* continent was undoubtedly necessary, or there would have been no mammalia at all in Madagascar; and the nature of its fauna on the whole, no less than the moderate depth of the intervening strait and the comparative approximation of the opposite shores, clearly indicate that the connection was with Africa." While not explaining precisely how a united Africa and Madagascar might have come to be separated, Wallace felt sure that this is what happened. Even if he was right, however, he still needed to explain how lemurine mammals came to exist in such widely separated parts of the world.

Then he had an insight. He came at it in part by considering a highly unusual creature known as the solenodon. This foot-long, shrewlike mammal is native to Cuba and Hispaniola (the island now comprised of Haiti and the Dominican Republic). Yet its closest living relatives, indeed the only animals in the world to share with it the family Centetidae, are the tenrecs of Madagascar. The solenodon, Wallace noted, further added "to our embarrassment in seeking for the original home of the Madagascar fauna." Must Madagascar, he wondered, have been formerly attached to these West Indian islands in order to account for this family's highly divergent distribution? "We might as reasonably suppose a land connection across the Pacific to account for the camels of Asia having their nearest existing allies in the llamas and alpacas of the Peruvian Andes," he wrote, "and another between Sumatra and Brazil, in order that the ancestral tapir of one country might have passed over to the other."

Wallace had a better idea. The Centetidae, the Strepsirhini, and the Viverridae (which includes the fossa, ring-tailed mongoose, and other endemic carnivores of Madagascar) had very ancient evolutionary pedigrees. Fossils of tenrecian, lemurine, and viverridian animals had been unearthed that dated back to Eocene times (65 to 57 million years ago). Moreover, those three groups, whose greatest concentration of survivors can be found today on Madagascar, once had a far greater distribution, because fossils of all three groups had been dug up in Europe. Wallace hypothesized that

> all extensive groups have a wide range at the period of their maximum development; but as they decay their area of distribution diminishes or breaks up into detached fragments, which one after another disappear till the group becomes extinct. Those

animal forms which we now find isolated in Madagascar and other remote portions of the globe all belong to ancient groups which are in a decaying or nearly extinct condition, while those which are absent from it belong to more recent and more highly developed types, which range over extensive and continuous areas, but have had no opportunity of reaching the more ancient continental islands.

Wallace had hit upon an idea that would come to gain great currency with biologists and remains to this day a cornerstone of their understanding of species distribution: vicariance. The statement above is a fairly cogent description of the theory, which holds that seemingly incongruous distributions of closely related animals, such as the lemurs and lorises, or the tenrecs and the solenodon, can best be explained by supposing that these animals formerly lived across a far greater portion of the planet than they do now and, in the fullness of time and in most cases because of overwhelming competition from more highly evolved species, died out in intervening areas, leaving the widely dispersed survivors we see today.

Vicariance was a key insight for Wallace, but the one idea that might have put it all together for him was that of continental drift. Others had proposed that the world's continents may have once been linked. For centuries people had remarked on the seemingly close fit, for example, between the east coast of South America and the west coast of Africa. Yet no one had any reasonable idea of how such huge and far-flung landmasses could have been attached.

Until the meteorologist Alfred Wegener, that is. He based his theory on the newly established principle of isostasy, which holds that the Earth's crust lies on material that behaves like a fluid. Isostasy gave continents an "ocean" on which to drift. Wegener first proposed his theory in 1915, but it wasn't until 1924, when he published *The Origin of Continents and Oceans*, that he built a strong argument that, for the first time, made the theory sound plausible. He marshaled an army of evidence, including ancient climates, distributions of fossil and living plants and animals, and that age-old, enigmatic "fit" between widely separated lands.

Like most great ideas, Wegener's hypothesis was too earth-shattering, in this case literally, to take hold immediately. In fact, it was not until the 1960s, when the study of plate tectonics satisfactorily explained how continents drift at the continual urging of volcanic upwellings along the midocean ridges, that Wegener's theory found general acceptance.

In the meantime, Lemurian theories persisted. In 1887, the pale-ontologist Melchior Neumayr, based on the distribution of fossil ammonites—flat, spiral shells of extinct mollusks—had suggested that not only did Lemuria exist, but so did a "Brazilian-Ethiopian continent" stretching right across the Atlantic Ocean. Neumayr was one of the most influential of the "lost-continent" adherents. "Such was the need felt for land links to account for a whole range of faunal and floral similarities," wrote the geologist Anthony Hallam in the 1960s, "that Neumayr's interpretation was widely accepted by Euro-pean geologists during the succeeding few decades, including some of the leading figures of the time."

Forty-five years later, the paleontologist Charles Schuchert decid-ed that ancient biogeography fully indicated that land bridges crossed both the Atlantic and Indian Oceans until the end of the Cretaceous. But he couldn't ignore the isostatic problem, so he proposed much smaller land bridges across both oceans. This proposal was not "suffi-ciently radical" (in Hallam's view) for the geologist Bailey Willis, who took issue with Schuchert's claim that these bridges were made of continental crust. The same year, 1932, Willis proposed "isthmian links" between Brazil and Africa, and between Africa, Madagascar, and India, that were not leftovers of continents but rather thrust up by volcanism. These links, he felt, had subsided by the end of the Mesozoic era (245 to 65 million years ago). Willis's view gained popu-larity among biogeographers of the day, but subsequent research indi-cated that the submerged ridges and volcanic islands that he invoked as surviving evidence of these links are, in fact, much younger in age, indeed, entirely post-Mesozoic.

As plate-tectonic theory advanced, however, so did biogeographic possibilities for how lemurs and Malagasy animals might have made the passage, sans a lost continent. In an argument published in 1952 that put one more nail in Lemuria's coffin, the paleontologist and evo-lutionary biologist George Gaylord Simpson strenuously invoked the idea of a "sweepstakes" crossing of mammals between Africa and Madagascar, in which only a few mammals made it across the ever-widening Mozambique Channel:

> Available evidence indicates that there have been about a dozen or fewer successful colonizations of mammals in Madagascar by dispersal from Africa during a span of some 75 million years and that these colonizations have been scattered at random through-out that time. Probability of such dispersal must then have been

exceedingly low, or, in other words, there must continuously have been a highly effective barrier to [the] spread of mammals between Africa and Madagascar. Existence of a land connection is almost impossible in view of these considerations, and it can be taken as practically proved that a sea barrier existed throughout the Cenozoic. Such dispersal as did occur, because of its low probability and random timing, is positive against, not for, a land dispersal route. Existence of such a route would have made dispersal of many groups practically certain and determinate, which is contrary to the given biogeographic data.

By the mid-1960s, evidence from oceanic rift structures, paleomagnetism, and other sources finally gave Wegener's theory the upper hand, and the idea of lost continents was buried for good. Scientists today generally concur that in ancient times, a single continent known as Pangea existed on Earth. In time, that broke into two supercontinents, a northern one known as Laurasia and a southern one called Gondwana.* Again, in time, both those supercontinents broke up. Laurasia divided into the Northern Hemisphere continents of North America, Europe, and Asia. Gondwana fractured into the Southern Hemisphere continents of South America, Africa, Australia/New Zealand, and Antarctica. India, now a part of Asia, was a member of the Gondwana club until it drifted north and rammed into what is now Tibet about 50 million years ago. In the ancient supercontinent, Madagascar lay sandwiched between Africa and India before it, too, went its separate way.

Scholars have debated particulars of the breakup of Gondwana since Wegener, who felt that Madagascar and Africa must have been connected in the middle of the Tertiary period (65 to 1.6 million years ago), when lemurs evolved, in order for them to make it onto the island before it drifted away. (Like Wallace, Wegener did not believe that mammals, even ones as compact as lemurs, could disperse across oceanic barriers on floating mats of vegetation.) Theories that Madagascar was closer to Africa in the early Cenozoic than it is today, or even attached to it, have been proposed as recently as the 1970s.

A bevy of recent studies relying on advanced methods and sophisticated instruments have drawn a widely corroborative picture of the weaning of Madagascar from mother Africa. These studies have

* Ironically, the name "Gondwana" was proposed around the turn of the century by a leading proponent of lost continents, the geologist Edward Seuss.

brought to the fore a mountain of evidence, including data on seafloor spreading, pole positions, the distribution of early plants and marine invertebrates, and petrology and geochemistry. The portrait shows that Madagascar abutted present-day Somalia, Kenya, and Tanzania before tectonic forces began wrenching it away in the mid- to late Jurassic, perhaps 165 million years ago. For tens of millions of years, it drifted southeast attached to India, in a minicontinent known as Greater India. Sometime in the Cretaceous period—current suggested dates range from 124.5 to 133 million years ago—Madagascar reached its current position. It remained attached to India for tens of millions of years before the subcontinent split off about 88 million years ago, to begin its long drift north to Asia. So for at least 88 million years, Madagascar has been left entirely alone, a chunk of Gondwana preserved down to the present day.

5.

IN THE TWO DAYS since Raxworthy introduced us to our study subjects on the beach, I've started to get a feel for our small slice of that chunk. I've assisted the Earthwatch volunteers in checking traplines, searching randomly for herps, and laying down a 50-by-50-meter (165-by-165-foot) grid in the forest for a population study of stumptailed chameleons that Raxworthy plans to begin tonight. Until now, we've traveled largely as a group, including during a hike this morning to the reserve's 1,500-foot summit. Because the volunteers have *paid* for the privilege of doing this, I've let them do most of the work while I've stood by and watched.

But I've decided that if I am to have any hope of finding "butchery," I have to learn the ropes and familiarize myself with this rare, captivating forest. So this afternoon I head off alone. What could be more enticing than having four or five hours to explore a largely unexplored rain forest? I concur wholeheartedly with the great early botanist of Madagascar, Richard Baron, who opined that "there is in traveling in the forest a strange and fascinating illusion, a vague feeling of expectancy, which persistently recurs, in spite of disappointment, that somewhere on in front something of exceptional interest will be found."

I begin by checking the four traplines, which a previous Earthwatch team laid down on a series of lateral ridges rising like waves on the flank of the peak we climbed this morning, high above our camp. This involves walking along a 100-meter (330-foot) transect, along which the volunteers have sunk plastic buckets to their brims every 10 meters (33 feet). A calf-high fence of black-rubber webbing stretches the length of the line, bisecting the top of each bucket. Tied to wooden stakes, with its bottom edge forced several inches deep into the ground, the fence is a formidable barrier for a herp. In theory, an unsuspecting creature coming up to the fence will work its way along it until the animal stumbles into the bucket's maw. It will be trapped, unable to escape either by scrambling up the bucket's smooth sides or jumping the foot or so to its brim.

The work leaves me sweating like a marathoner on a humid day. I'm deep in the rain forest, but on these humpback ridges the trees thin out just enough to let a fair wallop of tropical sun through. The sweat on my nape is gooey with twiggy bits from working through the low-slung vegetation. You could easily put an eye out in here. Looking down the steep hill through the leafy canopy, I can see tiny puzzle pieces of blue sea shimmering in the sun. Oh, for a swim.

At each bucket I drop to my knees in the soft soil and scrape through the leaf litter that invariably collects in the bottom of each bucket, searching for prisoners. At first you hope you'll find something unusual, perhaps a new species or, yes, even *Z. boettgeri.* As you come upon one empty bucket after the next, though, you begin wishing for anything at all. As the late naturalist Gerald Durrell wrote about working his own traplines in Madagascar, "one's hopes are only buoyed up by the thought that surely there will be *something* in the next trap." But I find little of consequence in the traplines, so for the rest of the afternoon I do "opportunistic searching," as Raxworthy calls it. That is, I wander randomly and see what turns up.

I put all of our tools to use. When I spy a smooth-skinned skink sunning itself on a log, I stand stock-still, gently remove a fat, five-inch-long rubber band coiled around my wrist, and shoot it at the culprit, hoping to stun it long enough to catch it. No luck; it vanishes beneath the log. So I dive at the log, slashing it with my "stumpripper," a herping tool like a ski pole with its tip bent ninety degrees. Skinks and blind snakes will head straight down beneath such refuges, and to catch them you have to destroy the log and dig furiously downward in a matter of seconds, otherwise they're gone—as they all were for me, of course. I have no success with the stumpripper, and the only thing I catch with the rubber band is my eye when I release it wrong-end first.

Even my supposedly superior brains don't seem to help. At one point I come upon a bright-green day gecko on a vine. I have it seemingly trapped on lianas: Every time it goes up or down I block its path. I think, *You're mine.* It does seem to have few options. I watch as the animal works it out, well aware of its predicament. Its amber eyes look this way and that. Then, just when I am about to make my move, it leaps onto my shoulder and from there onto a nearby tree, whence it escapes.

Duped again, by a lizard.

When I do succeed in catching something, I'm to place it either in a cloth sack (for snakes, chameleons, and other reptiles) or a plastic bag with a splash of water (for frogs), and bring it back to Raxworthy. I'm also supposed to record date, time, habitat, and other circumstances surrounding the capture. Back at our makeshift lab, Raxworthy photographs the live specimens before immobilizing them with anesthetic, which subdues and eventually kills them. He then fixes them in 10 percent buffered formalin, which is formaldehyde gas in solution in water. Once injected into the body, formalin triggers cross-chains to form between proteins, causing the corpse to go stiff as wood. Later, researchers can study the arrangement of organs, examine the animal's last meal, check on its reproductive condition, and so on.

Next, Raxworthy ties a numbered tag onto the hind leg of each specimen. The tag will allow researchers for years to come to consult numbered entries in his field notebooks, which he will deposit along with the specimens at his target repositories. He also removes liver and muscle tissue from representatives of most species and freezes them in liquid nitrogen for later molecular work to study evolutionary relationships. Because variations occur between individuals of a given species' population, just as they do among people, Raxworthy tries to collect between five and ten specimens of each species. A single specimen may not be representative of the population as a whole, what with young and old, males and females, and plain old variety.

After an hour, I sit down for a rest. The reddened soil is dry and hard, a good seat. In the rainy season, this whole hillside must be as slippery as ice. Mopping my brow with the filthy sleeve of my T-shirt, I take a swig from my water bottle. The water is warm but still refreshing. You have to drink continuously in this environment not to get dehydrated. I sit perfectly still, soaking up the charged, pleasantly oppressive silence. Neither the faint *whoosh* of the sea far below, nor the miscellaneous insect and bird calls breach the bastion of stillness around me. The closeness of this forest seems womblike.

There's something about such private, unpeopled places that triggers a spiritual rush in me. I feel more in touch with the natural world in these moments than in any other. In my brain, I keep a neural photo album of such spots to be flipped through at will: a clear pool in which I took a swim one hot afternoon in an Indian rain forest; a stretch of old logging road along a trout stream in northern Pennsylvania; a tiny waterfall among sun-warmed boulders high in the Himalayas. I'll have to add a snapshot of this silent scene to that album.

All at once I realize it's not silent at all. In fact, it's deafening. I turn my head this way and that, gauging the enormity of the sound. It's like the whirr of an engine, one simultaneously right by your ear and incongruously far away for so insistent a sound. It's the cicadas. The sound, according to Baron, "is not a buzz-z exactly, and it is not a hum-m. It is a deafening, unceasing, rasping, irritating monotone." The insects start so quietly that you don't notice them and increase in volume so smoothly that you don't catch on for some time. In the midst of luxuriating in the silence, I've found myself thinking, all in the space of seconds, Hmm, it's so quiet, it's really *quiet,* how come it's so quiet, why is it so damn LOUD?, before realizing that the cicadas have got me again. You forget about them until suddenly you realize the whole forest is shaking with the intensity of their noise, and your head with it.

Then, just as suddenly, they fall silent. You remember what true silence is. And you begin luxuriating in it anew until you're caught off guard again. "Don't be like the cicada, whose voice fills the whole valley, though the creature itself is not a mouthful" goes a Malagasy saying. Sometimes the surprise is physical, as the cicadas, perhaps spooked by something, suddenly take off en masse, screaming like banshees, only to settle back down and fall silent moments later.

Madagascar's locusts, which also occur in this forest, are even more infamous swarmers. When it comes to the migratory locust, *Aedipoda migratoria,* the air can suddenly be so filled with them that, as an early observer once put it, "one cannot see one's way." An English missionary described another such onslaught: "When preaching at Ananalamahitsy, the western door, facing the pulpit, was wide open, and swarms of locusts swept by the church; and as the sun was shining brightly at the time, the light glancing on their outspread wings gave them exactly the appearance of a heavy fall of snow, except that the ground was not white." After they have swept through, it is not green either. "[T]hey eat all the herbage where they are deposited, and do not leave any verdure on the land, which they leave as if the fire had passed there," wrote another witness. Locust plagues can destroy crops for tens of miles around within a matter of hours.

The Malagasy make the best of such bad situations—by eating their attackers. "[L]ocusts are esteemed a great delicacy by the natives, indeed what a frog is said to be to a Frenchman, so is a locust to a Malagasy," observed an English visitor in the nineteenth century. "We have not yet tried the dainty dish, and don't know when we shall make the experiment. Only fancy, fried or stewed locusts! well, well, we have no need." One writer for the *Madagascar Times*, a newspaper published in Tana in the last century, did make the experiment. "If you cook them in lard, they mostly taste of lard; if you cook them in butter, they taste of butter," he reported. "I cooked mine without anything, and they tasted of that."

Cicadas and locusts may be bothersome, but they're harmless. And that's true of Malagasy wildlife in general. Madagascar harbors scorpions and black widow spiders, but its isolation has ensured few other dangerous animals. Except for crocodiles, all of the potentially lethal beasts one equates with nearby Africa—large flesh-eaters like lions, excitable herbivores such as elephants or rhinos, and poisonous snakes—evolved too late to make it onto the island.* About the most threatening critter in the Malagasy menagerie is the mosquito, which can give you malaria. In every rain forest I've ever explored there was always *something* to be afraid of. Not in Lokobe. Ever since I arrived, I have felt unexpectedly, delightfully above the natural law. You can grab at anything!

Well, almost anything. As I soak up the sun filtering through the canopy overhead, I suddenly laugh out loud, remembering Garfield Dean's story of his close encounter with the one animal at Lokobe you don't want a close encounter with. He was using a World War II entrenching tool that Ron Nussbaum, the Michigan herpetologist who is visiting our camp for a few days, loaned him. Dean was wielding the device like a stumpripper, tearing apart a termite mound high in the reserve, looking for skinks and burrowing snakes.

"At the top of the nest there were very few termites, but toward the bottom it was just raining termites," he told me. "Right at the bottom I saw this scorpion. I thought I'm damned if I'm going near that. I sat back to have a rest, and all these *Zonosaurus* lizards came in from

*Several Malagasy snakes, such as the odd-looking leaf-nosed snake, are mildly venomous, but the fangs of those species are tiny and face rearward at the back of the throat.

all directions to mop up the termites. It was incredible: You'd only see one *Z. madagascariensis* every half hour or so, and yet there were about ten in front of my eyes.

"Then I saw one that looked as though it was eating a worm. I thought, Damn, that might be a skink. So I went in and picked the thing up. What I'd picked up was the scorpion's tail. Felt a complete sod, really."

The scorpion stung him twice in the thumb. Ever the amateur scientist, Dean paused long enough to poke the creature into a plastic bag for later identification. Then he raced back to camp, where he learned that he had been lucky. He'd been stung by a fat-tailed scorpion; in rare cases, people have died from the sting of the thin-tailed variety. The Malagasy in camp told him there were three things he could do to stanch the pain. He could put the scorpion in alcohol, then apply the brew of dissolved scorpion juices to the wound. He could mix local Malagasy rum with local tobacco and rub that in. Or he could take his live scorpion, boil it for twenty minutes, and guzzle the juice. Dean settled for a beer.

Now, as I resume my search, I suddenly catch sight of the last thing I expect to see. No, it's not *Zonosaurus boettgeri,* alas. It's a leaf-tailed gecko. These guys are almost impossible to discern during the daytime, when they blend into the bark of the tree they're sleeping on like snow on a glacier. The only reason I see this one is that I notice its body bulging from the smooth trunk as I step past.

Even though its eyes are wide open—leaf-tailed geckos have no eyelids—it appears to be asleep. At least it hasn't moved a muscle since I swung into view. Cherishing the chance to marvel at this famously elusive character before capturing it for Raxworthy, I bring my face slowly toward it, until I'm just inches away. It is while I'm admiring the leaf-shaped tail that I catch a slight movement in my peripheral vision: The head is rising off the bark. Slowly, ever so slowly. The animal, I realize, has likely been watching me from the moment I first appeared. Understanding now that its camouflage has failed, it begins plotting its next move.

I don't give it the opportunity. With the memory of too many lost lizards in my head, I fall back on the one strategy that has worked well for me: lunge and pray. In a flash the gecko is in my hand, its mouth agape, its clawed feet raking my fingers. After taking a moment to complete the admiration that the animal's awakening interrupted, I gently peel him off my hand and stuff him into my cloth sack.

On the way back to camp, the gecko and several other reptiles thumping gently against my side, I feel a pang of guilt, knowing that these beguiling creatures will soon be dead. I hate to kill anything, but here I am consigning my favorite animal in Madagascar to death by lethal injection. And what if I find "butchery"? It is possibly the rarest lizard in the world. Can we really justify trading its life for science?

I feel mixed emotions as I add my three cloth sacks to the string of sacks dangling in the field lab like hanged men. Why does Raxworthy take so long to process his herps? Some of these sacks, their bulges hinting at the size of the creatures within, have hung here for days; one of them holds Achille's "butchery." I look at the watercooler that contains Raxworthy's specimens; it is packed to the brim with "fixed" herps. Doesn't he have enough already?

After a swim in the ocean to scour off hours of sweat, I join Raxworthy on the blue tarp that serves for meeting, dining, napping, you name it. Night falls quickly in the tropics, and soon stars appear high over the Mozambique Channel. Raxworthy fires up the kerosene lantern, which casts its garish glow through the trees and out to the wavelets tipping softly onto the shore not thirty feet away.

"Chris, how can you justify killing that *Zonosaurus boettgeri*?" I ask. I think I know what he's going to say, but I have to ask anyway. "Why can't you just take a photograph of it and let it go? I mean, what if it's the last one alive?"

Raxworthy darts his walleye at me without turning his head. He seems surprised at my question.

"Well, the first thing, Peter, is the population. Lokobe is over 2,400 acres of primary rain forest. It's inconceivable that you're dealing with the last animal in the reserve. At the absolute minimum, the population will be in the hundreds, if not thousands. If it was so low as to be one or even a handful of individuals, then the species is almost certainly doomed to extinction anyway. With reptiles and amphibians, there are almost no cases known in the world where populations have gotten to that size, except for really tiny, rocky islets. And with a normal population, the turnover from predation and potentially disease is typically very high, so we've really had no impact on the natural population."

Clearly he's answered this kind of question before. He settles back on his hands and looks up at the overhanging trees. A few volunteers have joined us on the tarp.

"So," Raxworthy continues, "assuming that taking the animal isn't going to harm the natural population, what's the justification for doing it? For one thing, it provides evidence that the species does still exist. There's no question about interpretations or results; we've actually got the specimen. Second, it will be valuable for research work. We can look at its gut contents and aspects of its morphology, we can compare it with other species of *Zonosaurus* to work out relationships. And that specimen will be available for study for hundreds of years."

"If it's not lost," I say.

"Right. Finally, conservation. We can firmly say, yes, this is the only reserve in Madagascar with *Z. boettgeri* in it. It is an extremely unusual lizard, it is arboreal, it is completely unlike the other plated lizards in Madagascar. By taking that specimen, we can provide good propaganda, if you like, to justify protecting the whole population."

He aims that errant eye directly at me.

"So for the specimen itself it is bad news but, for the population as a whole, taking that specimen will probably assure its long-term survival by protecting its forest."

We're interrupted by dinner, a tasty fish-and-rice concoction whipped up by Angeluc and Angelin Razafimanantsoa, the twenty-something identical twins who serve as the team's guides and cooks. Raxworthy met the twins several years ago while surveying herps on Montagne d'Ambre, the rain-forest national park in the far north where they live. Now, wherever he goes in Madagascar, he takes them along. Slight and boyishly handsome, the brothers are a perfect blend of African and Indonesian. They have identical, highly infectious laughs that rend the air every few minutes from dawn till late at night. The only way I have any hope of telling them apart—and I'm often wrong even then—is to look at their teeth: smoking has slightly yellowed Angeluc's.

Suddenly I feel better. The hot meal and two bottles of Three Horses, the national brew, surely have something to do with it. But mostly it's what Raxworthy said. His logic is incontrovertible. I certainly have selfish reasons for wanting to find the second *Z. boettgeri*—the thrill of discovery, the wish to please our leader, pride—but I've known in the back of my mind that it might also help safeguard this gem of a place. In fact, finding "butchery" might be more important than finding any other species at Lokobe.

As we ready ourselves for our after-dinner project—the chameleon population study—I realize my resolve to secure Number Two has been redoubled. And the fact that we'll be working tonight in the very patch of forest where Achille found Number One only heightens my eagerness.

We'll be doing the third type of research Raxworthy brought us here to do: work the grid. We laid down the grid two days ago on a narrow, forested plain just along the beach from our camp. Here Raxworthy will study *Brookesia stumpffi,* a chameleon no bigger than your thumb. The word "chameleon" comes from the Greek *chamai leon,* dwarf lion, but this species dwarfs the dwarfs. *B. stumpffi* is an ideal subject for the study, because it is plentiful and easily found where it roosts on shrubbery a few feet off the ground. By finding virtually every *B. stumpffi* within the grid, Raxworthy hopes to estimate the population size of this species of stump-tailed chameleon within Lokobe Reserve. It will be only a ballpark estimate, because different parts of the forest, including areas at higher altitudes, likely bear more *B. stumpffi* than others. Nevertheless, Raxworthy can use the data, which are difficult to come by for species more cryptic or less numerous, to compare Lokobe to other rain-forest sites on the mainland. The findings will give him a rough idea of what healthy populations of at least this one species require in the way of space, habitat, food, and so on. The data will also serve as a baseline against which to measure future changes in the reserve's population of *B. stumpffi.* Conservationists and wildlife managers use such comparable information to guess by extrapolation what other rain-forest species might need as well.

"Aleffa!" cry the twins. "Let's go!" We switch on our headlamps and head along the beach toward the grid. With the beams slashing the darkness like headlights in fog, we enter the forest through a narrow opening, where an erstwhile stream pours out onto the beach during the rainy season. We aren't but a dozen steps in when Angeluc (or is it Angelin?) exclaims, "Manditra!" He has found one of Madagascar's three species of boa, coiled on a low-hanging branch over the now-dry streambed. An auspicious sign.

When we get to the grid, ten of us line up along one edge of it, an invisible line running through the jungle over uneven ground. It is pitch-black save for the personal sphere of light given off by our headlamps, which blanch the green leaves pushing up into our faces. The only sound is the hushed *thwump* of waves hitting the beach half a mile away through the trees. The air smells of living forest, dead leaves, salt-sea air, body odor. With an image in my mind of a little white dinosaur on a twig—I'm trying my damndest not to think of *Z. boettgeri*—I step into the grid.

In theory, working the grid is simple. You move from one side of the marked-off square to the other, leaving no leaf unturned in your search for *Brookesias*. In practice, it is exhausting. To ensure that I don't miss any *Brookesia* on my watch, I make my way as deliberately as a sloth. The chameleons, I know, rest just a foot or two off the ground, so I bend over double as I maneuver gingerly through the tangled branches. After a long day already spent fighting through dense foliage, that quickly becomes painful, and I keep tripping on unseen vines running along the ground like wires on a dark stage. I only see them, wrapped tightly around my ankles, when I pick myself up off the ground after a fall. So I kneel down and hobble along on my knees. The close, humid air under the canopy is sauna-hot, and I'm soon drenched in sweat.

"*Brookesia!*"

Another chameleon, the eighth or tenth of the night. Isabelle Constable, a graduate student at the University of Michigan who serves as our logistics coordinator, makes her way over to record the animal in her notebook and, before releasing it, mark it with a dash of Wite-Out. That way, when we repeat the sweep tomorrow night for accuracy, we'll know if we've already counted that chameleon.

Every few minutes, a different voice yells out "*Brookesia!*" It has become both a mantra and a cry of victory, though I do wonder what local Sakalava might think of our catching *Brookesias*. Malagasy throughout the island harbor an intense fear of these harmless creatures. "It would be better to trample a divinity than trample a *Brookesia*," a proverb warns. The Malagasy naturalist Guy Ramanantsoa believes that "the somewhat exaggerated fear of [the] *Brookesia* results from the risk of crushing it inadvertently, and thus from unconsciously harming it." Since Malagasy ethics forbid harming any animal without justification, this has resulted in, he says, "a kind of collective psychosis." The Bara call the tiny reptiles *andro*, which is a synonym for bad luck. The Betsimisaraka name for them is *Ramilaheloka*, which means "He who seeks to condemn, or he who seeks to make guilty." Here on Nosy Be the Sakalava know them as *tsiny*, which is also the word for a forest genie. At the same time, the Sakalava also hold the *Brookesia* in high regard, Ramanantsoa writes, for legend says that the chameleon once saved a Sakalava ancestor from the attack of a "cannibal monster."

Looking off through the trees to get my bearings, I can see beams sweeping the darkness like the flashlights of a search party, which we are, after all. In between calls of "*Brookesia!*" come the simultaneous

guffaws of the twins whenever Achille cracks another joke in Malagasy. As I search, I can't help but secretly hope *Z. boettgeri* will pop
out of the underbrush. The first one was found during the day and
high up a tree, but maybe they come down at night to feed or sleep.
No one has any idea, of course; the species' behavior is completely
unknown.

"*Pardalis!*" cries Angeluc (or is it Angelin?). The twin has found a
hefty green chameleon. It is the size of a guinea pig, and it is truly
outlandish, with its curved scaly back, coiled tail, and swiveling, gun-
turret eyes. The first Europeans to lay eyes on this curious creature
harbored strange ideas of what it could do. Etienne Flacourt, author
of the famous 1661 *Histoire de la grande Isle Madagascar*, wrote that "if a
native happens to approach the tree where it hangs, it instantly leaps
upon his naked breast and sticks so firmly that in order to remove it
[he is] obliged, with a razor, to cut away the skin also." Not remotely
true, nor is the long-held notion that it lives on air. William Finch, an
English merchant, solved the mystery by carefully watching one he
brought aboard his ship during a stop at Saint Augustine's Bay in
1608: "On perceiving a fly sitting, he suddenly darts out something
from his mouth, perhaps his tongue, very loathsome to behold . . .
with which he catches and eats the flies with such speed, even in the
twinkling of an eye, that one can hardly discern the action."

Throughout this long night, many other herps end up in our
sacks. I find many myself: a small leaf-tailed gecko, *Uroplatus ebenaui*,
a space alien on all fours. Its famous screaming cousin, *Uroplatus
henkeli*, standing bowlegged on a tree trunk, tail up, bug eyes staring
me down (I leave it alone). And a fish-scale gecko. When I catch sight
of this critter scooting among the rocks of the dried-up streambed, I
quickly grab it. Which is when I learn the reason for its name. In one
hand I see a sticky mess of silver-brown scales; in the other lies the
gecko, a naked, gooey, red-orange blob.

Long after midnight, I finally crawl into my tent. Shoving aside
my junk—daypack, camera tripod, roll of toilet paper, books (*Faith
and the Good Thing* by Charles Johnson, *Muddling Through in Madagascar* by Dervla Murphy, a guidebook)—I collapse onto my foam pad.
What a day—a good day. I now feel comfortable with the tools and
techniques of herping, and I'm finally beginning to feel at home in
the forest. I could leave Lokobe right now with a range of memo-
rable experiences. There's just one thing left to do—*find that damn
"butchery"!*

6.

NO ONE KNOWS WHEN the ancestors of *Z. boettgeri* or any of Madagascar's other native wildlife reached the island. All that can be safely said is that different groups arrived at different times — and suffered different fates. Some creatures, such as the boas and perhaps the chameleons, likely never had to touch water, but rather already lived on the chunk of Africa that slipped away and became Madagascar. The boas may not even have evolved much from the time the two lands were united in Gondwana more than 165 million years ago to the present day, while the chameleons may look appreciably different from their Gondwana forebears.

Then there are the later arrivals, those whose ancestors must have arrived long after the island rafted to its present spot. The lemurs are among that group. Paleontologists know the prosimians couldn't have hitched a ride on Madagascar when it was weaned from mother Africa in the Jurassic period, because they didn't exist then. Indeed, the fossil record indicates that even when Madagascar had finally reached its present position, about 100 million years ago, another 40 million years still needed to elapse before primates of the same evolutionary level as lemurs would even appear on Earth. Did the ancestors of Madagascar's prosimians arrive in a single event, say a cyclone that washed a stand of African rain forest out to sea carrying a troop of lemurine creatures? Or did such animals come on two or more occasions? Though biologists consider it unlikely, it is theoretically possible that all sixty-seven types of extinct and extant lemurs so far identified on Madagascar arose from a single pregnant female that crawled ashore 40 or 50 million years ago.

Like digging into memory, the farther back in time one goes, the less precise one can be about when the plants and animals found on Madagascar today first got there. We know that eucalyptus and Mexican pine appeared around the turn of this century, when they were introduced for reforestation purposes. *Raketa*, as the Malagasy know the prickly pear that infests the south, is also exotic to Madagascar. Among the animals, it is known that humans, who have only lived on the island for two millennia, brought over cats, dogs, sheep, goats, and cattle. Perhaps unintentionally, they introduced rats, the palm civet, and at least one species of shrew. They probably were responsible for the bush pig, though it's possible that it got there on its own from Africa. Finally, by taxonomic reasoning, paleontologists figure the pygmy hippopotamus, which is closely related to African hippos,

must have come in the late Cenozoic, probably by swimming
the Mozambique Channel. Perhaps, like the first lemurs, it inad
tently got washed out to sea in a storm and pushed by winds or cu
rents to Madagascar, where it lumbered ashore and got on with its life
in an unfamiliar land.

Without a good fossil record, scientists must base arguments on
arrival on the relatedness of Madagascar's creatures with those elsewhere, both those alive today and those no longer with us. Most
authors believe that the island's carnivores and rodents arrived in the
mid-Cenozoic, while the lemurs, tenrecs, elephant birds, and the
aardvark-like mammal *Plesiorycteropus* probably came in the early
Cenozoic, if not, for some of the groups, the late Cretaceous. In
minority opinions, it has been suggested both that the carnivores
arrived with the earliest animals and that the Malagasy rodents may
be survivors of a diverse rodent fauna that made it onto Madagascar
at the beginning of the Age of Mammals some 64 million years ago.

Where Madagascar's wildlife came from is as enigmatic as when it
came. One look at a map will have you thinking that most or all of the
island's plants and animals must have arrived from nearby Africa
rather than from India or Indonesia, more than 2,000 and well over
3,500 miles away, respectively. And indeed, despite the imbalance in
species, genera, and even families between the island and the continent, most scholars concur that the bulk of Madagascar's flora and
fauna originally did come via Africa, through either a vicariant or a
dispersive process.

Yet there are those curious presences to contend with. The Asian
connection is very strong. Did the Malagasy bats, frogs, mollusks,
and birds with Indo-Malayan affinities arrive directly from there? Or
did they come by way of Africa? *Pteropus*, the genus of the flying fox,
a giant bat seen by visitors to Madagascar's Berenty Private Reserve
winging silently over the trees at twilight, clearly flew from Asia.
Species of *Pteropus* have successively occupied western Indian Ocean
islands, including the Seychelles, Aldabra, the Mascarenes, and the
Comoros Islands (though somehow *Pteropus* never reached mainland
Africa).

The Asian connection that has baffled biogeographers the longest
concerns the lorises, the lemurlike mammals of India and Southeast
Asia. Africa, as we've seen, also has lemurine animals, the pottos and
bushbabies. But does that mean that Madagascar's lemurs originated
there? Might they have arisen in Asia? The herpetologist Jean-
Claude Rage has suggested that adapids, now-extinct close relatives

lemurs themselves, might have come through India
here they reached a cul-de-sac. Adapid fossils have
ous places in Laurasia, the northern supercontinent,
contested fossils of adapids have ever been found in
es one to consider the possibility, Rage contends, that
the lem... e it to Madagascar from Laurasia via India, though he
offers no means as to how they could have made the traverse.

The American connection is as striking as the Asian. The boas,
the iguanas, the rere turtle, the traveler's tree, the *Rhipsalis* cactus:
Did these animals and plants come from the Americas? If so, how?
Did they travel to Madagascar through Africa from South America
when all were connected as part of Gondwana? Or did they come
through Laurasia with the lemurs? Rage includes the boas and iguanas in his India-to-Madagascar argument, with the same caveat of not
knowing how they might have made the crossing.

Then there is the global connection. The most striking example of
this is Madagascar's extinct elephant birds. They have surviving
flightless relatives in South America (rheas), Africa (ostriches), New
Guinea (cassowaries), Australia (emus), and New Zealand (kiwis).
Until a few hundred years ago, they had another cousin in New
Zealand's moas. Where did the ratites, the group to which all these
birds belong, originate? Alfred Russel Wallace addressed this question in his *Geographical Distribution of Animals*. While he had no firm
answer, he described two possible ways that ratites might have gotten
to where they are today. Their distant ancestors may have been present worldwide during supercontinent times and simply walked to the
lands where they exist today, evolving into different forms as these
lands became isolated from one another. Or their ancestor might have
been capable of flight and flown to its respective new homes, where,
perhaps for lack of predators and long isolation (in the case of Madagascar and New Zealand, anyway), it lost the ability to fly and
diverged into the nonflying forms we see today.

Supercontinents may hold the answer to where many of Madagascar's relict denizens originated. Yet the picture is far from clear. In
spite of its ancient heritage, for example, Madagascar has neither
marsupials (the opossums of the Americas, the kangaroos of Australia, for example) nor monotremes (the echidnas of Australia).
Why? The most helpful way to look at both when the island's plants
and animals arrived and where they came from is to look at *how* they
might have come—a question that lies at the heart of current studies
on Madagascar of biogeographic origins.

7.

BIOGEOGRAPHIC ORIGINS are arguably the last thing on our minds as we head off to a slide show being given one night in an abandoned soap factory next to our Lokobe camp. The talk will be all about the present and future, not the distant past. The speaker is Josephine Andrews, a slim, bright-eyed Scottish anthropologist who, since 1991, has run the only ongoing research program at Lokobe Reserve: the Black Lemur Forest Project. She has had more contact with people living around Lokobe than any other vazaha. As the slide projector beam casts the factory's ruined hulk in an eerie, submarine light, Andrews enlightens us about the complexities surrounding Lokobe's preservation. Like so many of the world's fragile ecosystems, the reserve is caught between encroaching development and the people who call it home.

Nosy Be is Madagascar's premier tourist destination, coveted for the beautiful beaches lining its west coast. But while the strip of modern hotels there rake in income, Andrews tells us, outlying villages receive only the scraps. The cruise ships that dock regularly at Hellville disgorge hundreds of wealthy passengers, who are often woefully uninformed of local culture. Since Lokobe is closer to Hellville than the beaches are, the tourists inevitably end up sweeping through nearby villages on their way to the reserve, snapping pictures and leaving little behind except bad feelings. Children miss school to beg, young women offer themselves as prostitutes, and old people just sit and stare, saying the exploitation reminds them of French colonial days, which effectively ended only in the 1970s. And the tourist influx is likely only to increase. The island currently has 220 hotel rooms of international standard; by the year 2000, officials plan ten times that number. Workers recently expanded the Nosy Be airport to take 747s, and the first international flight is due in next month from Johannesburg.

"What do these very, very poor people think when they see all these rich people coming in?" Andrews asks rhetorically, her face lit cadaverously from below by the projector's bare bulb.

Andrews first came to Lokobe in 1988 to do a two-month behavioral study of the black lemur, a handsome species found in Lokobe and in a few nearby places on the mainland. But a series of talks she gave at local schools convinced her to greatly expand her efforts.

"When I showed slides of lemurs in other parts of Madagascar, they'd say, 'My auntie came from there' and 'Oh, they've got those things there?' And I'd say, 'Yes, and they *only* come from there.' They knew more about African and European animals than their own."

Most rural schools still use textbooks the French brought from Europe years ago, and there is little money to pay teachers. At Marodokana, the village closest to our camp, 140 schoolchildren go to class in shifts, Andrews says, because there are only two teachers.

What do local people think, then, of Lokobe Reserve? To find out, the next morning I pay a visit to Julien Mohamed Jules, a young Malagasy who lives with Andrews in a two-room house next to the soap factory. Ocean breezes blow through the cloth-covered doorway, and several puppies play on the concrete floor. Julien sports dreadlocks and a muscle T-shirt with SAFARI emblazoned on it. A mechanic from the mainland, he has lived on Nosy Be for six years. A few years back, he gave up a job in Hellville to assist Andrews on her project and to make one of the first local attempts to serve as a tourist guide. Most locals remain "a little frightened of vazahas," Andrews says, but Julien has learned enough English to take the few visitors who come specifically to see Lokobe—even from the proscribed distance—around the periphery in pirogues.

Julien feels more comfortable speaking in French than English, so Andrews translates for us. Younger people, Julien confirms, are "taking the mentality" of Westerners by ignoring taboos and other traditions. He says local Malagasy know the value of the forest "in general"—it is one of their main sources of water, for one thing—but not such that they would keep it at any cost. Educating children about Lokobe is the way to encourage conservationist thinking, Julien says, so they might grow up to become forest guides.

Andrews concurs. "It's not a deathly grim conservation message at that level, but just one of instilling pride in the local environment," she says. "If the kids are really into it, then the adults will switch on as well."

But education must be managed carefully, Julien adds, to avoid a glut of guides and ensure that earnings get funneled back into the community.

Lokobe is currently classified as a strict nature reserve, which means it is only open to scientific research teams (and journalists they kindly invite along). But plans are afoot in the Malagasy government to change Lokobe into a national park, which allows for greater access—and much-needed tourist revenue—but must be managed properly. Visitors can put undue pressure on wildlife and forests, and Lokobe already contends with exploitation by Malagasy living nearby. In my peregrinations near the edge of the forest, I have come upon stumps of freshly cut trees, which locals take to build houses and dugout canoes. Furthermore, neighborhood boys readily admit to collecting the leaf-tailed gecko and other herps for the international pet trade. When Raxworthy pressed them on how many leaf-tailed geckos they thought they had collected, they guessed 40,000 over the past six years. Even allowing for gross exaggeration, the strain on the *Uroplatus* population has to be tremendous. For each leaf-tailed gecko, Raxworthy told me, foreign middlemen pay the boys a few hundred Malagasy francs—less than twenty-five cents—then sell them in the States for up to $250 a pair, a three-orders-of-magnitude markup.

After my first few days at Lokobe, I had begun to wonder how anyone familiar with this natural museum could even consider opening it up to hordes of tourists. Blinded by Lokobe's beauty, I had stepped into a trap that well-meaning Westerners are prone to fall into: pondering conservation out of context.

After talking with Andrews, Julien, and others about Lokobe, however, I have had to reexamine my position. Thinking back on the hardscrabble villages I saw through the smudged windows of the Land Rover on the drive in from the airport, I realize that Nosy Be's villagers have only what little an annual per capita income of just over $200 can provide. For the first time, it strikes me that it might be better if Lokobe were to become a national park, despite the increased impact. Tourist revenue would potentially soar, and if the park was managed well, local people could reap sustained economic benefits.

Because many people consider the reclassification a fait accompli, Andrews has worked hard to help locals prepare for it, primarily by promoting village cooperatives.

"We've had a great deal of enthusiasm," she says. "It hinges on two things: trust among the village communities, and educating tourists on responsible behavior."

To build the former, Andrews has helped create village committees bearing, she says, "a voice from all corners of the community"—local politicians, elders, mothers, children. To begin educating visitors, she has printed flyers describing local customs and is trying to raise money from the U.S. Agency for International Development and other funders to build a tourist center staffed by local people.

Andrews's tactics mimic those increasingly used by Malagasy and Western aid and environmental groups. They have a buzzword for it: integrated conservation and development. Since 1992, the National Association for the Management of Protected Areas, or ANGAP, a

Malagasy nongovernmental organization, has set aside half of all park entrance fees for local people's use. "This would apply to the people around Lokobe if it became a national park," Sheila O'Connor, the World Wildlife Fund's chief representative in Madagascar, told me. "Other revenue-sharing schemes and direct economic benefits would depend on the local people's willingness to be involved, and on the operator involved in preparing and executing a management plan."

As I leave them, Andrews says, with an almost visible shudder, "It's a weird time, *before* all the tourists come."

8.

THERE ARE ONLY TWO ways that the ancestors of the flora and fauna Andrews would like to protect at Lokobe and elsewhere got to Madagascar. Either they were there already when the island drifted away from Africa. Or they arrived later. It's that simple.

If only it were that simple.

Vicariance may suffice as an argument for certain relict species — the boas, the iguanas, possibly the elephant birds. But vicariance doesn't cut it for Madagascar's mammals and many other vertebrates, for the simple reason that they hadn't evolved yet. The latest thinking is that Madagascar broke free from Africa at least 150 million years ago and has rested in its present spot for at least 124 million years. Yet mammals only took off after the dinosaurs disappeared 65 million years ago, and animals akin to lemurs did not arise until about 58 million years ago. How, then, did everything that didn't yet exist when Madagascar went its separate way get to the island?

By dispersal. They may have come to the island actively, by flying or swimming, or passively, by being carried unwittingly in mud stuck to the feet of birds (as the seeds of certain Malagasy plants likely did) or on rafts of vegetation washed out to sea from African rivers during storms (as many Malagasy plants and smaller animals probably did). Humans may have brought some of them, either intentionally or not. And some very ancient lineages might have made it to the island by land.

Paradoxically, animals could even have *walked* to Madagascar from Africa long after the two split apart. It would have been a long walk, but in the immensity of time it might have happened. For the sake of argument, let's say Madagascar parted ways with Africa 150 million years ago. (This is a conservative estimate; others range up to 165 million years.) Until 130 or possibly even 120 million years ago, however, Madagascar remained linked to Africa via the rest of Gondwana. For during this 20- to 30-million-year period, the east coast of Madagascar was attached to the southwest coast of India. The east coast of India was, in turn, connected to Antarctica, which linked to the southern tip of South America through the Antarctic Peninsula. The part of South America that we know as Brazil, in turn, was jammed into the west coast of Africa centered on what is today Cameroon. Thus, African animals and plants that weren't on Madagascar when its west coast cleaved from Africa could still theoretically have reached the future island via its east coast.

It's possible that the elephant birds, for example, reached Madagascar along this route. The ratites may have arisen in South America, where today we find their descendant the rhea. They may then have walked or, if their ancestors were not flightless, flown in one direction to Africa, where they are represented today by the ostrich, and in another direction to Madagascar, where they became the now-extinct elephant birds, as well as to the chunks of Gondwana that became Australia (the emu), New Guinea (the cassowary), and New Zealand (the kiwi and extinct moa). No one knows where the ratites arose or how they dispersed, but this is one possible scenario.

In fact, Madagascar would remain attached to India until 88 million years ago. So the island's connection to its Gondwanan roots was open, albeit to an ever diminishing degree, via a direct land route for a full 62 million years after Madagascar diverged from Africa. Of course, this doesn't answer for most of Madagascar's wildlife and all of its mammals, whose ancestors had not yet appeared on Earth. For them, one must look at a more challenging route—across the Mozambique Channel.

Plate tectonic theory may have doomed the notion of a lost continent between Africa and Madagascar, but Lemurians have their modern counterparts. As recently as the late 1980s, Jean-Claude Rage proposed that during the Cretaceous period a land bridge might have existed between the two lands along the Davie Ridge. (Oriented north-south down the middle of the Mozambique Channel, the submerged Davie Ridge is the strike-slip fault along which Madagascar slid in its journey away from Africa.) In a 1988 paper, Rage argued that the Davie Ridge rose above the surface of the sea in the C̲ ceous, perhaps forming "a more or less continuous land b̲ recent studies of ancient sea levels indicate that the se̲ ranged between 330 and 1,320 feet above modern levels throughout the late Cretaceous and well into the Tertiary. Only about 10 million years ago did global sea levels drop below present-day levels. Other studies have shown that the channel has not changed since Madagascar stopped moving relative to Africa over 124 million years ago. In a 1997 paper, the paleontologist David Krause and two coauthors conclude that "no conceivable sea-level fall could expose a continuous path across the channel floor."

How about a discontinuous path then? Over the years, various scholars have theorized that ancestral Malagasy wildlife island-hopped across the channel. In the 1970s, for example, the geologist Joël Mahé suggested that during glacial periods, when a significantly greater portion of the Earth's water is locked up in ice caps and glaciers than during nonglacial periods, much of the submarine ridge that underlies the Comoros Islands might have been emergent. The Comoros would seem to provide ideal stepping-stones. The islands are close enough to Madagascar that on a clear day off Cape Delgado in the northwest, one can reportedly see the outline of 7,791-foot Mount Karthala on Grand Comore. But the Comoros are a volcanic chain that only emerged above the Indian Ocean about 8 million years ago. So lemurs didn't come that way, nor did many other creatures descended along antedeluvian lines.

It is possible, however, that several seamounts along the Davie Ridge might have been islands in the mid-Cenozoic. Currently the highest point on the ridge is about 1,115 feet beneath the waves. But in former times this feature, which lies between Madagascar's Cap Saint André and the city of Mozambique (today the shortest distance between the two lands), might have risen up. The same holds true for Mounts Antandroy and Betsileo, two seamounts south of it. Krause and his coauthors postulate that

> if both Mount Antandroy and Mount Betsileo were emergent at the same time in the Miocene, then the trip from Africa could be broken into three north-south stages of [183, 130, and 77 miles] in the direction of the prevailing currents. This island-hopping route seems worth investigating with respect to Madagascar's "mid-Cenozoic" carnivores and rodents (and possibly tenrecs); no other routes seem as favorable, and the route would have provided a short-lived opportunity, possibly at about the right time.

Most animals in the Malagasy menagerie had no need, however, ɔ await the opportune rise of a seamount to make the journey.

Observers have long propounded the notion, as I've intimated, that animals reach oceanic islands by unwittingly hitching a ride on buoyant bundles of vegetation wrenched free from their continental homes during storms and forced far out to sea. Carrying everything from mosses to palm trees, insects to small mammals, these rafts eventually wash up on distant shores.

Such rafts have been recorded periodically. In *Island Life,* Alfred Russel Wallace reports that "rafts with trees growing on them have been seen after hurricanes; and it is easy to understand how, if the sea were tolerably calm, such a raft might be carried along by a current, aided by the wind acting on the trees, till after a passage of several weeks it might arrive safely on the shores of some land hundreds of miles away from its starting point." Madagascar is just such a land. In 1972, the scholar Jacques Millot wrote that "a number of great water courses flowing to the African coast, such as the Rovuma and the Zambesi, were able to carry out to sea through their estuaries floating masses of vegetation, carrying a varying number of animals, and send them in the direction of Madagascar." Even the boa, that paradigm of vicariance, might have reached Madagascar this way. Wallace mentions a boa constrictor that apparently rafted from the South American coast to the West Indian island of Saint Vincent almost 200 miles away. The snake, he wrote, was "twisted round the trunk of a cedar tree, and was so little injured by its voyage that it captured some sheep before it was killed."

Rafting might be a tidy way to account for dispersal, but scholars must account for a number of constraints. First, colonization through rafting and successful establishment is obviously an extremely rare event. Not many storms can rip large plant masses away from a mainland, not many combinations of currents and winds can succeed in moving such masses intact over large distances, not many creatures can survive weeks at sea. If dispersal were a more regular occurrence, Raxworthy notes, the natural world would be a far less interesting place. "Think of the fauna of, say, North America or Australia," he says. "They're very distinct. If dispersal was so common, what we should see is a really active mixing of all these different animals so that eventually you get a very homogeneous global fauna. But we don't see that." Not surprisingly, he cites the herps of Madagascar and Africa. "The huge difference we see between the herpetofaunas of the two places is clearly telling us that dispersal has been very rare."

For one thing, currents and winds and even the pattern of cyclones that periodically strike Madagascar have mitigated against dispersal from Africa to the island. All three—the South Indian Drift surface current, the southeast trade winds, and cyclone tracks—are part of a counterclockwise gyral circulation that operates in the Indian Ocean. The currents may have helped bring the first Malagasy to the island from Indonesia; the trade winds are responsible for the strip of rain forest up and down the east coast; and the cyclones typically hit from the east. There are exceptions, however. During the peak of the winter monsoon, both winds and currents can sweep down the Somali Basin and curve southeastward toward northwestern Madagascar. And the rare cyclone can act somewhat similarly, swinging back toward Madagascar after having passed over it from the east. Cyclone Georgette did so in 1968. After passing over northern Madagascar, it swept across the Mozambique Channel to Africa, where it hooked back toward Madagascar, crossing the southwestern part of the island ten days after leaving the African coast. Because Southern Hemisphere cyclones circulate in a clockwise direction, Georgette and storms like it would leave Africa with offshore winds and hit Madagascar with onshore winds, both of which favor transport of flotsam.

This system would seem to answer for a lot. But the monsoon system that generates those currents and cyclones is thought to have fully developed only about 8 million years ago. By the same token, Madagascar and Africa have been moving steadily north through the Cenozoic and, geologically speaking, have only recently come under the subequatorial influence of this monsoonal system. Yet judging from the relict nature of Madagascar's wildlife, the clock seems to have stopped there 30 or 40 million years ago. Since then, few new types of animals, save those introduced by people, have apparently made it onto the island. Why no successful rafting events for 30 million years? As one group of primatologists has written, "if crossings of the Mozambique Channel today are in theory no more difficult than they were back in the Eocene, how have the supposedly competitively inferior lemurs managed to hold the Madagascar fort against more advanced newcomers? Or was the crossing of the Mozambique Channel in fact achieved against such overwhelming odds that only once was it ever accomplished by primates—coincidentally earlier rather than later in the Age of Mammals?"

The extreme rarity of such events must play into it. Such chance migrations across the sea have been termed "sweepstakes" routes of dispersal. As Anthony Hallam has written, "only a small minority of the available fauna obtains winning tickets in the sweepstakes." Left out, of course, are the largest animals, whose weight natural rafts cannot support. Even lemurs appear to have made the crossing only once. Recent studies by the molecular systematist Anne Yoder indicate monophyly among all thirty-two extant species of lemurs. That is, they all seem to be more closely related to one another than to other lemurlike primates—the lorises of Asia and the galagos, bushbabies, and pottos of Africa.

Finally, dispersal is only half of any successful colonization. Once a plant or animal has arrived in a new land, it must then establish itself. No matter how many empty ecological niches might be available to new arrivals, a single male or female of any one species, or a group of all males or all females, is doomed from the start. Moreover, it isn't easy to survive in a strange land. Plants may come without their natural pollinators; animals may not find the food they're accustomed to eating or the habitat they prefer. Raxworthy told me about a study of plant colonization done on a volcanic island that emerged in the North Atlantic. One species in particular took hold with a clump of a few plants, which flowered and appeared ready to spread across the island. But then a small landslide wiped out that group, and it was years before the species colonized the island. "It was a nice example of how that very early point of colonization is really, really crucial," Raxworthy told me. "How vulnerable populations are at that stage, and how chance events can sometimes have a very big effect."

The general consensus is that, in lieu of a better explanation of how flora and especially fauna colonize distant islands, dispersal by rafting is the only reasonable explanation. It might be almost impossibly rare, but it must happen.

One aspect of such scenarios that is often overlooked is the sheer immensity of time involved. As George Gaylord Simpson has written, "Any event that is not absolutely impossible . . . becomes probable if enough time elapses." Simpson noted that even if the chances of an animal or plant making it across a wide barrier in any given year are one in a million, then in a million years, the probability is 63 percent, and in 10 million years, it is 99.995 percent. Ten million years is not so long to Madagascar, which has floated alone in the Indian Ocean for at least eight times that. Yet Raxworthy councils caution in relying too much on such thinking. "If you accept Simpson's argument," he told me, "then you can take the most ridiculous dispersal hypothesis and always say it's true, given enough time. In that case, we should expect lemurs on the moon."

9.

ON MY SECOND TO LAST morning at Lokobe, Garfield Dean confides something to me that seems as unbelievable as lemurs on the moon. He had a "butchery" in his hand and lost it. *In his hand!*

He had gone out early to work the traplines. Now, you don't have to design air-traffic control systems to work a pitfall trap; simply reach into the buried bucket and grab whatever's there. The fact that Dean *does* design air-traffic control systems somehow made it all the more incredible that he had let the world's least-seen lizard slip through his fingers.

"I somehow managed to miscatch hold of it," Dean tells me. "I ended up with a tail that wriggled in my hand for about a minute." He smiles. "If you'd stuck that thing on the end of a fishing rod, it would have made damn good bait."

I look at Dean. A thirty-three-year-old Briton living in Paris, where he runs a sweet shop in his spare time, Dean is out of shape and out of fashion for the forest. Yet despite his button-down shirts and a straw hat that can barely contain hair rising like bread in the tropical heat, he has become one of our most intrepid team members, and he knows what he missed.

"How did Raxworthy take it?" I ask.

"He played it with the ultimate in sangfroid," Dean says. "The look on his face said, 'Oh, well, it's just another herp that got lost.' "

But it wasn't. It was the century's second. How could he have let such a prize get away?

In disbelief, I set off on my day's mission. Several of us are hiking around Lokobe Peninsula by way of the beach to scout out a river that appears on Raxworthy's rumpled Department of Water and Forests topo map. He thinks it might make an ideal sampling site, offering insights into species living on the other side of the reserve. If it looks good, Raxworthy will set up a satellite camp there to do a day and night of thorough collecting.

Daypacks stuffed with food, water, and sampling bags, we head east along the beach right after breakfast. We cross a series of pretty coves, with the turquoise waters of the Mozambique Channel to our right and rain forest rising to our left. Stiff-legged mangrove trees march across the sand like floral brigades, red-clawed sand crabs disappearing beneath them into holes in the mud. At the end of each cove, we have to clamber over a mountain of slippery boulders—carefully, for the ocean pounds unforgivingly just below. As we come alongside Nosy Komba, lying a quarter mile away across a shallow strait, the coves give way to a narrow, rock-strewn passage, where the thick foliage threatens to shoulder us into the surf.

The hike gives me ample time to reflect on Dean's news. I'm surprised to realize that, once again, I've had a shift in thinking. Admittedly, his discovery has taken the edge off my search. He stole my goal: finding Number Two. Who knows? Maybe my vaunted quarry is actually common here. But my change in thinking is not purely selfish. With Andrews's and Julien's voices still in my head, I'm beginning to feel that perhaps it's better he didn't catch it. *Z. boettgeri* might have the power to become Nosy Be's spotted owl, causing economic hardship and bitterness among the Sakalava who have lived here for generations, relying on the forest and its bounty.

After an hour and a half traipsing under a hot sun, we finally reach the river. Bearing the typically unpronounceable name of Andranomainty, it is really a stream ten feet across, flowing lazily out of a solid wall of rain forest into a wide, sandy estuary. With the same excitement I felt when I first stepped into Lokobe, I immediately strike off with the others up the streambed. For once I'm not looking for *Z. boettgeri*. Instead of craning my neck to look up tree trunks, I keep my eyes aimed low, and with a week of solid searching under my belt, my eye catches the slightest movement.

It's a snake day. Under a rock by the brook, I grab a small, thin-bodied snake that leaves nine separate bite marks on the back of my hand. Later, a half hour apart, I catch two menaranas, the evil snake of the Malagasy Bible. Each is three feet long, and one I have to chase twenty-five feet through a field of boulders. I bag both of them for Raxworthy, and watch in amazement as one pushes its "sweet head" right through the thick plastic of a Ziploc bag.

With a horde of specimens in our sacks, we take a refreshing dip in a deep pool beneath a waterfall, then lie on rocks in the sun to dry off. It's a classic rain-forest setting, with moss-draped trees overhanging the pool and fleshy plants wiggling as heavy droplets flung from the falls hit their pendulous leaves. Perhaps I'll add this to my neural photo album. I finally feel completely relaxed and at home in Lokobe—now that I'm about to leave. Isn't that always the case?

Such feelings never last, and on the way back I begin to feel a rising frustration. How *could* he? Number Two! That was mine to find, dammit. It suddenly occurs to me that, depending on how you look at it, Dean didn't find the second one after all. I remember Sir Edmund Hillary's comment after a reporter asked him how he'd feel if it was ever shown that George Mallory and Andrew Irvine, who disappeared high on Mount Everest in 1924, had actually reached the summit almost three decades ahead of his historic first ascent in 1953. With a twinkle in his eye, Hillary said he felt that the *getting down* is important when talking about climbing a mountain. As is getting the other half of a creature in finding it.

In spite of myself, I start scanning the trees along the beach, using the search image that is now burnished into my brain: dark green skin with yellow spots, long toes, a ten-inch body with a tail half again as long.

Naturally, I see nothing of the sort, and as it turns out, I'm in for a surprise when I return to camp. As I approach the tents late in the afternoon, Dean shuffles out to meet me on the beach. He's beaming.

"What's up?" I ask.

Dean tells me what he's now told everyone else. He pulled a fast one on Raxworthy, along with the rest of us. He put the tail of the "butchery," which had broken off during capture, in one bag and the body—alive and well—in another. He managed to keep the lizard hidden long enough for Raxworthy to write in his notebook, "*Z. boettgeri* [lost]." Only then did he reveal his hoax.

"It's typical of the British sense of humor," Dean says. "You try and wind people up a bit."

I have to laugh. I'm happy for him, happy for Raxworthy, happy even for myself (I'm off the hook).

But we're just visitors to this precious place. Lokobe's fate lies in the hands of Malagasy officials and the international conservation agencies they choose to work with. It's now up to them to decide if it remains a strict nature reserve or opens its gates as Lokobe National Park. We can only hope they make their decision sensitively, with the interests of both natural and human worlds in mind.

Deep into a Primordial Land

THE SPINY DESERT

The ancestors come into our lives

like guests who need no invitation.

—*Malagasy proverb*

1.

"RIGHT. YOU'VE GOT to get off, Peter."

It's Garfield Dean, and he's just gotten word in French from a green-bereted Malagasy policeman who waved us down that I have to get off the roof of the "Brookesia," as Raxworthy has christened his Land Rover. I've been sitting up here with an Earthwatch volunteer amid a heap of tied-down luggage, enjoying the view. Seems it's illegal to have anyone traveling on the roof, and our Malagasy driver has been ticketed and summoned to appear in court tomorrow for the offense. The two of us stuff ourselves in with the rest of the team, and we drive off. A few miles farther on, the driver pulls over, flashes a smile, and we climb back onto the roof.

It's a year and a half after my trip to Nosy Be, and we're driving through an area that could not look less like the lush rain forest of Lokobe—the spiny desert of southwestern Madagascar. Here Raxworthy is conducting a herp survey of a 30,000-acre proposed reserve known as PK-32. It is named for the nearby POINT KILOMÉTRIQUE 32 sign north of Tuléar, a port that is the largest town in the southwest. A sweaty Dean met me and a new Earthwatch crew at the airport and is driving us north along the coast to the research site at Lake Ranobe (Big Water). With his easygoing manner and his fluency in French, Dean has graduated from Earthwatch volunteer to Raxworthy's logistics coordinator on this expedition. I'm delighted to see he hasn't changed a bit: all white button-down shirts, Yorkshire-pudding hair, and dry British wit.

On the roof, I crouch rather than sit to spare my bones. The road we're on is Route Nationale 9, but it is more like a dune-buggy trail than the American interstate its name might conjure up. For long stretches the Brookesia glides along a pair of smooth tracks through the pale-orange sand, and you forget that the road is unpaved. Then up ahead you see the oh-my-God-it's-going-to-tip-over lurch of a taxi-brousse, Madagascar's small, brightly colored buses, and you prepare yourself to be tossed about like a sausage on a griddle. The driver slams on the brakes, and the Brookesia pitches forward into a nest of broken-up rock, so out of place in this soft beach environment. I comfort myself with glimpses of the Mozambique Channel, which lies just over a ridge of scrub-covered dunes to the west.

Even as I get used to the lurching, I can't get used to what I'm seeing along the road. It would not surprise me to learn that the term "hardscrabble" was coined to describe the living people have to scratch out of this inhospitable coast. Walter Marcuse, an Englishman who passed through this region about eighty years ago, wrote that "every green plant, including the cacti such as prickly pears and Barbary figs, was withered and burnt to a sickly brown color, the earth was baked to the hardness of asphalt, and it was almost inconceivable that a living creature could exist amid such a scene of desolation." Yet living creatures do. People do.

A typical hamlet consists of three or four weathered huts spaced about thirty feet apart on a patch of bare sand alongside the road. The huts, blasted by airborne sand every time a vehicle swooshes past, are about ten feet by six feet, with a frame of stout sticks holding up walls and a peaked roof of sun-seared thatch. There is a door, either a plank or simply an opening, but neither windows nor chimneys. The cooking fire is kindled on the floor inside, and smoke finds its way out through the thatch above. Set amid a rain forest, these huts might look quaint to Western eyes. Here they couldn't appear more bleak. "A few very small tumble-down huts, cattle-pens in the midst of them, with here and there an old stunted tree on a burning-hot sandy hillock, form their village, and miserable and desolate did it look," wrote a visitor to Tuléar Bay in the last century. I could not agree more. I'm ashamed to admit that the thought of having to spend even one night in such a claustrophobic, smoke-filled, stinking-hot hut sounds like torture.

They are truly set amid a "scene of desolation." Most of their yards—if they can be called such, for no fence, hedge, or line separates one hut from the next—can boast not a single blade of grass, much less a lawn. Rarely one will feature a small plot of rosy periwinkle, the world-famous Malagasy plant used to cure childhood leukemia. Otherwise, save perhaps for a leafless tree or two, whose trunks provide a sliver of shade in this largely shadeless land, they are completely devoid of vegetation. Herbicides couldn't do a more effective job than nature has here. As if this weren't bleak enough, the naturally orange sand in these hamlets is often stained black from the charcoal locals produce from wood collected in the spiny desert. The charcoal, a chief source of income for these desperately poor people, is stuffed into tall burlap sacks and placed along the roadside for purchase by people traveling between Tuléar and Morondava, the next town up the coast. In larger hamlets, the black stain stretches right across the road between huts on both sides, making the whole look like the remains of a village burnt to the ground in wartime.

Beyond the huts empty sand gives way to the tangle of spiny plants and trees that give the desert its name. This thorny thicket is no place for kids to play. No place to hunt, now that locals have hunted out all the lemurs and many of the birds. No place to grow crops. In the arid southwest, it's not unusual for an entire year to pass without measurable rainfall, and a drought in the region in the early 1990s left many thousands dead. The only thing it's good for is wood, but if the market for fuelwood in Tuléar continues to grow as it has in recent years, the spiny desert won't be good for that either, as it will all be gone. The advance of the charcoal-cutters is the main reason Raxworthy wants to survey this area, to provide reliable information as quickly as possible on rare species that conservationists might use to secure formal protection for PK-32.

Forty-five minutes after I retook my position on the roof, we stop in a large hamlet to stretch our legs. On the drive, I saw few people, only the occasional face poking inquisitively out of the comparative cool of a hut, or a mother sitting in the shade of a tree, picking lice out of her child's hair. But now a sea of children suddenly washes around us, having materialized out of nowhere. Defying the harshness of their lives, they are smiling and laughing, their curiosity and safety in numbers just getting the better of their shyness. They are beautiful, with perfect white teeth and handsome, finely featured faces. Their crinkly hair and dark skin suggest a strongly African heritage.

These people are Vezo, an offshoot of the Sakalava, the same tribe we encountered at Nosy Be. (The Sakalava are the most widespread tribe in Madagascar, ranging along almost the entire west coast; north of Nosy Be, they give way to the Antankarana tribe, south of Tuléar to the Mahafaly.) The Vezo are not one of the eighteen officially recognized tribes of Madagascar, perhaps because they are defined less by ethnic differences than by way of life. More than any other tribe in Madagascar, they are a people of the sea. It may be that the only way they can tolerate their hardbitten life ashore is by spending most of their time out on the ocean, fishing along the region's extensive coral reefs from their outrigger canoes.

It seems inconceivable that the English would choose this region for their first attempt at colonizing the island, but they did, and all because of the tragic misinformation provided by the merchant Richard Boothby (he who dubbed Madagascar "the chiefest paradice [*sic*] this day upon Earth"), and by his shipmate Walter Hamond. The two Englishmen traveled to Madagascar on the same voyage with the East India Company in 1630, and their ship paused at Saint Augustine's Bay, just south of Tuléar, before proceeding on to India. Though they stayed only three or four months and never went inland, both Boothby and Hamond later published influential accounts that raved about the island.

Hamond, a surgeon by trade, wrote two pamphlets: *Madagascar, the Richest and most fruitfull island in the World*, and *A Paradox proving that the Inhabitants of the Isle called Madagascar or St. Laurence (in Temporall things) are the happiest People in the World*. Boothby was no less gushy, even though it's clear that a lot of his information came from hearsay or was entirely made up:

> The country about [Saint Augustine's Bay] is pleasant to the view, replenished with brave woods, rockie hils [*sic*] of white marble and low fertile grounds . . . [I]n the Island is a large plaine, or champion country of Meadow or Pasture ground as big as all England: which . . . must, by reasonable consequence afford multitude and variety of Fowles and Beasts and other

creatures for food, cloathing, necessary use and delight, and no doubt but such low grounds affordeth also store of large and small Rivers, Tanks and Ponds replenished with multitude of good Fish, water Fowles &. and it is apparently manifest or very probable by the quantity of brave fat Oxen, Cowes, Sheepe and Goats brought downe and sold unto us by the natives for refreshing so many people that the Country is very fertile. . . .

Most later visitors to the region, many acting on Hamond and Boothby's advice, found out quickly and tragically how unreliable such depictons were. They suffered terribly from hostile Malagasy, heat, fever, and a land that could scarcely be less fertile. They didn't realize that these two men arrived in winter, when the weather can actually be pleasant. They couldn't know that Hamond and Boothby had encountered only friendly Malagasy. In short, the pair's visit was freakishly free from trouble—they lost not a single man from the ship's company of 460 to illness, attacks, or any other cause—and everyone who followed them paid dearly for it.

After years of advocating for a settlement, Hamond and Boothby finally got their wish. In August 1644, five years before Robert Hunt would leave England on his ill-fated mission to colonize Nosy Be, John Smart sailed with 140 settlers to establish a colony at Saint Augustine's Bay. The trip began auspiciously. By the time they reached the Cape of Good Hope at the tip of South Africa in January 1645, they had lost but one person (ironically, the settlers' doctor had succumbed to illness). In fact, Smart was able to report optimistically, "Wee are increased in our number by the birth of foure brave boyes, besides expectation of others: which makes us conclude God goes along with us."

Smart would change his mind soon enough. He and his party arrived safely in March 1645 and immediately began setting up the colony on the south side of the bay. They sowed crops, built a fortified compound with "many and well-built" houses, and established trade relations with local Malagasy. But they had arrived at the end of the brief wet season and ended up planting at the wrong time. Their crops failed, and pasture for their cattle vanished. They soon realized that Boothby's fertile "champion country" was a myth. Their relations with locals turned sour, and settlers began dying from fever and dysentery. By August, Smart reported to his principals in London that only forty men remained who were fit to bear arms, the rest being "old, ignorant, weake fellowes." The women—"she-cattle," as he called them—were of "no other use but to destroy victuals." The Malagasy, for their part, were "of soe base and falce a condition that they have not their fellowes in the whole world."

Smart decided to sail one of his ships to Nosy Be to get supplies and see if he could transfer the colony there. But the ship hit a reef on the way and was forced to return to Saint Augustine's. By this time, the original party of 140 had dropped to 63, mostly from disease but lately at the hands of local Malagasy. On May 19, 1646, just over a year after Smart had arrived with such enthusiasm, he and the remaining 59 settlers sailed away from "this most accursed place." Some wound up in India and others in Sumatra (including Smart, who died there); fewer than a dozen of the original 140 men, women, and children managed to return to England. One of them was Powle Waldegrave, who wasted no time in publishing his seething retort to Boothby.

Though a massive colonization attempt was never made again in the region, Saint Augustine's Bay, and Tuléar Bay just to the north of it, remained important trading stops for ships rounding Africa on their way to India. Travelers could water up and trade for slaves, livestock, hides, and fresh fruit. In the first two centuries after European discovery, such trade was dicey, with Malagasy unsure of Europeans and vice versa. "After staring about us for a little time, about a dozen fellows with muskets made their appearance, and I must say I looked on them with much distrust," wrote one such visitor who came ashore at Tuléar Bay, only to retreat empty-handed when the Malagasy began clamoring for brandy. "We might, perhaps, by waiting, have got bullocks, sheep, goats, or our throats cut, as the humour happened to take them."

Over time, however, this southwest corner of Madagascar became one of the chief trading centers, and local Malagasy became less unsociable. The town of Tuléar came into existence in 1895 and is now the provincial capital. According to legend, Toliara (as it is also spelled) got its name when a European sailor asked where he could tie up his boat. *"Toly eroa,"* replied a local. "Tie up down there."

After traveling through a wide open landscape from Tuléar, with dunes to the left and low-slung spiny desert to the right, we turn off the national highway and enter a thick woodland. It's a relief to get out of the sun. We drive along a sandy road wide enough for one vehicle, and I have to be mindful of overhanging branches. The forest here is green but dry. Lianas drop from trees averaging no more than forty feet tall. There is plenty of shade beneath the thin trees and giant, oaklike tamarinds, but also open patches of red, sandy soil where the unforgiving sun beats down.

We come out on the grassy shore of Lake Ranobe ("*ron*-oo-bay"). It is large enough that you could not make out a taxi-brousse on the far side, and a forest of tall green reeds fills at least half of it. We drive along the lake to the camp, which Raxworthy has established just inside the forest, about fifty yards from the lake edge. I immediately recognize the accoutrements: blue tarps for eating and meeting spread in a clear space between dangling lianas, the cookstoves nearby; beyond, green expedition tents parked in clefts in the thick shrubbery; a makeshift laboratory with cloth sacks dangling from an overhanging tree. With grins and pats on the back, I greet Raxworthy, Achille, and Jean-Baptiste Ramanamanjata, a boyishly handsome Merina who is Raxworthy's other graduate student, along with the irrepressible twins Angeluc and Angelin.

There are also two smiling Sakalava women from a nearby village, whom Raxworthy has hired to help the twins with the cooking. They are tall and have the frizzy hair, high foreheads, and large, deep-set eyes characteristic of the Sakalava. They also have open, friendly faces, and appear to be so amused by our appearance or behavior that they can't stop laughing, their white teeth flashing in the dim light here in the shade. I'm sure they get along just fine with the twins, who can't stop laughing either.

For my part, I'm delighted at the prospect of perhaps finally getting to know the Sakalava. At Lokobe, I had little contact with them, since we spent the whole time in the reserve, and what little contact I did have, epitomized by my encounter with the shark fisherman, left me baffled and no closer to understanding the Sakalava mind-set. Here, our camp lies smack between two Sakalava villages, and Raxworthy has hired several villagers to help us as guides and cooks. While the rest of the team heads off to set up their tents, I sit on the blue tarp taking notes and occasionally stealing glances at the two women, trying to match them with what I've read about the Sakalava of this region.

The Sakalava were long reputed to be one of the most warlike of Malagasy tribes. Many carried both guns and spears, the latter being called "the wives of the gun." Early in the last century, their ferociously independent nature helped them stave off several military advances by the Merina king Radama I, who unified all of Madagascar save for this region of the deep south. Yet constant infighting among various subtribes also kept the Sakalava from ever achieving a true confederacy, and by the mid-nineteenth century the Merina had usurped the Sakalava's position as the island's most dominant tribe.

For the Sakalava, a focus on toughness begins in early childhood. During circumcision, which takes place when a boy is six or seven, "the father . . . takes a loaded musket, commands the boy to lie down with his face to the ground, places the musket across the child's back, and fires," reports Arne Walen, a Norwegian missionary who spent two years among the Sakalava of this region a century ago. "After this, all the company join in kicking the child on the hinder parts with the view of making him a strong and brave man when grown up, that is to say that he may become a true Sakalava." Because circumcision is performed with an ax or large knife, Walen says, it is "sometimes attended with very sad results, especially as drunkenness is one of its invariable concomitants." Sad results can attend even the simplest quarrels among boys, Walen notes:

> On one occasion a Sakalava boy said to his companion: "You are a disgrace to the free and brave Sakalava, for who does not know that you have married a slave woman." The other replied: "You ought to be more ashamed than anybody else, because you, proud as you are, have married the ugliest woman I ever saw." Then both cried out: "Let us fight," and in a moment the duel took place. They fought with spears and muskets, and they speared and shot at each other like two little savages . . . the result being that both were dreadfully wounded, and only narrowly escaped with life.

Even games can be rough. Marcuse describes a Sakalava pastime that "for sheer pluck . . . will compare favorably with any sport in which the white man indulges." In the Malagasy version of the running of the bulls at Pamplona, a young Sakalava dashes into a herd of 500 or 600 zebu, the lyre-horned cattle of Madagascar, and attempts to return with a bull barehanded. To do so, he throws his right arm around the beast's hump, grasps its muzzle in his left hand, and tries to lead it away from its companions. Invariably the bull struggles furiously, and not infrequently a stampede ensues. "Usually, in such a case, the game fellow is shockingly trampled; but this catastrophe is merely a signal for the next brave man to step out and try his fortune," declared Marcuse. "I have seen as many as a dozen men injured (some fatally) before the desired result was achieved. . . ."

The only thing to come in the way of a Sakalava's bravery is superstition. "They are not wanting in courage and prefer to die fighting than in any other way," wrote E. O. McMahon, an English missionary who worked among the Sakalava in the late nineteenth century, "but they are so full of superstition and fear of charms that directly one or two of a party are killed they generally give up the fight, not from fear of the enemy, as they will certainly return and fight them again, but because they think the gods are against them."

The Sakalava have charms against just about everything, including spears, poisons, and bullets. They have charms to secure a woman's love, to grow rich, even to make vazaha willing to give the wearer goods and money. (The Sakalava are not alone in this among Malagasy tribes. The botanist Richard Baron wrote that the Malagasy have "many herbal quacks who, for a consideration, are able . . . to furnish charms against fire or tempest, locusts or lightning, leprosy or lunacy, ghosts, crocodiles, or witches.") Sakalava fetishes often consist of a zebu horn containing a mixture of grease and various small objects. These might include a needle, bits of bone or sacred wood, crocodile teeth, a scorpion's tail—whatever the Sakalava sorcerer who creates it deems is necessary to make the fetish efficacious.

Sakalava take great care in making their charms. Jørgen Ruud, the ethnographer, describes the elaborate procedures that attend the making of a single charm, the poison fetish known as *Raiboboka,* "He who is the origin of swelling." Also known as *Fangotsohana,* "Pain in the body," this fetish, Ruud notes, aids robbers and assassins, bullfighters and cockfighters, and those who fight with their fists or feet. To make the charm, a sorcerer gathers the following nefarious creatures: a scorpion, a wasp, two kinds of venomous spider, a deadly water insect known as *tsingala,* and a poisonous beetle whose Malagasy name means "he is continuously hit." He mixes these with a powder made from seven plants bearing sharp thorns, stinging nettles, poisonous berries, or other menacing properties, and stuffs the blend into an ox horn.

After the sorcerer smokes the horn with incense and makes various incantations laced with words of hate and revenge, the fetish is ready for use. The user, who often smears the concoction on his spear or bull's horns in addition to hanging the horn from his person, must adhere to strict taboos: He must not eat meat that was roasted on a spit, or he might himself become spitted. He must not eat kidney, for the Malagasy word for that organ, *voa,* also means "to be hit." Finally, he is forbidden to kill any animal represented in the charm. "All the things in the sacred ox-horn are conscious beings, with wills of their own," Ruud writes, adding that each animal is thought to "cooperate with the man who fights, and with its fatal nature it will kill his adversary."

Ferocious in battle yet emasculated by charms—such contradictions in the Sakalava character abound, mitigating against any quick understanding by the curious vazaha. In olden days, the Sakalava put to death babies born on unlucky days, yet they consider childlessness as among the greatest of evils. The greatest evil of all, according to Walen, is for a decent man to fall from grace, for the Sakalava value goodness higher than all other attributes and believe that one can lose it only by some flagrant disregard for the *lilin-draza,* the "laws of the ancestors." Yet a good man might steal, indeed is encouraged to do so. Walen says Sakalava parents punish their children only for getting caught while stealing. "They are not punished for *the stealing;* that is not considered an evil deed at all, indeed it is considered a very good way of acquiring property," he notes. "But to steal so that the theft be discovered is such a foolish and stupid deed that it deserves contempt and punishment, which latter is actually carried out." One of the most striking contradictions is that, despite their seeming disregard for human life, the Sakalava cherish it. If any Sakalava are killed in battle, "it is immediately known on both sides, and all become sorrow-stricken," Walen reports. "The weeping women stand up to negotiate peace, and these seem, by their weeping, to be able to subdue the hate and thirst for revenge of the men; and so they cause the war to cease. . . ."

Looking at the Sakalava women chatting away across from me in the camp's "kitchen," I smile sympathetically. They're just like women everywhere, I think—warm, caring, baffled by the brutal nature of men. Suddenly one of the women catches me staring at her. Her companion notices her expression and turns to look in my direction as well. Instantly the broad smiles vanish from their faces, and they stare back defiantly, without a hint of shyness. Their boldness unnerves me, and I look away, embarrassed. Seeing their laughing faces on the way in, I'd thought here at last were some Sakalava who felt comfortable around vazaha and might share some of their secrets. But those brazen, who-the-hell-are-you expressions argue otherwise. In a trice they seem as remote and exotic as I can imagine. Do I have any hope of approaching this most arcane tribe?

I slip away to set up my tent in the forest. The thicket is so dense in places that I have to crouch to get through. Spiny vines hang everywhere; one resembling a rose stem grabs skin and clothing with equal tenacity. I find a flat space with a view through the trees to the lake. Perfect—or so it seems; this, after all, is Madagascar. Just as I'm pushing the last stake into the soft soil, I feel a sudden seering pain in my arm. I jump back, slapping and scraping in a panic. A Sakalava charm in action? In a flash I see the culprit flying away: a large yellow wasp. I don't stick around to find out if I've disturbed a nest, but in a rush yank my tent up by its stakes and shuffle off to find another spot. Hamond and Boothby: proven wrong again.

2.

THE ORIGINAL SPECIES TO descend on Madagascar—the distant ancestors of large yellow wasps and everything else in the island's glorious assortment of plants and animals—encountered feeble competition and few large predators, or so scholars have long presumed. Plenty of empty ecological niches existed, and the pressures of natural selection were low. In addition, these pioneers found a land with extremely diverse topography and climate. All these factors, the argument goes, worked together to spark intense speciation.

But how? What triggers one species to become two, let alone the preposterous number found on Madagascar? Was speciation of original species landing on the island as clear-cut as that argument suggests? What about all the animals and plants that came later? Raxworthy has begun racking his brains for answers.

A century and a half after *The Origin of Species,* biologists are still struggling to determine how one species grows out of another. Even the notion of how to define a species remains in dispute. The evolutionary biologist Ernst Mayr came up with a definition half a century ago that has long held sway among biologists: Individual species are "reproductively isolated" from one another. Even if they look identical and are genetically capable of reproducing, individuals of two different species do not because they have evolved different courtship displays or songs. In some cases, they might try to interbreed but fail because the sperm cannot fertilize the egg or the egg dies or any offspring are weaker than those produced within any single species and cannot compete. Even genera and families are amorphous, Raxworthy says, often growing out of the biases of scientists. Ask ornithologists how they came up with specific genera and they often can't provide a decent answer, he says. Perhaps the most obvious subjective placement is that of human beings. We have classed ourselves in our own family, because we don't wish to be included with the apes, even though we share over 98 percent of each other's DNA.

Another problem with studying speciation is that the process takes time—on the order of millions of years. No matter how fervent a biologist might be, he or she can observe only several generations in his or her lifetime. For this reason, biologists are forced to infer past evolutionary developments based on present-day species or on fossils. Madagascar's fossil record, as we've seen, is woefully threadbare. So models and theories about past scenarios—and future ones—take over.

Biologists define two principal types of speciation. In the first, known as allopatric speciation, a new species develops after being geographically isolated from the parental species. A storm blows a species of Darwin's finch over to a new island in the Galápagos. That species adapts to the new environment and, over time, becomes a new species that, even if it were brought together again with the first species, would unsuccessfully breed with it, if it bred at all.

In the second, sympatric speciation, species can arise with no geographic barrier but through other mechanisms, such as genetic or behavioral changes. This has long been observed in plants. One way is through polyploidy, in which the halving of chromosomes that normally takes place during cell replication fails to take place, and all chromosomes are passed on to the gametes. These gametes cannot reproduce with their parental forms, which have the usual number of chromosomes, but they can with other gametes like themselves that have twice the number of chromosomes. In addition to polyploids, which include most of the world's domestic crop varieties, botanists estimate that up to 70 percent of all flowering plants arise from hybrids. Hybridization occurs when two closely related species produce offspring. Despite this, evolutionary theory holds that hybrids will always be less fit than their parental forms and so will rarely if ever engender new species.

Sympatric speciation among animals has long been controversial. The controversy began in the 1850s, when Charles Darwin wrote in a manuscript for his posthumously published book *Natural Selection,* "I do not doubt that over the world far more species have been produced in continuous than in isolated areas." Unfortunately, he didn't spell out why he thought so. The first reported case of possible sympatry comes from 1862. That year, a local newspaper declared that the apple maggot *Rhagoletis*, whose natural host is the hawthorn tree, had turned up on apple trees in the Hudson Valley. An amateur entomologist named Benjamin Walsh proposed in 1864 that when the two groups of fruit flies began eating and laying their eggs on the different fruits—apples and hawthorn fruits—they sufficiently separated themselves as to become separate species. But Walsh couldn't prove his theory. The two fruit flies looked exactly alike and, for all anyone knew, were interbreeding.

The idea kicked about until 1947, when Mayr tackled it. He claimed that neither Darwin nor anyone else had solid evidence that sympatric speciation took place. There the notion sat until the 1960s, when one of Mayr's graduate students, a Harvard biologist named Guy Bush, decided to have another look at the apple maggot. Were the two groups of fruit flies reproductively isolated? And if so, what caused the isolation? After long study, he had an answer. It seems that the apple flies mated only on apples, and the hawthorn flies only on the small, red fruits of the hawthorn tree, but not on both. He had found his isolating mechanism for sympatry: The two flies had chosen the two fruits as their feeding, courtship, and mating grounds. A later study revealed that only 6 percent of apple and hawthorn flies were interbreeding with one another.

Mayr notwithstanding, the notion of sympatric speciation has gained supporters in recent years. Studies of cichlid fishes in newly formed crater lakes in Cameroon show that they are more closely related to one another than to fish in nearby lakes or rivers. This indicates that the former evolved within the lake rather than through immigration from other lakes or rivers. Similar findings have come from studies of stickleback fish in small lakes in Canada formed when glaciers retreated at the end of the last ice age.

So how might sympatric speciation have occurred on Madagascar? Regardless of the specific mechanism, biologists have long believed that the periodic mass extinction events that have struck the world's biota, wiping out as much as 97 percent of all extant species, have offered a grand stage on which sympatric speciation could take place. With all those empty niches to fill and with little competition to fill them, new species can arise sympatrically. But such thinking, Raxworthy feels, can be "a bit of a minefield." For one thing, biologists disagree on how to define an ecological niche. For another, even if species found a whole host of empty niches, it doesn't mean they automatically speciated like crazy. Raxworthy offers an alternative theory: "In those conditions you tend to find that a species becomes a generalist. If there are no other competitive species, it will start to occupy different niches as a single species. Only afterwards, when another competitor comes on board, say, feeding on the same resources, that's when you might see some kind of evolution."

Sympatric speciation remains highly controversial, and Raxworthy concedes he has found no solid evidence for it yet among the reptiles and amphibians of Madagascar.

Allopatric speciation, on the other hand, that arising through geographic separation, is widely believed to be the chief engine that drives the evolution of new species. (Mayr, for one, feels that it accounts for more than 99 percent of novel forms.) And the minicontinent of Madagascar, with its wide range of climatic regimes, landscapes, and ecosystems, offers prime turf for such speciation. Indeed, the primatologist Alison Jolly and two colleagues have written of the island's diversity of ecosystems that "the ring of wet and dry forest of Madagascar acts almost like an archipelago of islands, in which evolution runs more quickly than [on] either fully divided islands or continuous mainland. This is one clue to Madagascar's riches of lemurs as well as all its other burgeoning forms of life."

Alterations in the natural environment can also spur allopatric speciation. Until recently it was thought that Madagascar's climate changed little since the island became isolated in the Indian Ocean. But recent paleoecological work, led by David Burney of Fordham University, has shown that the island's climate has fluctuated more or less continuously over the ages, with consequent transformations of ecosystems and evolution of new species. The same goes for natural disasters. Volcanic eruptions, floods, cyclones, and the like can alter the topography, isolating animals or plants that, over time, become distinct species.

3.

I FEEL AS IF I'VE BECOME a new species—a poor, miserable, suffering one. This morning, my first at Ranobe, I woke up feeling as if someone had hit me on the head with a sledgehammer, poisoned me, and drained every drop of energy out of my body. I can hardly remember feeling sicker.

I don't know what I've got, but I do know what triggered it. Last night after we arrived at the camp six or eight of us sat on the blue tarp drinking *toaka gasy*, the local rum. Loren Caira, an emergency-room doctor in Los Angeles and a member of Raxworthy's team, broke out a bottle of Armagnac he'd brought along, and some us, or perhaps just one of us, stupidly began mixing that with the other. The talk dwindled to a frat-house level, and as the night wore on I noticed women volunteers vanish one by one. The conversation turned toward illness, with everyone trying to outdisgust everyone else. Caira, the ER doctor, led the way.

"You don't get sick with the common roundworm, *Ascaris lumbricoides*," he said. "You eat them, and they burrow through your intestinal wall into your perineal cavity, through your diaphragm into your lungs. You cough up the small immatures and swallow them again. They become adults in your intestine."

"Oh, my God," murmured a listener.

"Very pleasant," said another.

"They get to be eight inches long and very muscular," Caira continued. "They can come out your nose and your mouth."

"Oh, please."

Raxworthy didn't miss a beat.

"A friend of mine was brushing his teeth in the States after coming back from Egypt," he said, "and he looked into the bathroom mirror and saw the tip of a little worm doing this." He waves his fingers as if mimicking a fish swimming. "He coughed and managed to get the thing out, and he took it to his doctor: roundworm. The doctor gave him a tablet and said, 'When you take this, make sure you're close to a toilet.' My friend said, 'What do you mean by close?' The doctor said, 'Really close.' He took it and within about five minutes his whole gut began going through tremendous spasms. These spasms are so violent that they actually throw the scolix out."

Moans.

"Anyway," Raxworthy continued, clearly relishing the telling, "the thing was shimmering in the pan. This guy was into parasites, so he ran off to get some alcohol to preserve it. But his girlfriend flushed it down the toilet before he got back. He felt cheated; he really wanted it."

Such talk proved a harbinger. This morning, I woke feeling like death. Not true death, but a kind of Haitian zombie voodoo death. I was too sick to appreciate the irony of all that talk about illness. Lost my dinner, lost my breakfast. As the others headed off for a first day of checking the traplines in the forest, I lay in my tent, groaning. I was literally unable to rise. It felt too vicious to be simply a hangover. Before he headed off with the team, Raxworthy appeared at the door of my tent. "Try one of these," he said and thrust his arm through the open tent flap. In his hand was an old plastic medicine bottle with no label save a piece of white tape bearing a handwritten inscription: FOR FLU AND HANGOVER. I took one and dozed off the morning.

By God, it worked. Around noon, I went into remission. I ate lunch and actually went for a hike in the spiny desert. I don't know what Raxworthy gave me, but I feel as if it saved my life. I'm even considering a bit of rice for dinner.

Raxworthy knows something about illness in Madagascar. In his ten years on the island, he has had hepatitis A, numerous gastrointestinal disorders, and more bouts of malaria than he'd like to remember. In the late 1980s, he began having side effects from all the antimalaria medication he was taking, what with spending six months of every twelve in Madagascar. He suffered from blurred vision and upset stomachs, and doctors told him the medicine would ultimately do more damage to his liver than the disease. So he stopped taking the prophylaxis in 1989.

"Within two months, I got malaria," Raxworthy told us last night on the tarp. He collapsed and was rushed to a hospital in Fort Dauphin, a town on the southeast coast. A doctor took a blood test and within fifteen minutes found malaria. Raxworthy was pumped full of Fansidar, an antimalarial drug, and placed into a private ward. "The only thing to distinguish it from any other ward was that it had two surreal paintings on the wall," he said. "One was a knight in armor coming out of a chessboard, and the other was a woman merging into a horse. I began having these great hallucinations, sort of crawled into the paintings."

The next morning the fever broke, and he was released from the hospital.

"It was a really grim place," he recalled. "It stunk, and there were people just screaming. But they probably saved my life. I was running a temperature of 40.2 degrees C. Forty-one kills you." Since then, he said, he has come down with malaria ten or twenty times, though thankfully each new case has been milder than the last.

Early visitors to Madagascar built up no such resistance, and they died in droves, mostly from fevers contracted from exposure to what they termed "miasmatic vapours." A nineteenth-century observer vividly elucidated the distinct stages of such fevers:

> The first—the cold stage—is introduced by a general feeling of languor, with yawning and stretching, and before long intense cold is felt all over the body, with violent shivering, chattering of the teeth, and the knees knocking together. Pain in the limbs, especially in the back and loins, often accompanies this, and the respiration is short and quick. This lasts for a period varying from half an hour to two hours or more, when the second—the hot stage—commences, during which the body becomes of a dry and burning heat; there is often intense headache, aversion to food, but great thirst, sometimes severe and prostrating vomiting . . . This again lasts for a variable period . . . and is succeeded in its turn by the third—the sweating stage—the advent of which is welcomed with the greatest of joy by all who have any personal acquaintance with Malagasy fever. During this stage the skin becomes soft and the body moist with perspiration, the temperature and pulse fall to their normal condition, and the sufferer forgets the weariness and agony of the few preceding hours in quiet sleep, from which he awakes greatly refreshed, and, except that he is weaker, feels as well as before the attack came on.

Many others were not so lucky as to survive. Indeed, the swiftness with which "the intermittent fever of the ephemeral or quotidian type" could carry people off was breathtaking. An early visitor to Tamatave, a town on the east coast surrounded by mosquito-nurturing swamps, was astonished to find "men with whom you dine at the *cercle* in the evening are buried hurriedly the next morning."

Madagascar harbors most of the expected tropical diseases and afflictions, but it is those that are unique to the island that excite the most fear and awe. One of the most dangerous attacks one can have is from the tsingala, a poisonous black water beetle. Cattle drinking water bearing this insect typically die within twenty-four hours if an antidote is not administered. In the last century, one Rev. H. T. Johnson described a tsingala assault on one of his palanquin-bearers, who minutes before had drunk from a dirty pool while traveling near the town of Fianarantsoa:

> He stood stretching out both his arms and throwing back his head in a most frantic manner, at the same time shrieking most

hideously. My first thoughts were speedily seconded by the words of his companions, who said: 'He has swallowed a tsingala.' Of course, I immediately got out of my palanquin and went back to the poor fellow. He was now lying on the ground and writhing in agony. His abdomen had become very swollen, and his skin very hot, and I felt that unless something could be done, and that speedily, the man must die. My other bearers, seeing the extreme urgency of the case, called to passers-by, asking if anything could be done to cure their companion. . . [A] Betsileo . . . ran to procure some leaves, with which he returned in about ten minutes; he soaked them in water from a stream close by, and then gave the sufferer the infusion to drink. With almost the quickness of a flash of lightning the poor fellow shewed signs of relief and began to shiver violently. After drinking this infusion several times more, he said that he was free from pain, but felt very weak and faint. Soon we were able to proceed to the nearest village, and I left him there for the night, thankful that his life had been spared. It was some weeks before the man got thoroughly strong and able to carry the palanquin again.

In former times, Malagasy forced afflictions akin to that of the tsingala upon people suspected of crimes. The most infamous of these trials by ordeal was the *tangena*. Tangena is a tall shrub growing throughout eastern Madagascar. It produces a greenish-yellow fruit about the size of an apple, which contains an almond-shaped nut of extreme toxicity. During the ordeal, the accused was made to eat some rice and three pieces of fowl's skin, followed by two halves of two different tangena nuts (to increase chances that the poison would be of average strength). Then he was forced to drink copious amounts of tepid water. Violent and prolonged vomiting usually ensued, and if the accused brought up all three pieces of fowl's skin, he was declared innocent; if all three did not come up, he was guilty.

Regardless of innocence or guilt, however, the accused often died from the ordeal. If enough of the poison got into the bloodstream before it could be ejected, the victim was doomed. As if the mere administration of the tangena were not unjust enough, not infrequently the poison acted more as an enema than an emetic. Symptoms began with a numb, tingling sensation in the mouth and throat as it went down and then all over the body. Intense vomiting attended by great anxiety and debility ensued, with victims paralyzed and unable to stand. Some became delirious. Just before death, the fingers and toes twitched spasmodically. All told, the tangena proved fatal in one out of ten cases, and since it was often administered to entire villages at once, the number killed by the practice before the government officially abolished it in 1862 was stupendous.

Malagasy normally reserved tangena for the most heinous transgressions, such as witchcraft or treason. But since the Merina monarchy approved its use, it was often used for nefarious purposes. "It can easily be understood that state policy readily attained its crooked ends by the administration of the Tangena," wrote an English physician who described the affliction in 1890. "It was observed that those who might be called 'the opposition members' of the government seldom recovered from the ordeal." The notoriously cruel Ranavalona I even used it to "screen" people to work for her. According to Lt. Frederick Barnard, who traveled to Madagascar in the 1840s in an American attempt to suppress the slave trade, the queen once

> sent to Fort Dauphin for some singing girls, who had all to undergo this fiery ordeal, and one poor girl remained in a most dreadful state for a long time, and was left on the beach to die, her mother watching her from as near a spot as she dared; and at last the poor creature got rid of the fowl's-skin and begged her mother to run to the Governor, to say she was innocent, who immediately sent down soldiers to beat out her brains on the spot, as it had remained down so long.

Singing itself, along with dancing, lay at the heart of one of the most curious disorders known from Madagascar. This illness, known as Ramanenjana, reached epidemic proportions in the 1860s. Ramanenjana, which comes from a Malagasy root meaning "to make tense," was described by eyewitnesses as a kind of choreomania—spasmodic and severely prolonged dancing and singing. Similiar epidemics reported in Europe during the Middle Ages were known as "devil's dance" or "Saint Vitus's dance." Dr. Andrew Davidson, a physician who wrote a detailed description of Ramanenjana for *The Edinburgh Medical Journal* in 1867, considered it "a disease so strange that I might well have hesitated to record the facts, if they had not been witnessed by so many whose character and judgment place their evidence beyond question." Davidson noted key symptoms in victims:

> After complaining, it may be one, two, or three days, they became restless and nervous, and if excited in any way, more

especially if they happened to hear the sound of music or singing, they got perfectly uncontrollable and, bursting away from all restraint, escaped from their pursuers and joined the music, when they danced sometimes four hours at a stretch with amazing rapidity. They moved the head from side to side with a monotonous motion, and the hands in the same way, alternately up and down. The dancers never joined in the singing, but uttered frequently a deep sighing sound. Their eyes were wild, and the whole countenance assumed an indescribable abstracted expression, as if their attention was completely taken off what was going on around them. The dancing was regulated very much by the music, which was always the quickest possible — it never seemed to be quick enough. It often became more of a leaping than a dancing. They thus danced to the astonishment of all, as if possessed by some evil spirit, and with almost superhuman endurance, exhausting the patience of the musicians, who often relieved each other by turns, then fell down suddenly, as if dead; or, as often happened if the music was interrupted, they would suddenly rush off, as if seized by some new impulse, and continue running, until they fell down almost, or entirely, insensible.

Davidson felt that the disease, which rarely proved fatal, could be linked to religious and political upheaval then taking place in Madagascar. In 1861, Radama II assumed the throne and quickly reversed the xenophobic, anti-Christian policies of his mother, Ranavalona I. Foreigners were once again welcome in Madagascar, and freedom of religion was restored, with Christianity essentially becoming the country's official religion. Davidson noted that few Christians or city dwellers, who were generally in favor of the new policies, contracted the disease. Ramanenjana's victims were poor villagers in the countryside, where idolatry was still widespread.

Whatever the cause, foreigners called upon to treat the disease were at a loss as to what to do. After such fits, patients recalled that it felt as if a dead body were tied to them that they could not shake free, or as if some enormous weight continually pulled them downward or backward. They could not bear hats, pigs, or anything of a black color. Such anomalous symptoms left physicians bewildered enough to try anything, as this Englishman did on a Ramanenjana sufferer:

The strongest current of a four-cell battery did not seem to affect him, otherwise than to make him yell out at the top of his voice.

The hysterical attack still kept on, he holding intercourse with the old kings Radama and Andrianampoinimerina, as he afterwards told us in all seriousness. A douche of cold water stopped the attack for the time, and later on he sent his slave all the way to his house, about 12 miles away, to throw away a certain hat, which he believed was the cause of his illness.

Breaking fady could bring on similar fits of possession. Richard Baron, while traveling 1,200 miles in a palanquin through northern Madagascar in the 1880s, reported coming upon a woman who had just learned that her cook had used lard, a fady substance. "I found the woman on her knees on the floor, with open hands turned upwards, her eyes rolling, and her whole body in a tremor, while she gave vent to the strangest gibberings and jabberings I had ever heard," he wrote. "If ever woman was possessed, she was." At his wits' end, the woman's husband implored Baron to find some remedy. As much at a loss as those seeking to cure a fit of Ramanenjana, Baron approached the woman with a cup of cold water hidden behind his back. "Now, Ramatoa . . . I am going to cure you, don't be afraid, you'll soon be well," he said, and quickly dashed the water in her face. "Now, you are better, are you not?" She replied, *"Maiva, maiva* (Better, better)." "All right . . . now I think I can give you something that will quite complete the cure, but wait a little while, let the water act first." Five minutes later he slipped her a small piece of money, "more as a corrective to the first rather rude remedy than anything else," after which she replied, "I'm cured, God bless you."

The Sakalava have their own peculiar methods for curing illness, to which vazaha would not take kindly. Indeed, it is hard to know which form of illness, mental or physical, a Westerner would *less* rather have the Sakalava treat, for they have special "cures" for both. Arne Walen tells of a woman in Morondava, a town about 200 miles up the coast from Tuléar, who became possessed by a "demoniacal spirit" and ran about as if mad, ranting and raving. "The remedy for her illness was soon decided on—she was beaten; and she got indeed a sound drubbing," Walen wrote, adding that "radical evils require radical remedies."

Which is how one might describe the curious treatment of any Sakalava who falls sick: No matter what malady one is suffering from, one must eat copious amounts of food, on the principle that "food increases life," especially when one is weakened by illness. "This," Walen wrote, "is the only rule of diet which the Sakalava know with regard to sick people." He once visited the home of a "stout and strong" Sakalava boy suffering from gastritis, whose mother and relatives were force-feeding him maize from a large pot. ". . . I told his mother that her son would die solely on account of the food they had obliged him to eat," he recounted. "But nobody believed me, they rather laughed at my simplicity; and indeed the mother looked very pleased that her son had been able to take such a quantity of food." The youth died before sunset that very day.

For protracted illness, the Sakalava resort to an elaborately staged ritual known as *miantsa bilo*, "to sing the bilo." Dressed in their finest clothes, a sick person's friends and relatives, and not infrequently the entire village, gather around a tall, wooden platform bearing a ladder and, at its foot, a wooden effigy known as the *valy* bilo, "the bilo's consort," both specially created for the occasion. The sick person arrives, or if too ill to move, then a close relation serves as proxy. A few rifle shots open the ceremony, after which various family members and friends proceed to dance and sing.

Walter Marcuse, who took some of the first photographs of a bilo ceremony around 1910, describes what happens next:

> The patient, whose face has been rendered hideous by smearing it with white manioc powder, is carried up to [the platform]. A rope with a running noose is then slipped round a fat ox's neck, and the beast is dexterously drawn up to the stage by heaving the rope over a convenient tamarind bough. Then the animal is stunned by a blow from a club, and while it still breathes, the [medicine man] dispatches the poor beast by the barbarous method of cutting out its heart. Draining off the outpouring blood in gourds, the sorcerer raises them on long poles above the patient, and after an incantation pours the still warm liquid over him. During this ceremony, the male spectators create a terrific din by beating tom-toms, while the women clap their hands and sing dismal dirges unceasingly.
>
> Meanwhile, the ox is being cut up and roasted, hide, horns, and all, so that a sacrifice of chosen meats, generally the entrails, may be offered upon the platform or altar to the spirit presiding over the destiny of the sick person. A drinking bout of the wildest description follows these proceedings, and in it the patient is expected to join. He is also encouraged to participate in the subsequent dancing, a recreation which does not strike the onlooker as a particularly good "cure" for the invalid.

The ceremony reaches a fever pitch, with everyone imbibing great quantities of toaka gasy. "They load their muskets more and more heavily, and owing to their intoxication and consequent carelessness, often allow them to explode, sometimes causing serious accidents," writes Walen, who also witnessed bilo ceremonies. "[S]ome get their hands crushed, others are maimed in the face or chest, or some other part of the body." In the final act of the ritual, which Walen says typically lasts half a day, the villagers wash the patient in the sea, after which everyone returns home.

If all goes well, the patient recovers. A French businessman who worked in Tuléar in the 1930s described how a woman between fifty and sixty years of age, who had been seriously ill for some time and whom he had tried to cure in vain with Western medicines, had appeared "perfectly well and quite normal" after her bilo ceremony. If the patient dies, however, the medicine man may be held responsible. "This happened in one case that I know of," reports E. O. McMahon. "[A] chief's brother called Solifa was being treated in this way, and as the poor fellow was suffering from pneumonia, he died, as was to be expected, and Tsivatoa the medicine man was killed accordingly by Solifa's friends."

Thank God the Sakalava in camp never took notice of my sickness. I have no idea if they perform the bilo or force-feed the ill around here, but I certainly don't want to find out in my condition—which is now in its third day, by the way. Just when I thought I was over it after taking that pill out of Raxworthy's dubious-looking bottle, it came back with the vengeance of a Sakalava warrior. No dinner that evening, nor any food all day yesterday, save for some plain rice. Not a drop of energy, head spinning, diarrhea, vomiting.

Only last night did the three Fansidar (in case it was malaria), the three Tylenol with codeine (spaced over the day), and the Cipro (for gastrointestinal disorders) I took yesterday begin to gain any ground on this vicious affliction. I doubt a Sakalava poison fetish could have wrought more bodily harm.

4.

THE HERP COLLECTING THAT, God willing, I'll soon help Raxworthy undertake here is the essential first step in answering questions of speciation and endemism: First you have to document what lives where. Hence our use of pitfall traps and opportunistic searching and stumpripping, a protocol Raxworthy has followed in Madagascar for a decade.

Biologists like Raxworthy have strived to build scientific collections of Malagasy herps ever since an early collector first deposited a leaf-tailed gecko in the National Museum of Natural History in Paris in the late 1600s. While many early collections remain valuable to herpetologists today, Raxworthy cautions that data from the first centuries of herp gathering can be unreliable. In the early days, collectors made little or no reference to a specimen's habitat, behavior, or even place of recovery. Often the person who bagged an animal in Madagascar differed from the person who described and archived it back in European museums. A collector might sell a specimen to a third party, for example, who in turn would sell it to a museum in, say, London or Paris. As we've seen with *Zonosaurus boettgeri,* the chance for errors in identification and documentation was great.

Even in cases in which trained herpetologists collected, described, and archived their own material, problems have arisen. Some herpetologists have chosen to publish descriptions of new species privately rather than in peer-reviewed journals, or to keep specimens in personal collections rather than deposit them in museums or other repositories where scientists can have free access to them. In some cases, these hoarded exemplars have even included the type specimen—that is, the very animal used to describe the species. Such practices not only frustrate scholars who play by the rules but can lead to errors that remain in the literature for decades. Raxworthy says he has spent countless hours revising species descriptions, distributions, or localities that earlier investigators misidentified. Such work must be done, however, before scholars can properly address larger questions such as those concerning speciation.

Fortunately, the tide has begun to turn in recent years. Squirreling specimens away in private collections and publishing in nonrefereed outlets is no longer an accepted practice, and the collectors who first see and capture a species in the wild are often the same people who later describe, archive, and study it. People like Raxworthy, who

John Behler says has done "more than any other soul to gazette Madagascar's amphibians and reptiles and make sense of the chaotic state of the literature about them."

Following fieldwork, any new species need scientific descriptions, which Raxworthy and Ron Nussbaum are currently providing for upward of 150 distinct herps from Madagascar. The physical description of each animal that scholars publish in peer-reviewed scientific journals is almost obsessive in its specificity of detail. Take this characterization of just the head of a new species of dwarf chameleon, *Brookesia ambreensis,* which Raxworthy and Nussbaum recently described:

> Head with lateral, orbital and posterior crests that form a dorsal helmet; divided into three regions by a pair of longitudinal parasagittal crests that start above the eyes and begin to converge before terminating at the posterior crest; between the parasagittal crests there is a pair of short parallel longitudinal crests, with a short median crest anterior and posterior to the parallel pair; posterior crest of helmet notched between parasagittal crests; three similar-sized pointed tubercles on each side of posterior helmet crest, one at termination point of lateral crest, one at termination point of parasagittal crest, and one between parasagittal and lateral crests; two pointed tubercles on lateral surface of head, one just above posterior angle of mouth, and one below lateral crest in temporal region; orbital crest dentinculated; supraocular cone rounded, does not project forward of the nostril; supranasal cone does not project beyond snout tip; horizontal distance between snout tip and anterior border of eye 0.9 times eye diameter; head longer than wide; chin and throat with longitudinal rows of slightly enlarged tubercles.

And that's for a head the size of a pea.

The problem is, many of Madagascar's dwarf chameleons look superficially alike. To distinguish one from another, taxonomists like Raxworthy must make the most minute analysis of morphological characters, or body traits. This includes external features such as coloration and, say, the distance between the snout and the vent.

When the going gets rough, as it does among the dwarf chameleons, it may also include internal morphological characters. Among the *Brookesia,* for instance, the hemipenis holds diagnostic characters, ones that can be used to differentiate species.* Taxonomists separate Madagascar's dwarf chameleons, housed in the genus *Brookesia,* from Africa's dwarf chameleons, of the genus *Rhampholeon,* on hemipenis characters, as they do species within the *Brookesia* genus itself, for chameleons in this genus often look on the outside like the spitting image of one another.

Where possible, scholars match such descriptions with those of ecological and behavioral characters. How abundant is a given species? What altitude range does it prefer? Does it have a courtship display, and if so, how does it differ from that of other species? Where does it sleep? The idea is to give as complete a portrait of a species and its place in nature as possible. Only then can biologists like Raxworthy begin to address the bigger questions of where it originated and how its kind may have speciated so wildly.

5.

OUR WORK TO GATHER the information needed to describe such characters starts early. "I think we should go out and start sweating," Raxworthy says about eight o'clock on my fourth morning at Lake Ranobe, the first I've felt well enough to participate. "It's just about getting hot enough to start work."

I smile, thinking how oblivious Raxworthy acts toward any discomfort. At this time of day, you can be fooled into thinking the temperature will be pleasant; nighttime temperatures drop into the quite comfortable seventies. But as the sun rises higher, the heat quickly becomes appalling. "By reason of the increasing power of the sun beating down upon the unprotected mountain-sides," wrote a botanist working in this region in the last century, "we were obliged to abandon our botanical research in the coral-tree thickets . . . as early as half-past seven in the morning. . . ." But will a little heat stop Raxworthy—or, by extension, the rest of us?

Some members of the expedition head off to check the pitfall traps that Raxworthy and his Malagasy assistants planted in the gallery forest near camp before we arrived. But after losing three days to sickness, I opt to join four Malagasy on an all-day opportunis-

---

*Reptiles have retained a dual penal structure known as the hemipenis, while most mammals, including humans, have effected a fusion of the two parts.

tic search in the spiny desert. We hope to reach the edge of a plateau whose western edge rises up out of the sweltering plains about ten miles to the east. Cliffs there mark a change between the sea-level spiny desert and a lofty, calcareous plateau, and between Sakalava country and that of the Bara, who live in the south-central part of the island. Raxworthy wants to see if a certain species of rock-dwelling "iggy-wanna," as he pronounces it, lives in those cliffs. It should take us several hours out and as many back. My companions are the powerfully built yet soft-spoken Achille, carrying his regular herping gear of floppy sun hat, wrist rubber band, cloth sacks, and stumpripper; Angelin, ready as ever to guffaw at Achille's jokes; a quiet man from Tuléar whose role on Raxworthy's expedition I have yet to learn; and James Bond.

A local Sakalava man, James Bond gained his nickname when a volunteer learned that his Malagasy surname sounded like that of Ian Fleming's character. The name—we never call him Bond, always James Bond—has stuck partly out of relief, the relief Westerners feel when offered a way out of having to attempt pronunciation of Malagasy names. But it has remained mainly because of his personality. Even as Raxworthy introduced him to us that first afternoon, it was obvious the man was crafty. About five feet five, he stood defiantly with hand on hip and head cocked to one side, not smiling, his dark, staring eyes bespeaking a cold calculation. It was the first time I'd seen a Malagasy with attitude. He wore nothing but a pair of beat-up shorts—and that attitude. Shining but sweat-free in the tropical heat, his taut skin and fit, gangly body made it impossible to tell his age. Forties? Sixties? Just when I began wondering whether he ever smiled, something one of the other Malagasy said made him throw back his head, open a mouth with no front teeth, and squeal in an unsettlingly high-pitched laugh not unlike Bart Simpson's.

The warmth of most Malagasy makes you feel an immediate, well, bond with them. Over the coming days, however, I could sense that most of the volunteers, particularly the women, didn't want to get too close to James Bond, as if he were a homeless person, say, or a beggar. And, indeed, his seeming defiance, exacerbated in many eyes by his fondness for toaka gasy, might, if he were the only Sakalava you ever met, have left you suspecting more than a kernel of truth in the otherwise ludicrous characterization of the Sakalava by an early visitor to these parts: "They have . . . supercilious airs and bold appearance, having good mental power and strong and easily excited passions, making them rude, wild, and often raging in their conduct. On the whole, the Sakalava are a sly, perfidious, brutal, and arrogant people, given to stealing, drinking, fighting, and plundering at every place where they make their appearance." I have detected a certain shyness in James Bond, however, along with a profound sadness born likely of a life harder than most of us can even imagine, which has left me feeling strangely protective of him, protective against what I perceive as an open aversion by some of the volunteers. True to the contradictory nature inherent in his people, he evinces both strength and vulnerability.

An incident the day before yesterday, or rather two incidents, brought this dual nature out. Garfield Dean was relaxing on the blue tarp in camp, gazing up at the sky, when a drop of burning sap from a euphorbia bush growing overhead landed in his eye. (Poor Garfield: He seems to attract misfortune like Murphy of the infamous Law.) The milk-white latex from certain species of euphorbia can cause blindness, and we feared this might be one of them, for Dean's eye turned strawberry-red and swelled like a toad after the rains. Without a word, James Bond slipped into the forest and returned with a euphorb stem. To our untrained eyes, the plant looked exactly the same as the one rearing over the tarp. He snapped the stem in half and, with Dean's hesitant permission, dripped some of the milky fluid into his eye.

The next morning, Dean was back to normal—and ready to repay the favor. Since our first day here, James Bond had been complaining of a toothache. The Malagasy used to call toothache sufferers "poorly through the worm," as they believed the pain was caused by a small worm inside the tooth, and anyone looking into the mouth of one so afflicted would have the same thing befall him. But Dean, if he even knew of the superstition, did not hesitate when James Bond asked his new friend if he would pull it. As you know, Dean designs air-traffic-control systems, and he had never done such a thing in his life. But he went at the tooth with a pair of pliers from Raxworthy's toolkit, and out it came. James Bond never felt a thing, having steeled himself beforehand with an entire fifth of toaka gasy.

He has apparently steeled himself similarly for our hike today. It's a bit disconcerting to find your chief guide for a daylong trek across a largely trackless desert bombed at the start. Nor did I find it reassuring when James Bond, while leading us through Ranobe, the Sakalava village nearest our camp, needed a crippled five-year-old to hobble out of the village on his homemade crutch to show us where the trail to the cliff began.

Achille and Angelin seem unconcerned, and we press on into the otherworldly spiny desert. Looking around me, I soon forget all about James Bond. The landscape has all the expressiveness of a van Gogh painting. Between the red, sandy soil and the cobalt-blue sky grows an assortment of outlandish plants. Green "octopus trees" soar twenty or thirty feet into the air like fireworks caught in midburst. Their long, tentacle-like trunks bristle with four-inch spines like rows of shark's teeth, ready to impale you if you get too close. Along those dry gray trunks run lines of tiny, boxwoodlike leaves that, from afar, give the trees a deceptively furry appearance. Beneath the octopus trees huddle all manner of euphorbia, their rubbery, finger-thick branches intertwining with the geometric precision of a chemist's model. These plants have no leaves; their smooth green branches handle photosynthesis. I break apart a euphorb twig and watch the oozy white latex ball up at the broken ends. What would it feel like to have that in my eye?

Every now and then we come upon a loose cluster of baobabs. Baobabs are enormous—one in the western town of Majunga measures forty-six feet around at the base—and their smooth-skinned trunks are completely devoid of branches until the crown, which, in the tallest species, can be six stories off the ground. There lies a disheveled hairdo of scraggly branches, looking in some species entirely too few and too tiny for such a massive tree. These rootlike branches have lent the baobab the name "upside-down tree." One writer deemed the genus "a Caliban of a tree, a grizzled distorted old goblin with the girth of a giant, the hide of a rhinoceros, [and] twiggy fingers clutching at empty air."

For their part, the Sakalava call it *reniala*, the "forest's mother," because within the bloated, hard-skinned trunk they can find up to 200 gallons of fresh water stored in the tree's pulpy tissues. In times of drought, the Sakalava tap these trees for their water, and even in normal conditions will cut them down and slice them lengthwise to provide water-soaked pulp for their zebu. Life in this environment, it's clear, is one big struggle against water loss. The Tropic of Capricorn passes just south of Tuléar, yet this scorching, bone-dry desert seems about as far from the temperate zone as the Sahara is.

As is often painfully obvious in this open forest, the baobab's barrel-like trunk is just one of many adaptations that help spiny-desert plants cope with the harsh climate. Many trees here have swollen trunks, roots, or leaves. The "elephant's foot," for instance, a relative of the baobab, looks like just that. The octopus trees, on the other hand, have gone in the opposite direction, featuring trunks no thicker than my arm and tiny leaves whose small surface area helps reduce water loss. These trees, along with many others in the spiny desert, drop their leaves in the dry season—yet another strategy to retain water. It's now the end of the rainy season, though it can hardly be called that, and many plants have new green leaves, including the octopus trees and baobabs. Some plants have leaves with waxy or hairy coatings, others droop their leaves for most of the day to protect them from the direct rays of the sun, still others forgo leaves altogether in favor of spines or just branches as in the euphorbia.

These defenses can do nothing for them against the charcoal cutters, however. All too frequently, we cross an ashy spot along the trail where locals built a small fire to cook a pot of rice while collecting fuelwood. Sometimes we pass tidy bundles of wood stacked on the trailside, waiting to be hauled out to the road and burned to charcoal. We've come upon clearings where cutters have piled up octopus-tree trunks that will soon help shore up a house. The desert around us seems as empty as a graveyard, yet people obviously visit often. Indeed, they are quickly deforesting the spiny desert north of Tuléar, including PK-32. Biologists have declared the proposed reserve a Site of Biological Interest, but that scientific designation means nothing in terms of protection, and the cutters are free to do as they please.

We're keeping our eyes peeled for some of the creatures of "biological interest." One of them is the radiated tortoise, so-named for the striking black-and-yellow starburst patterns on its domed, basketball-sized shell. Another is *Phelsuma standingi,* a handsome gecko with a greenish-brown camouflage pattern that lives on baobabs. We're also looking for *Zonosaurus trilineatus,* a sleek brown plated lizard and cousin to our old friend *Z. boettgeri* from Nosy Be. Raxworthy suspects that *Z. trilineatus* ("three-lined") is actually the same species as *Z. quadralineatus* ("four-lined"). Both species live in the area, and he wants to collect enough specimens of each to prove that the difference in line number constitutes natural variation within the same species. I, and perhaps I alone, also keep an eye out for two birds that, since they're known in all the world only from this patch of spiny desert north of Tuléar, have been deemed "among the highest priorities for bird conservation in the Afrotropics": the subdesert mesite and the long-tailed ground-roller.

Aside from occasional murmurings between Achille and Angelin, no one has spoken since we left the village. We're on a mission, the rising heat suppresses superfluous chat, and what's there to say anyway? My Malagasy is limited to the odd phrase such as *mafana be* (very hot), which my companions have heard enough already from my lips on this trek, and these guys range from speaking English haltingly (Achille) to not at all (James Bond). We focus on the task at hand: opportunistic searching for herps. Every few steps a tiny yellow sand iguana with an arresting "third eye" atop its head scuttles across the sand. That species' ubiquity, though, contrasts with the near-total absence of other animals. It's getting so hot that even the lizards are hiding. We spot one of them—a skink—in a heap of bark sherds littering the base of a dead baobab; Achille and Angelin go after it in vain. Later, looking for a species of nocturnal gecko that spends the day sleeping on tree trunks, I peel back a strip of bark only to scare up a seething mass of tarantulas, scorpions, and giant hissing cockroaches. (Needless to say, I abandon the search.) I'm beginning to notice a strange imbalance in my senses: The desert has few sounds and is too dry for smells, but it makes up for it in sights and touch. I don't know how the Malagasy get by without sunglasses; my eyes hurt even with them on.

The going gets rougher. The "road" has petered out into a confusing network of trails. Increasingly we're forced to crouch down and push through dense brush. Beneath those soaring octopus trees lurks an army of plants just waiting to ambush me, it seems. (The others don't seem the least put out.) Shrubby acacias stab me with unseen thorns, and thin vines bearing fishhook barbs latch onto my bare legs, leaving stinging lacerations when I yank them free. I have to watch not to put an eye out on some five-inch thorn or broken twig, or to twist an ankle in the occasional thigh-deep hole some Sakalava has dug to retrieve a wild yam root. Toward noon, as I pick my way through the bush, I suddenly bolt like a horse, hooting and slapping, a victim yet again of a yellow wasp attack. I manage to get away with just one sting, but it *hurts*, dammit. From now on, whenever I hear anything even remotely resembling a buzz, I'm going to break into a sprint.

Through it all, the heat beats down as indomitably as a migraine, which I won't be surprised if I get before the day's out. Yesterday peaked at 102°F in the shade, which is common enough around here. "The heat was grilling and the path devoid of all shade, while, to make matters worse, a blinding glare rose from the white, salt-encrusted earth," wrote one early traveler through the spiny desert. Another opined that "the air seemed to scorch the nostrils and lips in breathing." Still another recalled "sweltering in a heat beyond description. The sun, a ball of molten brass set in a cloudless sky, beat down upon a silent stifled earth [and] even the rickshaw coolies, naked saved for their scanty breechclouts, lay stretched in the shade . . . panting like fish tossed upon a bank to die." All I can manage is an increasingly enfeebled "mafana be," which my companions have surely noticed no longer includes a sympathetic smile.

And now we're waiting for James Bond. He's *behind* us, doing God knows what. Perhaps he's looking for the proper trail, though heading due east will get us where we're going. Maybe he's lost, which is easy enough to become in this flat landscape with no geologic features to guide you. As the rest of us take to the shade of a baobab to await his return, I recall with a twinge the time a few days ago when one of the Earthwatch volunteers vanished in the desert. It was on that day when my illness briefly went into remission, so I was able to join the search. It was scary, let me tell you. The volunteer was an overweight, fifty-year-old lawyer from New Jersey who had recently undergone a triple bypass operation. Of the twenty or so people in our camp, he was the last one you'd want lost in the spiny desert. Yet somehow he became separated from his group early in a "long march" just like the one we're on now. Thinking he must have turned back, the group continued with its work, returning to the camp early in the afternoon, just when the heat had reached unbearable proportions. When they learned he hadn't come back, they were shocked. A sense of alarm shot through the camp like a bolt of lightning. It was the only time I've seen Raxworthy truly worried. Had the man collapsed along the trail? Was he dead? If not, how could he possibly survive?

Within minutes, Raxworthy had dispatched three separate search parties. I joined a group of three or four Malagasy led by a local forest ranger, who made us wait in Ranobe village while he went to his house and inexplicably returned with a loaded revolver. Did he suspect foul play? As we set off away from the lake, I half-expected to find the volunteer's body frying in the sun. But before we even got into the spiny desert proper, a beaming Angeluc appeared ahead of us on the trail, the lawyer at his side, exhausted but otherwise fine.

The man told us he had sat down to take a break when he noticed his heart rate had climbed to 165. The trail was wide and clear, but why he thought he could catch up with the group, all of whom were far fitter than he, I don't know. Anyway, he got lost. Remembering Raxworthy's warning that if you get lost, walk west, because you'd eventually reach either the lake or Route Nationale 9, he did so, forgoing the confusing trails to crash due west through brambles, compass in hand. Over the next six hours, with his water running out and thoughts of the worst coursing through his brain—"I imagined a headline in my local paper: LOCAL LAWYER DIES IN MADAGASCAR," he told me—he got stung by wasps, scratched by plants, and so startled by a zebu that he spent half an hour ten feet up a tree. The man didn't leave camp for the next three days.

The lawyer's ordeal was a stroll in the park compared to that under-gone by Robert Drury. In the early 1770s, this young Englishman spent *weeks* surviving alone in the spiny desert. If that had been his only misfortune, Drury would have jumped for joy, however, for it was only one small part of his fifteen years as a slave in Madagascar. Chronicled in his book *The Pleasant and Surprising Adventures of Robert Drury, During his Fifteen Years' Captivity on the Island of Madagascar*, Drury's story is one of the most remarkable ever to come out of the island. First published in 1729, his book was long considered the prin-cipal source work in English on Madagascar, as valuable as the Frenchman Etienne Flacourt's 1661 *Histoire*.

Almost as soon as it came off the press, the book fell under suspi-cion, however. Discerning readers noticed that certain passages appeared to have been lifted straight out of Flacourt's work, while some of Drury's philosophical musings bore an uncanny resemblance to those of Daniel Defoe, the prolific author of *Robinson Crusoe*, who shared a publisher with Drury. After two and a half centuries of debate, though, most scholars stand convinced that the work is gen-uine. Many details, including historical figures and events, can be independently corroborated, and the narrative rings true both to casual readers and to people intimately familiar with the region and its history, who claim that Drury's story could only have been recounted by someone who had spent considerable time in the area. An editor, perhaps Defoe himself, probably heavily embroidered Defoe's manuscript, but the basic story is true.

In 1701, when he was only thirteen years old, Drury convinced his father to let him try the life of the sea and shipped out on the *Degrave*, an East Indiaman bound for Calcutta. After nine months in Bengal, the *Degrave* set sail for home. Running aground shortly after leaving Calcutta, the ship was refloated and went on its way, leaking heavily yet apparently seaworthy. But as the *Degrave* approached the southern tip of Madagascar, order was given to abandon ship. One hundred and sixty Englishmen (Drury makes no mention of women or children) came ashore near Cap Sainte Marie, the island's southernmost point.

Here their nightmare began. The king of the local Antandroy, one of the most warlike of Malagasy tribes, welcomed them but made it clear he would never let them go, as the presence of Europeans increased his clout with neighboring tribes. Over the coming days, with forced marches in the desert under armed guard, it became obvious that the king was telling the truth. The Englishmen rioted, temporarily holding the king himself hostage, but were subdued. When the Antandroy discovered that several of their captives had slipped away in the night, they slaughtered the remaining Englishmen, all save for Drury and three other youths. One of these escaped, another died, the third was killed, and Robert Drury began fifteen years of slavery.

He became the slave of the grandson of Deaan Crindo,* the regional Antandroy king. Drury appears to have spent about ten years among the Antandroy, tending his master's zebu herds, learning the tribe's language and lifeways, even marrying. He told the bride's mother not to worry, that "I would take more care of her than of myself, and though I was not a black man, I had as tender a heart as any black man whatever, and designed to make her my wife, if she liked it." She did, and so Drury devised a simple marriage ceremony, and "we lay down and were as happy as our circumstances would admit for, notwithstanding we had no bridemen or maids, nor throwing of stockings."

One day an ambassador from the Fiherenana, a tribe living along the river of the same name north of modern-day Tuléar, arrived among the Antandroy to plan attacks against a common enemy, the Mahafaly. Another proud and hawkish tribe, the Mahafaly lived in the extreme southwest, between the Antandroy and the Fiherenana. The ambassador said if Drury could get to the Fiherenana, about twenty days' travel to the north, he would see that Drury got on the first English ship that put in at Saint Augustine's Bay. Not long after the ambassador's visit, an Antandroy *ombiasy*, or medicine man, predicted that Drury would escape to the north and eventually return to England. He cast a spell by making Drury eat scrapings of various

---

* Modern spelling: Andriankirindra. One of the most convincing aspects of Drury's chronicle is his transliteration of Malagasy names, which the English historian Mervyn Brown says is "just what one would expect from a Cockney with limited education and an imperfect ear for language."

roots that were supposed to make him ill if he tried to escape. Drury realized that if he ever did fall ill, his master would consider it proof of an attempt to escape and kill him.

He decided he must go immediately. He tried to convince his wife to join him—he was, he said, "downright in love" with her—but she was fearful of the ombiasy's spell. He slipped away to a northern Antandroy village, where he was led to believe he would find help in getting to the Fiherenana. But his contact there enslaved him as well. After six months, he managed to escape, taking more than three weeks to cross the spiny desert to the Onilahy, a river that flows west into Saint Augustine's Bay. One night on the way, he was attacked by what he calls a fox, but which must have been a fossa, the pumalike carnivore of Madagascar. He killed the animal, but not before it had bitten him in the foot, which swelled up and forced him to lay low for six days. He crossed the Onilahy on a makeshift raft, and several days' more on foot brought him to the land of the Fiherenana.

They greeted him warmly but gave him bad news. The tribe was at war not only with the Mahafaly to the south but with the Sakalava to the north. Continual fighting had impoverished the region to such a degree that European ships rarely put in at Saint Augustine's Bay to trade. He learned, too, that the Fiherenana were organizing another campaign against the Mahafaly, and another ombiasy claimed that it would only be successful if led by someone born outside Madagascar. Drury had no choice but to retrace his steps as the Fiherenana traveled to Antandroy country to unite with their allies. He met with both his former masters; one of them pleaded with him to return, saying his wife had been inconsolable since he'd fled. But he managed to return to Saint Augustine's Bay with the Fiherenana after the campaign, which proved brief.

Drury's life must have seemed to him like a recurring nightmare. Learning that the Mahafaly were launching an attack from the south, the Fiherenana retreated north, Drury with them, only to be ambushed and vanquished by a Sakalava army. Drury was made the slave of a six-foot, eight-inch tall Sakalava warrior, the grandson of the great Sakalava king Andriamanetriarivo. Rer Trimmonongarevo, as Drury called him, was perhaps the most powerful Sakalava ruler at the height of the Sakalava empire, and thus the most powerful man in Madagascar at the time. "He was an old man, not less, by what I could find, than fourscore years of age: yet of robust and hardy constitution," Drury wrote. "His color rather tawny, like an Indian, than black; his eyes fierce, and his whole appearance frightful, or his singular habit and character made me think so."

During his captivity with the Sakalava, Drury met a fellow Englishman, William Thornbury, at Yong-Owl (modern-day Morondava). Thornbury had been marooned as well and, like Drury, was hoping to get away on the next English ship—which soon arrived, as it happened. Hearing that Thornbury had boarded the vessel, Drury tried to send a message to the ship written on a leaf. But his illiterate Malagasy messenger lost it and substituted another leaf, which obviously had no effect. The ship sailed. Drury sank into a deep depression and fell seriously ill with yaws. When he recovered, he became the slave of one of Andriamanetriarivo's sons, who reproved his nephew—Drury's former master—for treating him so poorly. Wasn't it trade with the English that had made the Sakalava strong? For two years, Drury then lived in reasonable comfort, even overseeing his master's armory of more than 100 muskets.

Then he heard that two English ships had put in at Yong-Owl. Shortly thereafter, two seamen arrived in his village with a message addressed to "Robert Drury, on the island of Madagascar." It was from his father, who had learned of his fate from Thornbury. To Drury's astonishment, his master let him go. After fifteen years in the island, he had forgotten how to speak English and his hair was so long and matted that the ship's crew deemed him a "wild man." But he was soon cleaned up and on his way. The ship sailed to England by way of the Caribbean, arriving in the Downs on September 9, 1717. It had been more than sixteen years since he'd left on the *Degrave,* and he was now almost thirty years old.

It was a disappointing return. Both Drury's parents had died, and his surviving relatives barely remembered him after such a long absence. He tried to work as a bookkeeper and clerk, but "things did not answer my expectation." How could they? Within a year, he went to sea again—ironically, on a slave-trader—and even visited his final Malagasy master, who had become king of Menabe, the southern half of the vast Sakalava kingdom. Drury's tale ends with his return from this voyage, and almost nothing is known of the rest of his life, though he lived more than two decades more, dying sometime between the third edition of his book in 1743 and the fourth in 1750. "In his last years," writes Mervyn Brown, "he was to be found at Old Tom's Coffee House in Birchin Lane, where he was ever ready to regale gentlemen visitors with authentic tales of adventure among the wild tribes of Madagascar."

Like Robert Drury on one of his long marches, James Bond finally drags himself back among us, trying to look as bedraggled and spent as he can. In fact, his shirt—a yellow T with THE ORIGINAL JEANS 701 printed above two blond American teenagers—is spanking clean and dry. (Mine, at this point, is scuffed, filthy, and more liquid than solid.) He complains to Achille that the "road" doesn't lead to the cliffs or it's too far or he just wants to get home and have a drink. Whatever, we turn around and head back. The closest thing to a cliff we've seen is a few towering termite mounds. Raxworthy's "iggy-wannas" will just have to wait.

Achille, even though his upcoming doctoral thesis is on *Zono-saurus*, apparently gives up on trying to find the "three-lined," because we spend the hike back fighting through brush to each baobab we see, looking for *Phelsuma standingi*. This hefty, bulgy-eyed gecko is so prized by reptile collectors the world over that it has been all but hunted out locally. When Raxworthy pressed him back at camp, James Bond admitted he does a bit of collecting himself now and then (which probably means regularly). *Phelsuma*s are on Appendix 2 of the Convention on International Trade in Endangered Species (CITES), which means that collectors must have permits and adhere to limits on how many they can take. But there's no regulation, Raxworthy says. "Anybody asks you, you show them the paperwork and say 'I caught 10' or 'I caught 20,' and you leave with 150," he told me. The markup is extreme. For a healthy *P. standingi*, middlemen currently pay local collectors like James Bond 5,000 Malagasy francs, or about $1.20.* In the States, Raxworthy says, a pair will sell for anywhere between $80 and $200.

The markup for the radiated tortoise, which we've also been hoping to find, is astronomical. Local Malagasy sell these strikingly marked tortoises beginning at 15,000 Malagasy francs (about $3.60). The most highly prized specimens—those bearing the most distinctively marked shells and adult females that can reproduce—sell in the States for up to $10,000. With a markup like that, it's little wonder we haven't seen any.† And it's little wonder that the species' CITES sta-

---

* The exchange rate has increased to about 4,000 Malagasy francs to the dollar since my visit to Lokobe.
†A *New York Times* reporter accompanying a local collector near Tuléar recently watched incredulously as the man and his partner, in an hour and a half's work in the spiny desert, gathered fifty-four radiated tortoises.

tus on Appendix 1, which forbids trade, means nothing. Middlemen simply avoid Malagasy authorities by arriving at night near Tuléar in powerful speedboats and whisking out their contraband before daylight.

Perhaps confirming my suspicions about the degree to which he collects, James Bond is the one who spies a *P. standingi* high on a baobab. The banana-sized gecko suction-grips the smooth bark just below the rootlike branches splaying out at the crown. It must be thirty feet off the ground, but that doesn't stop Achille. He can't use his rubber band; even he can't shoot that straight. Instead, he finds two thin saplings about three times his length and expertly lashes them together with vines. Holding the base of his swaying tool with both hands to steady it, he tries to scrape the gecko off the trunk. But it is savvy and simply crawls higher, out of reach. Perhaps it has seen too many of its brethren kidnapped.

Achille, in the wonderful way most Malagasy have of making the best of every situation, simply grins and drops the stick in the sand. Angelin darts his smiling eyes at me and laughs. I smile back and shrug, as if to say, "What can we do? Good for the gecko!" He shrugs back, eyebrows raised, and laughs again. Achille, still grinning, starts shrugging, and so does the quiet man from Tuléar. Soon we're all laughing and shrugging. All except James Bond: He stands mulelike, arms crossed. Then, without warning, he throws back his head and squeals with that Bart Simpson laugh, his toothless gums glistening in the sun.

6.

HOW DID A SPECIES SUCH as *P. standingi* evolve? What is its evolutionary relationship with other *Phelsuma* species? When did it go its own way on the family tree? Many biologists today rely on a method known as cladistics to work out questions of evolution—the next step on the ladder to possible answers regarding speciation.

Developed in the 1950s by the German entomologist Willi Hennig, cladistics is a means to establish the evolutionary relationships of organisms, either living or extinct. In the past two decades, the method has come to dominate evolutionary studies, ranging from the origin of birds to the distribution of present-day animals. Cladisticians build family trees, or phylogenies, based on features called "shared derived characters." Vertebrates, for example, all have backbones, which they inherited from a common ancestor. For extinct animals, paleontologists rely solely on morphological characters, but for living animals, biologists can turn to all manner of characters, whether morphological or behavioral, genetic or chemical. Cladistic studies of human evolution, for example, can look at the development of physical characters such as bipedalism and sexual dimorphism as well as emotionally freighted traits such as sexual identity and intergroup conflict. The goal is to construct a family tree using as many well-established characters as possible.

After a decade of surveying and describing Madagascar's herps, Raxworthy has just begun to examine the evolutionary relationships between them using a cladistical approach. For his purposes, a robust family tree of, say, the chameleons, should go a long way toward answering questions of when and where they arose.

That question—whether chameleons evolved on Madagascar and spread elsewhere, or vice versa—fascinates Raxworthy, who is trying to answer it with a second method of studying evolutionary relationships. He is sequencing the mitochondrial DNA from all those samples of liver and muscle that he has been carefully removing from specimens in the field and placing in tanks of liquid nitrogen.

Not long ago geneticists discovered that mitochondrial DNA in animal cells can serve as a valuable evolutionary clock. Mitochondrial DNA can do this for several reasons. It is inherited only from the mother, it retains its integrity even through multiple sexual recombinations, it apparently mutates at a regular rate, and those mutations appear to have no consequences for a species' fitness. Biologists can consult this molecular clock within closely related species to develop a well-supported hypothesis about when they diverged from one another. Raxworthy has just begun to do this with the chameleons. "It's still in the early stages of the analysis, so I'm a little cautious as to the interpretation," he told me. "A year or so ago, I would have said that the probability is that the chameleons are Gondwanan in origin. But now I'm leaning toward the thought that they actually evolved in Madagascar."

He has also begun to use biogeography to answer questions of endemism and speciation in Madagascar's herps. Biogeography is the study of how plants and animals got to be where they are today. The field took off after the publication in 1967 of the book *Island Biogeography* by the ecologist Robert MacArthur and the myrmecologist Edward Wilson. Biogeographers have indeed conducted much of their work on islands, which are thought, because of their relative isolation and limited number of species, to serve as simpler places than continents in which to investigate the steady march of evolution. The most famous islands subject to island biogeographic study are, of course, the Galápagos.

Island biogeographers typically build models based on two processes: immigration and extinction. New species arrive on an island by dispersal and, if they are fitter than resident species, may cause one or more of those residents to go extinct. But standard models of island biogeography, Raxworthy says, usually leave out a third factor: speciation. "That seems a little strange, because you think of islands as having lots of endemic species," he says. The reason is that many biogeographers consider active speciation a continental phenomenon. Since Madagascar is arguably more continental in nature than islandlike, and since the island's long isolation has caused both immigration and consequently extinction to be minimal, speciation must be factored in. Indeed, while admitting that evidence is not yet at hand to prove which of the three processes is most responsible for the high species diversity, Raxworthy says that "if somebody pushed me to speculate wildly," he'd lay his bet on speciation.

If there ever was a place to examine whether island biogeographers should add speciation to their models, not to mention how speciation actually occurs, it is Madagascar. Despite the enormous biogeographic interest in the island, however, scientists have undertaken only a handful of detailed studies of biogeographic patterns among Malagasy vertebrates. Based on the distributions of lemur species, the anthropologist R. D. Martin divided the island into seven areas of endemism. The herpetologist Mathias Lang described patterns of endemism among cordylid lizards (a family of armor-plated reptiles). And Raxworthy and Nussbaum proposed new biogeographic divisions in northern Madagascar based on the distributions of *Brookesia* chameleons.

The latter study, published a few months after we returned from Lokobe, was based in part on our work there. And it is one of the first studies ever undertaken on Madagascar that examines how speciation might have run its course, at least in this one area of the island.

If you're going to study endemism and speciation on Madagascar, dwarf chameleons are a good place to start. With twenty-three species identified so far, chameleons of the genus *Brookesia* are more speciose than either of the other two Malagasy chameleon genera, *Calumma* and *Furcifer*.* After years of fieldwork, Raxworthy and

*Among Malagasy herps, the *Brookesia* are second in diversity only to skinks of the genus *Amphiglossus*.

Nussbaum have determined that fifteen of those twenty-three species live in northern Madagascar, a region defined as north of 16° south latitude, which cuts across the island roughly from the Masoala Peninsula on the east coast to Majunga on the west coast. This is an extremely high percentage considering that the northern rain forests represent only about 5 percent of the island's total area. "Clearly," Raxworthy has written, "speciation in *Brookesia* has been disproportionately active in the rain forests of northern Madagascar." Moreover, all but three of those fifteen species exist *only* in northern Madagascar, many of them in but one locality. This is a high level of endemicity even for rain forest.

How did such active speciation and endemism come about? Raxworthy and Nussbaum approached the question by looking at distributions of the dwarf chameleons. The distribution data showed where those fifteen species resided geographically, and what habitat and altitude they preferred. The pair found dwarf chameleons throughout the northern rain-forest belt, which stretches from Nosy Be in the northwest to the Masoala Peninsula in the northeast and includes an isolated patch at Montagne d'Ambre, a massif near the northern tip of Madagascar. Except for one species found in the dry-forest limestone canyons of the Ankarana, near Montagne d'Ambre, no species turned up outside of rain forest. (Dry forest takes up about half of northern Madagascar.) All fifteen species have restricted elevational ranges as well, the researchers discovered. Low- and mid-altitude species live in rain forest below 1,650 feet and between 1,650 and 3,960 feet, respectively, while high-altitude specialists occur in lichen forests above 3,960 feet.

Based on these findings, Raxworthy and Nussbaum divided northern Madagascar into five new biogeographic zones: northwest, northeast, east, Montagne d'Ambre, and Tsaratanana. The latter two turned out to be the key to hypothesizing how such intense speciation and high endemism occurred here. Montagne d'Ambre lies by itself far in the north. Today, almost seventy miles of dry forest lie between it and the nearest patch of lowland rain forest to the south. Yet Montagne d'Ambre holds the three *Brookesia* species we found at Nosy Be, in the heart of the northwest region. How could this be? Those species are what Raxworthy calls "obligate rain-forest specialists," meaning they can't live outside rain forest.

Through vicariance, Raxworthy and Nussbaum theorized. Those three species could not cross seventy miles of dry forest. The only rational explanation is that lowland rain forest once connected

Montagne d'Ambre and the northwest region. Climate change must have caused a drying out, forcing rain forest in the region to become dry forest and cutting off those two parts of the same population. There is evidence for this in East Africa, Raxworthy says, where pollen analysis has revealed significant expansion and contraction of rain forest during the late Pleistocene epoch (1.6 million to 10,000 years ago). Since Pleistocene climatic changes left their mark globally, it is reasonable to assume—in the face of a lack of hard evidence in Madagascar—that they also had a strong impact on Malagasy flora.*

There was another telltale element. Montagne d'Ambre has an endemic species of *Brookesia* that appears to be the sister species of one found in the northwest region. (Sister species are more closely related to each other than to any other species.) By the same token, two species living either to the east or west of the Tsaratanana Mountains, but not on both, also appear to be sister species. Today, dry forest lies to the north and south of the Tsaratanana, forming an effective barrier to any crossing, along with the mountains themselves, which are the island's highest.

Both cases, Raxworthy and Nussbaum suggest, appear to be classic examples of vicariant, allopatric speciation. One species evolved into two after a significant barrier—in this case, dry forest—arose to separate two segments of the same population. Climate change may also be responsible for the high regional endemicity, they say. During Pleistocene glacial periods, when a far higher percentage of the world's water was locked up in glaciers and ice sheets than during interglacial periods, lowered rainfall would have caused the rain forest in the complex mountain systems of northern Madagascar to contract, leaving only small refugia. *Brookesia* would have had to follow suit, with the result that considerable allopatric speciation took place.

7.

RAXWORTHY IS NOW DRIVING a group of us to the cliff face James Bond failed to get us to the other day, to try to resolve a case of species distribution even more striking than that of the *Brookesia*. Previous investigators have reported *Oplurus saxicola*, that rock-dwelling "iggy-

---

*Published palynological studies of Madagascar's early environment currently reach no farther back than about 36,000 years.

wanna," from this region. But Raxworthy has only seen it in the south-east near Fort Dauphin. Does the same species occur in both places? Or are the two sister species like the *Brookesia*s on either side of the Tsaratanana? Or was the species reported here simply misidentified?

Raxworthy believes the latter is most likely. Over 200 miles of spiny desert separate the two regions, and a barrier as formidable as the Tsaratanana exists just west of Fort Dauphin: a climatic fault line where, in a matter of miles, average annual rainfall rises from 20 inches to 140 inches. But he has to make sure, so we're now driving in the Land Rover to the hot, boulder-strewn cliff — ideal habitat for *O. saxi-cola*. As this jaunt makes clear, Raxworthy is not above setting aside an entire day and a good bit of his resources — three Earthwatch volunteers, five Malagasy, and two Land Rovers — just to answer a single herpetological question.

Moreover, he is following protocol. Throughout Madagascar, when anyone enters a village unannounced or even the forest around it, he must immediately present himself to the president of the *fokontany*, the local committee in charge of community affairs, and explain his reason for being there. Failure to do so is at the least impolite and at the worst cause for punitive action. In at least one instance, a pair of Western scientists working in a remote region died at the hands of local tribesmen, who apparently feared they were heart-stealers.* Since we would be driving across sacred land, Raxworthy not only presented himself (and a bottle of coveted toaka gasy) to the president of Ranobe's fokontany, but invited him along for the ride. The president, in turn, asked two of his fellow Sakalava along.

Three Sakalava, including what passes these days for a chief, right here in our Land Rover. Will I succeed with these guys where I have failed with James Bond and our Sakalava cooks, namely, in sim-

---

*One possible clue to the origins of the bizarre belief that white men are out to consume the vital organs of Malagasy — a belief that remains widespread in rural Madagascar — comes from the writings of the early missionary E. O. McMahon. Writing in 1892 about the slave trade, which only ended in Madagascar with the French occupation three years later, McMahon recalled that "[o]ne African who is now free told us how, when the dhow they were on was becalmed near the Madagascar coast and [a European] man-of-war boat was in chase, the Arabs called up the strongest of the slaves to row the dhow, and to make them work harder told them that those in pursuit were cannibals, and were only chasing them to catch them for food; 'and we worked harder and harder,' he said, 'till the dhow grounded, and then ran off into the woods out of the way of the Europeans, and were caught again by the Arabs and sold to the Sakalava.' "

ply getting to know them? Or will they and their tribe remain forever aloof and mysterious?

Half an hour after leaving Ranobe on a narrow dirt track, the branches of spiny desert trees lashing at the open windows on both sides, Raxworthy turns off onto an open area near a stand of leafless acacias and turns off the engine. We're nowhere near the cliff, which I can see rising in the distance above low, scraggly trees. Why the stop? Raxworthy soon explains: Ramanoel, the village president, wants to consult with his ancestors.

We step out into the hot sun and follow Ramanoel to a spot near the clump of acacias, where he abruptly drops into a cross-legged position in the sand. The rest of us, including the two men who came with the president, follow suit, fanning out behind him. Facing east, a sacred direction, Ramanoel begins speaking quietly in Malagasy. His pink flowered shirt and pink-and-green straw hat belie his air of authority. Out of respect, all of us remain silent while Ramanoel performs the ceremony, occasionally flicking droplets of toaka gasy to the east.

Raxworthy, who speaks a little Malagasy and has attended numerous ceremonies like this one, leans over and translates in a whisper. Apparently Ranobe's origins can be traced to this very plain, which has become sacred ground. Ramanoel's *razana* or ancestors once lived here, and their tombs lie in the surrounding hills.

"So whenever they want to do anything with the razana, this is where they come," Raxworthy whispers, shading his eyes against the bright sun as he scans the horizon. "He formally introduced us to the razana, explaining what we're doing here, that we don't mean any harm, that we're helping to protect the forest. And he asked for their support."

After five or six minutes, Ramanoel suddenly stops talking, gets up, and smiles. One of his men, a thin, middle-aged man with long front teeth, moves around the group, giving each of us a shot of toaka gasy. It's too early to drink, and I have painful memories of the local hooch. But the man squats down in front of me, fills a small plastic cup with the rum, and grins. What can I do? I drink.

In contrast to the somber air of the ceremony, a heated discussion soon begins among the Malagasy, led by Ramanoel. He sweeps his arm across the plain, pointing to the cliff, whose trees appear bleached by the sun. I have no idea what they're talking about, so I walk over to Raxworthy, who's now standing off by himself.

"Got three satellites, that's good," he says.

It takes me a second to realize he has his Global Positioning System unit in his hand.

"Already have our position: 23°S, 43°E," he continues. "We're 9.86 kilometers as the crow flies from Ranobe, and that's our bearing to the camp." He points back the way we came.

So incongruous: the most venerable of ceremonies, the most modern of technologies, employed simultaneously beside this stand of leafless acacias in the middle of nowhere, southwestern Madagascar.

Raxworthy explains that the Malagasy are debating the finer points of a battle that apparently took place here between the Sakalava and the Merina king Radama I. In the 1830s, Radama conquered most of Madagascar but failed to bring the widespread Sakalava under his control.

"It's quite rare that you hear about history that goes back that far," Raxworthy says, stuffing his GPS unit into his rucksack. "We're talking about the 1830s, and they're still arguing about it over 150 years later."

As we climb back into the Land Rover, I take one last look around. Who would have thought this flat, featureless plain would have held so much history, so much sanctity?

Ancestor ceremonies are as common in Madagascar as rice growing or cattle herding. As the ethnographer John Mack has written, "Even to discuss questions of tradition and history, things 'of the ancestors,' with village elders in any relatively formal setting may well involve some form of invocation or ceremony, however truncated. At the very least," he adds, "a communal sharing of toaka gasy is likely to be indispensable. . . ."

The very concept of the razana is central to Malagasy thought. Most Malagasy, particularly in rural areas, would not think of doing anything of importance—building a house or marrying off a child or even leading a few vazaha on a lizard hunt—without consulting their ancestors first. To the Malagasy, the razana are like a committee of elders, whose blessing and sage advice is sought on all matters of significance. To go against the wishes of one's ancestors is at best foolhardy and at worst tantamount to suicide. For the razana are the repositories of all things good and sound. As Mack has written, "the idea of the ancestors . . . encompasses and expresses all that is considered morally desirable or appropriate in social relations." It is they who dictate the fady that govern the average Malagasy's every move.

The razana are not some dimly remembered people from long ago, but rather the recent dead, one's grandparents and great-grandparents and great-great-grandparents. After a few generations, the particulars of the deceased may cease to be part of the collective memory of a family or clan. These unnamed ancestors are still revered, but it's the recent dead, particularly those who were deemed most wise or noted during their lifetimes, whom the Malagasy consult, just as Christians consult God or Muslims Allah. When a person dies in Madagascar, he or she instantly assumes the status of a spirit, like the ancestors they have joined. The razana are not the only spiritual beings in the Malagasy cosmology—Andriamanitra ("the Perfumed Lord"), Zanahary ("the creator"), and the Christian God also play central roles—but they are the most prevalent.

A sign of how much the people of Madagascar cherish the notion of one day joining their ancestors is that, for many Malagasy, the worst possible fate is to be excluded from the family sepulcher, either intentionally or unintentionally. It is truly a fate worse than death, the equivalent, Mack notes, "to being condemned to eternal oblivion." When a Merina dies overseas, his relatives will stop at nothing to retrieve his body and place it in the family tomb. Those tribes that practice individual burial will act similarly, frantic to ensure that the deceased rests forever in the land of the ancestors. "The soul of the deceased can be content only when he is buried in the ancestors' soil (*ny tanindrazana*)," writes Jørgen Ruud, "because then he has come home, in the fullest meaning of the word."

Burial is so important that in many parts of Madagascar, people spend far more time, effort, and money on building tombs than on building homes. Indeed, Malagasy tombs, wrote one visitor in the last century, "may justly be reckoned among the most remarkable and impressive antiquities of the country." Westerners touring the south of Madagascar are often baffled to see the small, makeshift thatch houses of the Mahafaly and Antandroy compared to their massive stone sarcophagi. The amount spent often far exceeds one's annual income. On one Mahafaly tomb I saw in the southwest, beneath the name and dates of the twenty-four-year-old woman it contained, lay the amount spent on her funeral and tomb construction: 1,301,250 Malagasy francs, or about $325. As one bewildered vazaha once put it, "the living are impoverished to aggrandize the dead." But to the Malagasy such expense is only sensible. "The building of the tomb may impoverish the man and his family for years," Ruud writes, "but he does not feel this as a special burden, because he is confident that he will have his reward through blessings and prosperity in the time to come." As the Malagasy say, "One is willing to spend most on the house in which one will dwell the longest." King Andrianampoinimerina summed up his people's thinking on the subject when he stated, in effect, that a house is but a voyage while a tomb is for eternity.

Burial practices differ from tribe to tribe. The Sakalava don't invest much in graves, which are little more than piles of stones. For them, *tromba* or spirit possession is a more important way of assuring ancestral blessing. The Bara, according to one observer, "dig no graves; the corpse is put on the ground naked, and stones are piled around and over it, making an oblong structure from a foot to three or four high." The Betsimisaraka, the island's second-largest tribe, bury their dead in coffins laid on the ground out in the open, though often the cemetery is hidden away in a copse of sacred trees. The Merina build solid family tombs out of locally quarried blue basalt. The tombs are largely undecorated, and no effort is made to conceal them. The Betsileo, who live in the Highlands immediately south of the Merina, construct elaborate underground mausoleums. These can be as much as sixty feet below the surface, with long tunnels leading from an opening some forty or fifty feet from the tomb.

The most imposing tombs are those of the Mahafaly and Antandroy. While Highland tribes bury entire extended families in their tombs, these southern tribes build elaborate stone structures for single individuals. A typical Mahafaly stone tomb I saw in the southwest was 5 feet tall and perhaps 1,600 square feet. Its smooth, whitewashed sides bore a portrait of the deceased and a series of life-size paintings of events from his life. One depicted him doing a leaping kick, another disarming a policeman with a karate chop. These reveal that the man had been part of the kung fu craze that swept Madagascar in the mid-1980s, causing riots around the country before the government banned kung fu clubs. Another painting showed a jet in flight, indicating that the deceased may have once flown on an airplane, a status symbol among the desperately poor Mahafaly. On the roof of the tomb rested objects from his life, including a suitcase and a chair, and the horned skulls of the dozens of zebu sacrificed at his funeral. There were also six of the carved wooden totems known as *aloalo*. Each six-foot-tall aloalo was a pole of geometric patterns topped, again, by elements from his life: a house, a man tending a zebu, and a man being arrested by two gendarmes, his arms tied behind his back. This enormous tomb was not for a venerable elder, but for a man who died at age twenty-seven.

While burial practices differ across Madagascar, one feature most hold in common is second burial. Shortly after someone's death, relatives bury the deceased in a temporary grave; only later, after decomposition has left just bones, do they retrieve and place the body in the family tomb. The idea is to separate wet, polluting flesh from dry, sacred bones. At the end of the nineteenth century, the great French naturalist Alfred Grandidier, the father of Malagasy studies, described one such second burial among the Vazimba, a tribe living in western Madagascar:

> After having washed the corpse and clothed it in its finest garments, they place it in a squatting posture upon the *kibany* (a bed or couch), and as if it were still living; and the relatives of friends attend it night and day, talking to it, putting into its hand a spoon full of rice or any other kind of food, etc. Formerly the liquids produced by the decomposition of the flesh were taken to a special place, which was sprinkled with the blood of an ox in order to nourish the *fananina* or snake, which they believe to be produced from these putrid liquids. Since the conquest of the country by the Sakalava king Lahifotsy, these customs have been to some extent abandoned, and as soon as the effluvium becomes too offensive, the corpse is buried. But, at the end of about a year, they take it out of the ground and wash the bones, which are placed in a new coffin, and are then buried for good and all.

It's no surprise that Europeans found aspects of second burial abhorrent. "The first ceremony of drying the corpse is succeeded by a most revolting custom," wrote the prolific writer-missionary James Sibree about burial practices among the Antankarana. "[F]or at the end of several days decomposition produces a putrefying liquid, which is received into vessels placed under the framework on which the corpse is deposited. Then each one of those present holds his hand so as to receive a portion of the horrible liquid, with which he rubs his whole body!"

The Merina and Betsileo have taken the practice of second burial to its greatest extreme in the ceremony known as *famadihana* ("turning the bones"). Among the Merina, famadihana take place most often when the body of a person who died far away is brought home for interment in the family tomb, while the Betsileo typically hold famadihana when temporarily buried corpses are incorporated into a newly built family tomb. Descendants may also hold bone turnings simply to honor ancestors buried in the sepulcher.

The practice may be somber, but more often it is a celebration with music, singing, drinking, and dancing. During a famadihana, family members remove the body or bodies from the tomb, place them on a mat, and rewrap them in new *lamba mena,* the silk shrouds used throughout the island to clothe the deceased. If old shrouds remain, the new ones are simply wound around them; if only bones are left, they are washed and enfolded in a new shroud. Family members often carry the ancestors being feted around the village, showing them new buildings and other recent developments. Finally, sometime before dusk, they return the remains to the tomb, which is sealed anew until the next turning.

Such elaborate funerary procedures ensure that burial of the deceased fuses into a more general relationship with the ancestors. The Western practice of burying one's dead and then largely forgetting about them is as baffling to Malagasy as famadihana and second burial in general are to vazaha.* To ensure the livelihood and good fortune of the family, clan, and even community, one must take care of one's ancestors. As Ruud notes, "The whole family would feel it a burdensome disgrace should the dead in the tomb be forgotten and left uncared for in their mouldering lamba. On the other hand, it brings great honor to hold such a feast." For all Malagasy, burial is just the beginning of one's relationship with the ancestors. Whatever was fady to one's ancestors remains fady to all his or her descendants. If it was fady for your great-grandmother to sit in a doorway when she was pregnant, so as not to block the passage of the baby, then any pregnant women in your family must not sit in doorways either.

The Sakalava feel such respect for their ancestors that when a

---

*Victorian-era missionaries were particularly appalled by famadihana, none more so than the Englishman John Haile. "It is a source of vileness which pollutes the stream of national life, a vestige of foul heathenism which withers and blasts all that is fresh and pure and strong, augmenting and reinforcing an element in the moral atmosphere which renders virtue impossible and vice prolific," he decried in an 1892 article in the *Antananarivo Annual and Madagascar Magazine.* "Health contracts disease by contact with mortal corruption, and imperishable natures are contaminated by vapours of infamy which inflame with fires of hell. It is high time that such a custom with its inherent evils was suppressed, as inimical to the physical permanence and the moral soundness of the Malagasy people."

Sakalava dies, words associated with his or her name may become fady to family members; when kings died in the past, such words became forbidden throughout the entire kingdom. Walen reports that after the death of a king named Marentoetsa, two words in the Sakalava dialect fell into disuse: *mare* (true) and *toetsa* (condition). "Those who use such prohibited words are looked upon not only as extremely rude, but even as criminal offenders," Walen wrote. Even to mention a deceased king's name became fady among the Sakalava, who supplied new praise names (*fitahina*) for the purpose. Thus, after the death of a Sakalava chief named Tsiaraso II, survivors gave him the fitahina Andriamandefitriarivo, which means "the ruler who is tolerant of many."

Raxworthy is displaying a tolerance of his own, for we're now waiting in a cornfield beneath a fat-trunked baobab for a *second* ancestor ceremony. After the first, we drove along a disused nineteenth-century road to the edge of the cliff, which rises up a half mile to the east. Here, Ramanoel holds no sway. There is another village in the vicinity, Ankena, and one of Ramanoel's men has run off to find the president of its fokontany to perform a similar razana ceremony. The man's already been gone forty-five minutes.

"You can't really rush these things," Raxworthy says, standing near the baobab in his too-short fatigues, torn-up Keds, and—I can't help but notice—one black sock. "Did you read that book *The Tao of Pooh*? The idea of the 'uncarved block'? A block of wood is beautiful, you don't need to mess with it and carve it. Go with it as it is. It's a laid-back attitude, and it works very well in Madagascar." He smiles. "Control freaks have a really hard time here. I could be ranting and raving, we have to get to this site and do this and this and this. But this way, we're going with the flow. Any other approach would just lead to aggravation."

He pauses.

"I'll be interested to know how the mountain group got on, because I'm not sure if Loren has that particular philosophy."

He's referring to Loren Caira, the ER doctor, who has joined four others from our team—Dean, Jean-Baptiste, Angeluc, and an Earthwatch volunteer named Michael O'Dea—on a trip to investigate a purported patch of rain forest on a mountaintop about fifty miles northeast of here. They've been gone two or three days already. Of all the members of Raxworthy's team, Caira is that ill-fated control freak.

"He's probably been hacked to death with machetes," I say. But Raxworthy has already moved off to chat with Ramanoel.

Looking up at the cliff, I think of the time I had a close run-in myself with a Malagasy on another cliff, this one in the Isalo canyonlands in south-central Madagascar. The experience burnished into my memory how important the ancestors and their resting places are to the Malagasy.

I had been hiking by myself on the outer edge of the Isalo massif, which lies in the heart of Bara country. The Bara are notorious as the fiercest warriors of Madagascar. Legend has it that a Bara is so ready for battle that he will only wash one side of his face at once, leaving one eye open and a hand grasping his spear or gun. I read this in a vivid little book entitled *Lights and Shadows; or, Chequered Experiences Among Some of the Heathen Tribes of Madagascar* by one John Richardson, an English preacher who in 1875 became the first vazaha to travel through the Isalo region. In the book, Richardson describes how, in his purblind view at least, your typical Bara thinks:

> Give me my gun, my powder and balls, my spear; leave me my toaka, my wives, my oxen, and my king; let me rob, plunder, kill, and destroy anything or anybody I please. Let me despoil every man, and carry away any man's cattle, his wives, his children, his slaves, to my heart's content. Let no man molest me; and then, who cares who governs the country!

Richardson went on to record the rather stiff fines Bara imposed for seemingly moderate offenses:

> For sitting or reclining on another person's bed: a fine of one ox or to be shot.
> For striding over a person, or for striding over the foot even: the same.
> For brushing a person's face or any part of his body even with any part of your clothing: the same.
> For using spoons, plates, drinking vessels belonging to another person: the same.

On my hike at Isalo, I had reached the very edge of the massif, a point where it dropped sheer for hundreds of feet into a great plain that stretched into the hazy distance. The view was stunning, and the sky a drama of white puffy clouds scudding across a brilliant blue sky. I sat there on the brink for an hour.

Far below, a Bara walked across a field toward his thatch hut, which stood in a clearing of packed earth. On the way, he stopped to talk to another man coming the other way. Then they walked together to the hut. Even though they were mere dots far below, I wondered if they could see me up there, silhouetted against the sky. For another hour, I wandered along the edge of that escarpment, taking photographs and investigating some graves I found in some rocky clefts. When I returned to the brink at one point, I realized I hadn't seen the men for awhile. Where had they gone?

Just when it dawned on me that they probably *had* seen me up there, a Bara tribesman suddenly appeared at the edge of the cliff. He was carrying an eight-foot spear and looked none too happy. A Bara chief once said, "Our gun and our spear are our court of justice." I thought, *Oh, shit.*

In the split second I had to think, I had a brainstorm. It may very well have prevented things from becoming nasty. After all, I was several miles from the road along which I was staying in a small hotel, and I had told no one where I was going. It occurred to me that my body might never even be found. As it was, I was lucky I had my camera around my neck. I immediately pointed the camera at a plant at my feet and made signs that I was so *thrilled* to find this plant and to photograph it. That I was just one more dumb-fuck vazaha doing something stupid—wasting film on weeds—but at least I wasn't doing anything near the graves. (Which, as it happens, I was just about to photograph when he appeared.)

By God, it worked. The muscles in the man's face instantly relaxed. He didn't smile, but I could tell I was absolved. He said *veloma* (goodbye) and walked back toward the cliff. It was then that I realized he had a companion, no doubt the other man I'd seen on the plain, who stood concealed just out of sight in case he was needed.

As I made my way back to the road, I kept well away from any graves and tried to look as nonchalant as possible, in case they were watching me. And I used up an entire roll of film on plants.

Now, after an hour and a half of waiting in the shade of the giant baobab, Ramanoel's man returns with the news that he can't find anyone, much less the president of Ankena. Ramanoel shrugs his shoulders and soon we're back in the Land Rover, driving to the escarpment.

All the solemnity of the first ceremony, and now this one evaporates like dew. I don't get it. If the land you're on belongs to somebody else's forebears, are you off the hook? If not, would an ancestor-worshiping Sakalava risk insulting his neighbor's progenitors? Alas, the Sakalava continue to surprise me.

I'm not surprised, though, that our efforts to find Raxworthy's "iggy-wanna" are in vain. After all the time we've taken to appease the ancestors, or to wait to appease the ancestors, we only have an hour or so of concentrated searching among the cliff's jagged rocks before we have to leave. In whatever patch of shade we can find, we inhale a late lunch of biscuits, oranges, chocolate, and that most precious commodity, water, and climb back into the Land Rover. Raxworthy, however, says he's fairly certain that *O. saxicola* doesn't exist in this region, that indeed someone misidentified it. One more problem of species distribution apparently solved.

8.

LIKE OUR NONDISCOVERY of *O. saxicola,* Raxworthy and Nussbaum's biogeographic study of a single genus of dwarf chameleons might sound like a drop in the bucket, and it was. But it was the first attempt made on the world's greatest natural ark to account for the island's intense speciation and endemism. And the Michigan herpetologists were not content to stop there. The next step was to see whether patterns of endemism were congruent between different vertebrate groups. Okay, so the *Brookesia* speciated like mad in northern Madagascar. Did other animals, such as mammals or birds, do the same, and in the same way? If so, that might suggest a common response to historical events. That common response, in turn, would help nail down what those historical events might have been.

For their study, Raxworthy and Nussbaum selected reptiles, amphibians, nonflying mammals, and birds of the eastern region of Madagascar. They chose these eastern animals because they showed great species diversity in all groups and high local endemism, and because recent inventory and taxonomic data on these animals existed, whether collected by themselves or others. Sites ranged from Lokobe Reserve in the northwest to the Fort Dauphin region in the extreme southeast—basically the entire rain-forest belt of Madagascar. The herpetologists used a method known as parsimony analysis of endemism. PAE resembles cladistic analysis, except that geographical regions are used instead of species, and species distributions serve as characters.

The team discovered that patterns of endemism between these groups appear to be largely congruent. There were differences, of course. The birds, because of their greater dispersal ability, did not indicate the same adherence to the more detailed patterns exhibited by the herps, for instance. And Tsaratanana was "problematic," they noted, having species shared among several other areas. But in essence, Madagascar's rain forest appears to contain centers of endemism, where species of reptiles, amphibians, mammals, and birds existing only there concentrate, having responded in a similar way to some event or series of events in the past.

What, then, were those events? Raxworthy and Nussbaum are currently under way with another PAE study—only the second one ever done on Madagascar—that is giving some intriguing clues. This one focuses on the reptiles of eastern Madagascar. The island's reptiles are perfect for a study of this kind, because of their high levels of diversity and endemism. Moreover, their elevational and regional distributions are extremely limited. Preliminary analysis of the reptile data indicates high levels of regional endemism in expected areas, Raxworthy says, with patterns largely similar to those found in the other PAE study.

Yet there are curious disjunctions in the distributions of certain reptiles. One of these occurs between two high massifs in central Madagascar, Ankaratra and Andringitra. Both mountain ranges feature montane heath vegetation above about 5,000 feet, but are separated from each other by lower-lying areas of less than 3,300 feet, which feature moist montane rain forest. Both Ankaratra and Andringitra bear five montane specialist reptiles in common: two chameleons, two skinks, and a day gecko. Raxworthy and Nussbaum have not encountered these species below 5,000 feet, and dispersal between the two ranges, which are about 200 miles apart, seems unlikely given what lies between them. Today, that is. In former times, they conclude, montane heath vegetation must once have occurred at much lower altitudes, perhaps as low as 3,300 feet.

There are also disjunct distributions among rain-forest specialists. Despite its isolation, Montagne d'Ambre has twenty-three species of rain-forest reptiles that also live at other sites, again at least seventy miles away through hot, dusty dry forest. This evidence supports that found in the *Brookesia* study suggesting that rain forest once reached from Montagne d'Ambre right down to Tsaratanana and the north-west region. Even more strikingly, the Isalo massif, the spectacular sandstone canyonland in south-central Madagascar, contains relict patches of rain forest deep in its canyons bearing a plated lizard and four frogs that are rain-forest specialists. A vast, semiarid, treeless plain separates Isalo from the nearest patches of eastern rain forest. The only way to explain this disjunction, Raxworthy says, is through vicariance: Rain forest once must have stretched from Isalo to the eastern escarpment, connecting these five species in a single popula-tion or contiguous populations.

9.

OUR TEAM APPEARS TO have just discovered a third such disjunction in rain-forest reptiles. The mountain crew—the one Raxworthy sent off to search for a lost rain forest on a mountaintop in the heart of the spiny desert—has just returned. Within the first few breaths, in among the welcome-backs and how-did-it-goes, it comes out that, yes, they did find the relict rain forest, atop a mountain known as Analavelona.

Even more exciting, they discovered there two species, a skink and a snake, that are rain-forest specialists. Previously their kind had been known only from the eastern rain forests. Today, the closest patch of eastern rain forest lies more than 150 miles away over some of the most barren landscape to be found in all Madagascar. It seems unbelievable considering the landscape today, but Raxworthy says that those two species provide evidence of wetter climates in the southwest sometime in the past—so wet, in fact, that rain forest reached all the way down here.

To think that two humble creatures could suddenly make what seems impossible probable.

We all settle down on the blue tarp under the trees to hear them out. Darkness has fallen, but glare from the kerosene lantern shows them up clearly. All five of them look like escaped convicts. Their hair is matted, their eyes bloodshot, their clothes filthy, their every action bespeaking severe exhaustion. Suddenly I don't feel so bad about not having been able to go along. Yet the flash of white teeth in the night from all five indicate this is a happy bunch. Mission accomplished.

And what a mission. Dropped off in Tuléar five days ago, they hired a taxi-brousse to take them as close to the mountain as possible. After three breakdowns, they got out at a village called Mahaboboka. Here they hired a zebu cart, the only means of transportation, to take them across a tree-studded plain to the foot of the mountain. The "chariot"—as Michael O'Dea, a strapping, hail-fellow-well-met Irishman from Dublin, describes it—was barely large enough to hold their gear and one or two passengers on top. So they walked. And walked. And walked. Around midnight, they stopped for dinner, then pressed on, taking turns riding on the cart. Since daytime temperatures reached into the high nineties, they decided to continue all night. At dawn, they paused to sleep for an hour and a half, then they were off again. At three in the afternoon, they reached Mitia, a hamlet below the mountain. They had covered more than forty miles in twenty-eight hours.

"I felt that if we had stepped off a spaceship, the villagers would not have been more surprised to see us," says O'Dea in his thick Irish accent as he stretches out on the camp's hammock.

A resident of Mitia told Jean-Baptiste that it was only the fourth time in living memory that vazaha had passed through the area. At first, the villagers were highly suspicious. They couldn't comprehend why this ragtag group of Malagasy and vazaha would come so far to collect *bibi keli* (small animals). We can look to the journals of Alfred Grandidier to see what was going on in their minds. After encountering difficulties working among the Sakalava of this region in 1866, Grandidier wrote that "everyone is a sorcerer who distinguishes himself from other people by his actions or his words; and the traveler who passes his time in collecting information . . . managing a crowd of instruments each more extraordinary than the other . . . and putting reptiles into spirits . . . is naturally in their eyes one of those monsters one cannot fear enough, and against whom it is well to take every precaution."*

Eventually, the village president himself agreed to serve as guide, along with his son. Only one resident signed on as porter. Two weeks before, the villagers had tracked down and killed two cattle rustlers, which may have had something to do with it, O'Dea says.

It was by then late afternoon, but they set off at once. The way led through rocky ground where only spiny-desert plants grew.

"I wondered how even a small patch of rain forest could possibly

*Even Raxworthy's onetime Malagasy girlfriend admitted to him that for a long time she suspected he had really come to Madagascar to search for a cure for AIDS.

be anywhere nearby," O'Dea says. Dean and Caira, normally voluble, seem content to let O'Dea tell the story, and Jean-Baptiste and Angeluc are too polite to break in.

The Analavelona massif reaches nearly 4,450 feet at its highest point, so they had a hike ahead of them. And at the president's request, they took the long route to avoid coming upon the rustlers' bodies. When they reached the ridge of the mountain, they all looked expectantly down the other side. No forest, only more rocky grassland. Were their guides hopeless? Did this fabled patch of forest, which no research team had visited in years, still exist?

"It seemed to me afterward that I took just one step and moved from dry grassland to thick jungle in an instant," O'Dea says. They had found the rain forest. Hungry, sweat-soaked, and utterly spent, they fought their way downhill through thick vegetation, swarms of insects getting into every orifice. They decided to camp by a brook, but the president said no, a dead rustler lay just upstream.

"When I heard this," O'Dea says, "I must admit I found myself thinking, *What am I doing here?*"

Eventually they found a suitable spot to camp. They herped all the next day, collecting chameleons, frogs, and skinks. A snake that disappeared up Jean-Baptiste's pant leg when he tried to catch it provided the first sign that this forest was important: The snake was that "obligate rain-forest specialist," known only from humid jungles 150 miles to the east. The skink turned up soon after.

Suddenly O'Dea becomes quiet. The surround-sound *whrrs* and *ch-ch-chs* and *tock-tock-tocks* of forest insects seem suddenly cacophonous.

"We had a remarkable experience at the camp that night," O'Dea says, looking up from his momentary reverie. "We were out chameleon spotting. I was probably the nearest one to the camp, and suddenly I started hearing this very, very, very loud talk. At first I thought it was just talk, but then I thought, *That doesn't sound right.* I walked back a bit, and the nearer I got, the more I realized one of the Malagasy was raving and chanting. It was the president's son. He repeated a phrase again and again and again and again, speaking really fast in a *really* loud voice, the way people, especially Malagasy, don't do in the jungle. He was shouting in a chantlike way. Garfield came back as well. I was at the fire, and Garfield shouted over, 'Is that the camp?' He thought it was zebu thieves or God knows what. It was just so weird."

O'Dea looks off abstractedly, his tired eyes glistening in the light thrown off from the kerosene lantern.

"Then, just like that"—he snaps his fingers—"he stopped and broke down crying and weeping. His father the president was there, too, and he then said some prayers over his son and anointed him with rum, pouring it over his head and rubbing it into his hair. He did a little ceremony over him. Then the son calmed down and sat there really calmly for the rest of the night."

He runs his hand through his hair and sighs.

"We had no idea what was going on, because neither Jean-Ba nor Angeluc was there," he continues. "So we sat there saying, 'What's wrong? Is there anything we can do?' Of course, they spoke no French, so we just had to sit and wait until one of the guys came back to ask them what on earth was going on."

Jean-Baptiste and Angeluc finally showed up and, after speaking with the president, explained what had happened. The son had been visited by the razana in a tromba. Apparently the ancestors were angry that the vazaha were there and, through the medium of the president's son, asked that they leave immediately. The president explained what they were doing and, since it was about ten o'clock at night, asked if they could leave in the morning. The ancestors assented, after which the son fell silent.

"I've never heard before of razana actually asking one to leave the forest," Raxworthy says suddenly. It's the first voice other than O'Dea's I've heard in a quarter of an hour. "The fact that the president tried to and successfully did change their minds also is quite interesting. It was like he was talking to a committee of elders."

Spirit possession of this sort is widespread among the Sakalava all up and down the west coast of Madagascar. Originally, tromba were the spirits of dead Sakalava royalty, but in recent times, tromba has come to be associated with any kind of spirit possession. Sakalava may be possessed by their own ancestors, as the president's son was, or by the ghosts (*lolo*) of Malagasy who died far from home and were never buried in the family tomb. (Razana are one's own ancestors, lolo are the ancestors of strangers.) There are nature spirits, which are usually associated with certain places in the forest, and evil spirits, such as *njarinintsy*—the name derives from *manintsy,* which means "coldness"—a type of possession sickness characterized by confusion, memory loss, and uncontrollable screaming and crying. The most recent form of tromba possession involves Christian figures. In 1988, according to the anthropologist Lesley Sharp, who conducted an in-depth study of tromba in the Sakalava town of Ambanja, John the Baptist appeared through a medium during a Catholic service. "This spirit stood and began to evangelize in the style of a Protestant pastor, infuriating the officiating priest," Sharp writes.

To Western eyes, tromba might seem bizarre, even contrived. But the institution is so widespread and so deeply ensconced in the private and public lives of the Sakalava that one cannot fully understand what it means to be Sakalava without understanding tromba. It has been called "the quintessence of Sakalava religious experience," and it offers clues to interpreting the tribe's worldview. As Sharp notes, "Discussions of the causes of spirit possession are best argued when rooted in the internal logic of the culture in which they are found. The interpretation becomes flawed and misleading if it draws exclusively from Western conceptualizations of the body and mind as distinct categories, a notion that only distorts indigenous notions of possession."

O'Dea doesn't even try to interpret.

"It's a very difficult thing to understand," he says now, staring at the ground before him. "I don't know what was going on in their minds at the time, but it was a very powerful thing for them. Whether we believe it or not, whether we believe it's all just nonsense, it was not nonsense for them. There was no bullshit about it; it was the real thing."

The next morning the president and his son acted as if nothing out of the ordinary had happened, and the team hiked back down to Mitia without incident. By now the villagers had accepted them, and they were able to hire a second "chariot" for the two-day ramble back to the road. They even became close with one of the drivers, who showed them a neat trick. Having run out of food, the team, like born Sakalava, had taken to eating the fruit of the ubiquitous raketa, the prickly pear. As our friends discovered, the fruit is delicious, but its skin bears thousands of fine spines that get stuck in your skin at the merest touch. The driver caught a fly off the belly of a zebu, held it by the wings, and let its scrabbling feet pull out the cactus spines.

O'Dea shakes his head.

"I have no idea how the Sakalava discovered this trick, but we were immensely grateful for it."

Later, as I flop down in my tent, exhausted from another long day yet wired from O'Dea's vivid tale, I decide that it's time to give up and admit that the Sakalava remain as distant and enigmatic to me as they were the day I met that fisherman on the beach at Lokobe. A week at Raxworthy's camp has brought me no closer to an understanding of this largest of Malagasy tribes, and with just one day remaining here, my chances of reaching even a semblance of that understanding are as slim as my chances of holding down a meal a few days ago. Perhaps that's what's most intriguing about the Sakalava: They remain unknowable.

10.

BY CONTRAST, AFTER A decade of fieldwork on the island, Raxworthy, aided by Nussbaum and building on the work of herpetologists who came before him, can safely say that the herps of Madagascar are not unknowable. He's starting to get a good handle on species, distributions, abundances, and even behaviors, and he has nearly finished surveying the protected areas of Madagascar, updating their often lamentably inadequate species lists. He says he feels confident now that, if he were taken to any part of the island and asked which species of reptiles and amphibians likely existed there, he could probably accurately guess about two-thirds of them. With Nussbaum's help, he is working his way through descriptions of those 150 new herps. And through his initial cladistic and biogeographic studies, he's beginning to see some patterns in how the island's generous complement of creatures might have evolved.

Raxworthy is the first to say that much remains to be done, however—very much. Despite his efforts and those of other biologists and systematists working in other fields, a surprising lack of information on the island's biodiversity yet exists. "I find when I talk to people that they're flabbergasted to learn that in a place like Madagascar, there's still so much that needs to be discovered," he told me. More fieldwork like that we've done at Lokobe Reserve and PK-32 is needed, mostly in places outside of protected areas, just to determine what lives where. Many of these places may hold rare or endemic species found nowhere else on the island. Certain special locales, such as the relict rain forest atop Analavelona Mountain, cry out for research work, he says.

Fieldwork produces findings that are also fuel for thought about bigger questions of evolution and speciation. Take our preliminary findings from around Lake Ranobe. We have identified more than fifty species of reptiles alone, Raxworthy says. That represents an extremely high diversity, even higher than rain forest. We have also confirmed that many species are restricted to this subregion near

Tuléar, with several confined to extremely tight home ranges. For example, the Earthwatch team immediately prior to ours, which worked to the south near Saint Augustine's Bay, rediscovered a species of gecko, *Paragehyra petiti*, that was known from a single specimen collected more than fifty years ago. The team found twenty-five of them, showing that the species not only is extant but has a reasonably good foothold in the rocky cliffs where it lives along the Onilahy River. In and around the Fiherenana River, the next one up from the Onilahy, our own team turned up a large green chameleon with a prominent white strip down its side. This species, *Chamaeleo belalandaensis*, was also previously known from but a single specimen.*

How did such a high diversity of reptiles, including a handful of extremely rare, locally endemic species, come about? Is it because the forests here are transitional between spiny desert and western deciduous forest? Because ancient rivers like the Onilahy and Fiherenana isolated species in small areas from which they could not spread? Now that he knows which species exist here and what their distributions are, Raxworthy can begin to address such questions.

Results from such fieldwork are also useful, finally, for conservation planning. As Nussbaum wrote in a letter to *The Sciences*, "Unless regional biodiversity is documented with sound systematic studies—backed up with museum voucher specimens that not only reveal the characters of the species but also verify the precise localities where the animals were found—conservation efforts based on the idea of preserving maximum chunks of biodiversity are badly flawed. . . ." Raxworthy concurs. "If somebody says, 'This forest doesn't seem very important, why bother saving it?' we can say, 'We've got all these species here, including *x* number of local endemics, including species that have not reliably been found anywhere else in Madagascar.' Our data are especially important, because PK-32 is not yet a reserve, and it appears the process is stalling. The fact that it hasn't happened means that any new data will help get the process under way."

The Malagasy government is about to implement the second phase of its twenty-year Environmental Action Plan, with grant monies on the way from the World Bank–funded Global Environ-

---

*After I left Lokobe, the crew turned up a third specimen of *Z. boettgeri*, which should serve as "good propaganda," as Raxworthy noted, for continuing to protect Lokobe as a strict nature reserve.

ment Facility. A cornerstone of the plan is a Geographical Information System database that will concentrate data from all fields to help establish conservation priorities in the country. One such priority—the island's last sizable rain-forest wilderness, on the Masoala Peninsula in the northeast—will soon become Madagascar's fifth national park. Perhaps some of that money will be used to formally designate PK-32 as a protected area. Such efforts would come none too soon. "We're the last generation that has the freedom to implement conservation policy on Madagascar," Raxworthy says. "After about the next twenty years, what is left outside of reserves probably won't be worth saving."

11.

ON THE LAST NIGHT I spend with Raxworthy in Madagascar, I sit with him on the blue tarp, chatting. It is late afternoon, and we are waiting for the twins to stop giggling and serve up dinner. Twilight has settled on the ground, but sunlight still plays off leaves in the canopy overhead. Looking west through the trees, I can see the lake, its surface burned amber by the sun. I have spent a good deal of time with Raxworthy, but I am merely a blip on his radar screen. He has been at it ten years, and he'll likely be at it ten years more.

He is telling me about a research trip he hopes to mount into a remote canyon in northwestern Madagascar. The canyon wiggles through a crater thirty or forty miles wide and contains forests that no scientist has ever seen.

"I'm calling it Expedition White Zone," he says, leaning back against the log frames that serve as benches.

If one were to overlap distribution maps of every animal ever recorded on Madagascar, he says, that "white zone" would pop up, unknown and enticing. Apparently no researcher in any field has ever worked in the region.

"It's understandable," Raxworthy says. "The area is hot, dry, and inhospitable. It's two to three days from the nearest road. And it's full of cattle thieves."

One of his eyes swivels my way.

"I would just love to go there."

Search for the Pygmy Hippo

Bullock in a crocodile's jaws:

Willing, or unwilling, he must go.

—*Malagasy proverb*

1.

"Whoaaa!"

"That was a goody!"

"We have to slow him down," says David Burney, leader of this journey across western Madagascar. "This is really dangerous driving."

Four of us—Burney, two other scientists, and myself—are crammed in the way back of a Toyota Land Cruiser, hurtling down the highway to Majunga, a port town on the west coast of Madagascar. It's August 1996, over a year since my trip to Ranobe. Our truck is one of three on this paleontological expedition to a large cave north of Majunga. There are six more people in the middle and front seats, but they don't hit the ceiling like we do every time Leva Rakotovao, our quiet, avuncular Malagasy driver, hits a pothole.

"Leva! Leva! *Moramora, si'l vous plait,*" Burney yells toward the front, mixing Malagasy and French, the country's dominant languages. (Moramora means "take it easy.") The racing engine immediately slows.

"*Nous avons les enfants precieux ici,*" he adds, still yelling. Burney turns to me. "You don't mind being an *enfant* just once more in your life, do you?" he says more quietly, though it's obvious he was referring to his fourteen-year-old daughter Mara, who is sitting in the front passenger seat.

We've been driving west for an hour already, and it's only just now getting light. By five o'clock this morning, seventeen people had stuffed themselves into the three trucks. We sped through Tana's darkened streets, weaving past the eighteen-wheelers that use the free roads of the night to get through town. Malagasy wearing scarves and straw hats and wrapped to the eyes in *lambas,* the togalike garment of Madagascar, walked along the road or warmed themselves around fires set in ashcans. Tiny pickup trucks packed to the roof with loaves of French bread made deliveries, while commuter buses vied with bullock carts for space, even at that early hour.

The sun has just now begun to peek through the eastern hills behind us. The landscape here is typical of the plateau region around Tana. Low, rounded hills bear only small stands of trees, often in the spring-fed clefts of the deep erosion gullies that are sadly so ubiquitous in the Highlands. All valley bottoms hold rice paddies, their growth mere stubble now in midwinter, like the beards of old men. There are fields terraced into the hillsides, too, though they're not as steeply pitched as those in the rugged Betsileo country to the south. The road cuts through hills baked brick-hard by the sun—no FALLING ROCK signs needed here.

Compared to the environs of Tana, this western part of the central plateau is lightly populated. Occasionally we pass through a village or small town, but mostly the landscape is open countryside, with houses dotting the pastureland as they do in the American Midwest. The houses are red-brick, two-story structures, with brick columns rising to a pitched roof and, on the second floor, solidly built balconies, often decorated with window boxes. On the whole they look like fancy pueblos with a French cast to them. This baked-brick architecture lasts: Everywhere you look you see the ruins of abandoned houses, their red, rain-eroded walls seeming half-melted by the sun.

There are also many tombs. In contrast to the colorful red houses, these are squat, square structures of gray stone. Most are unadorned, though some have been whitewashed and bear Christian crosses of white wood. (All over the island, the Malagasy incorporate Christian beliefs into the predominant ancestor cult.) These tombs are well maintained. Other tombs, however, are grassed over and probably no longer cared for. Perhaps they are so old that no one remembers who is buried in them.

Leva, inscrutable in the rearview mirror behind his dark shades, slows and goes smoothly over a speed bump, the first we've come to on this highway.

"He would have gone over that twice as fast," I offer.

"I'm not concerned about that," Burney says, his dark eyes unblinking. "I'm concerned about him hitting one of these buses head on. Or a cow. It's cow-hitting time."

Burney is a paleoecologist — one who studies past environments — at Fordham University in New York City. He's leading an expedition to an enormous cavern complex known as Anjohibe or "Big Cave." In Anjohibe ("on-*zoo*-he-bay"), he and his multidisciplinary team will be looking for remains of some of the extinct large animals, or "megafauna," that used to live on Madagascar — the giant lemurs, elephant birds, and pygmy hippopotami, among others. Some of these creatures fell into or got trapped inside the cave and died over the past several thousand years, and their bones have gone to subfossil, a kind of fossil in the making. By radiocarbon dating and other means, Burney and his team will continue trying to solve the greatest paleontologic mystery on Madagascar: Who or what killed off the megafauna sometime in the past 2,000 years?

So far, paleontologists have identified more than two dozen species of native animal on Madagascar that they deem to be megafaunal — that is, weighing more than twenty-five pounds. Except for the Nile crocodile and the bush pig, every one of them is now extinct. They went extinct over a very short period, possibly as quickly as a few hundred years. And they disappeared wrenchingly recently, in some cases perhaps only a few hundred years ago. How did animals as diverse as half-ton birds and giant tortoises and lemurs as big as Olympic weightlifters become extinguished, overnight in geologic terms? These animals likely had completely different habitats, home ranges, and feeding strategies. How this came to pass constitutes an enigma that has proved insoluble since scientists and others first began asking why over a century ago.

Purported causes of the mass extinction have ranged from climate change to the introduction of exotic diseases. Save for the former, all theories have one element in common: the human factor. People did it, these theorists maintain. The most prevalent theory, first advanced in the 1920s, holds that early Malagasy slashed-and-burned their way through a vast forest that once blanketed Madagascar, leaving the species-poor grassland that now covers most of the island. The forests, this theory stresses, couldn't take the fire, to which they were unaccustomed, and the megafauna couldn't take the loss of their habitat. That humans did it does seem self-evident: They first arrived on the island about the time of Christ, and all twenty-some species of megafauna went extinct sometime between then and now. (Go figure, as Burney likes to say.) Moreover, striking examples of similar mass extinctions that occurred in the wake of man's advent exist elsewhere—those in Australia between 30,000 and 15,000 years ago, say, or in North America beginning about 12,000 years ago.

Strangely, though, precious little evidence exists that people did it. For one thing, the number of bones of extinct megafauna that show signs of butchery constitute, Burney told me, a double handful. There are no bison-kill sites as in the American West, or heaps of moa bones as in New Zealand.* For another, Burney's paleoecological research has turned up compelling evidence that, long before the first person ever set foot on the island, fires sparked by lightning occurred regularly and, furthermore, some of the widespread grasslands thought to have been created by man's actions formed naturally. As research on Madagascar continues, the mystery only seems to deepen.

The four of us way in the back of the Land Cruiser have been discussing the mystery off and on ever since we left Tana. The talk has made it easier to take the unpleasant traveling conditions: the incessant jouncing, punctuated by periodic head slams into the ceiling; the engine's roar, with the ever-present chirpy rattle of the dust-caked rear window; and the discomfort of having your legs jammed between two people you've just met. Those two people are Helen James, a paleontologist at the Smithsonian Institution who specializes in the extinct birds of Madagascar, and Trevor Worthy, a biologist from New Zealand who studies the ecology of his country's native large birds, such as the kiwi and the extinct moa.

"If humans did it, how did they do it?" Burney asks now, leaning forward in his seat beside me to keep his head as far from the sloping ceiling as possible. "That's the interesting question, not whether they did it. Of course they did it. But how? There's really not much evidence for *a way* that they did it."

"It's a little more difficult here, because we don't have big moa oven sites," says James.

*A relative of Madagascar's elephant birds, the moa was hunted to extinction less than a thousand years ago.

"Yeah, we don't have all your big piles of bones," says Burney, glancing at Worthy. "We just talked about all of them." He laughs. "Nine bones."

"There's no long-term exploitation, no moa-hunter culture, nothing like that," continues James, a fresh-faced, frizzy-haired woman in her early forties. "And there isn't evidence of a huge population that could have exhausted species." She looks at me. "You've just driven across the plateau, and here we are in the west. Are you really impressed with the high population?"

I smile. We've seen only scattered settlements, with most of the landscape appearing as unpeopled as the Australian outback.

"What really drives it home is when you look at the land from the air," Burney says. "If you fly over this, you don't see villages except fairly close to the road. It's a vast empty space out there."

"So it's particularly difficult to figure this one out," continues James, still looking at me. "If you've got any ideas, let us know."

"It's a puzzle," Burney concludes. "Part of the solution to finding out what happened is admitting that we don't know. That's been an obstacle here for a hundred years or more. The French went ahead and created a myth"—that Malagasy deforested Madagascar, forcing the megafauna into extinction—"then went around looking for evidence to prove that. Not a hypothesis to test, but a myth that needed support."

Burney is not a man for myths. For fifteen years he has been meticulously re-creating past environments on Madagascar, using studies of fossil pollen, diatoms, and other microscopic remains to paint portraits of long-ago landscapes and how natural and human forces might have altered them. He calls what he does landscape paleoecology. "That is, I not only want to re-create the vegetation and the climate, the things I'm trained to do, but I want to try to reconstruct, at different time horizons, all the other things going on," he told me. "I'm very interested, for instance, in fire ecology and grazing ecology and trophic ecology." In other words, he tries to bring that past environment back to life in all its particulars, with an eye toward solving that great enigma of the extinctions.

Anjohibe Cave is a good place to practice landscape paleoecology—and to try to solve the extinction puzzle. The cave's more than three miles of passages have preserved and concentrated myriad bones of the extinct megafauna. In 1992, during Burney's first expedition to Anjohibe, James and Smithsonian colleague Fred Grady stumbled upon the rotting bones of an entire herd of pygmy hippos that had perished in a dark pool deep inside the cave. In another part of the cave, a cathedral-like cavern known as the Grande Avenue, the 1992 team unearthed a trove of megafaunal subfossils hidden beneath a giant, toppled speleothem. (Speleothem is the general term for any cave feature.) These included bones of two extinct giant lemurs and the bizarre, aardvark-like *Plesiorycteropus*. They also found remains of now-vanished giant tortoises within archeological sites inside the cave—the first evidence within Anjohibe that early Malagasy hunted some of the extinct megafauna.

For his part, Burney has used pollen grains that wafted into the cave over the millennia and became preserved within dripstone and other cave formations to begin to assemble a 40,000-year record of vegetation that grew in the vicinity of the cave. Preliminary analysis of this record reveals that a similar wooded savanna of satra palms and other fire-tolerant trees existed in the area long before humans arrived, which implies that climate change may have had little impact on the extinctions, at least in this one region of Madagascar. Since climate and habitat types vary enormously across the island, Burney has worked hard to collect such regional records of human and ecological history. The only way to come at an understanding of what happened to the megafauna, he believes, is to collate data from as many sites as possible around the country and from as many fields as possible, including paleontology, archeology, and paleoecology.

His Anjohibe team is just such a multidisciplinary group. James, an ornithologist who specializes in extinct birds, will focus her attention on recovering fossils of the several species of smaller birds known to have gone extinct on Madagascar. Worthy, the paleontologist and moa expert, is interested in the elephant birds. He is also an expert caver, as is fellow New Zealander Alan Cooper. The two of them plan to make in-depth explorations into the cave, including perhaps rafting down an underground river to see where it leads and search for undiscovered fossil deposits. Based at Oxford University, Cooper is a molecular evolutionist. He studies fossil DNA to glean clues about evolution. At Anjohibe, he told me, he would like nothing more than to find viable DNA from elephant birds. Along with DNA from other ratites, including moas and kiwis and ostriches, for example, it might help him find an answer to the nagging question of where the ratites first arose and how they spread around the world.

Cooper, in turn, will help Mary Egan, one of three Burney graduate students on this trip, who hopes to collect DNA from subfossil

hippos in Anjohibe. Using gene-sequencing techniques she lea
New York's Memorial Sloan-Kettering Cancer Center, she will s
the systematics of the extinct hippos, of which there are thought
have been two species. She will also test Burney's hypothesis that a
different genetic signal should turn up if the extinctions occurred
rapidly—say, by humans quickly hunting them to extinction—than if
they occurred more gradually—in the wake of a climate change, for
instance. The two other graduate students, both Malagasy, will also
focus on hippos. Toussaint Rakotondrazafy and Jean-Gervais
Rafamantanantsoa, the world's only Malagasy paleoecologists, are
"the people I'm most proud of here," Burney told me. Trained by him
over the years, they are specialists in pollen analysis, sediment coring,
and other paleoecological techniques. But at Anjohibe, they plan to
return to the hippo herd site to collect more specimens, both for Egan
and to paint a more complete picture of hippo ecology. Finally,
William Jungers, a primate paleontologist and expert on primate
locomotion at the State University of New York at Stony Brook, will
continue the search for bones of the extinct giant lemurs, while
Ramilisonina,* one of Madagascar's leading archeologists and ethnog-
raphers, will investigate the human presence in Anjohibe over the
centuries.

"To use a corny old analogy, these guys are like the blind men
feeling the elephant," Burney tells me in the car. In this case, the elephant is life as it was thousands of years ago—and what might have
happened to it. "One feels the leg and says an elephant is like a tree.
Another feels the side and says an elephant is like a wall. All these
people feel a different part. I have the advantage of hearing about all
of it, and I might actually be able to visualize the elephant."

I try to respond, but just then we hit a pothole of existential proportions and my voice fails to come. When Leva steps on the brakes,
I've learned to lean in advance and brace myself for the strike, but
again and again I've found myself moving my lips with nothing coming out. A bigger problem, though, is my legs. After hours entangled
among bodies and baggage, they are dead asleep.

Leva suddenly slams on the brakes and comes to an abrupt stop.

"We've got a flat," Cooper says from the middle seat.

"Flats we don't mind," Burney says, preparing to climb out the
rear door. "Flat *spares* is what we all hate."

I'm eager for a break. Even though he got no sleep two nights ago

*Like many Malagasy, Ramilisonina uses only one name.

n three hours of sleep last night, Burney has been
r hours. Before we left Tana, he smilingly told me,
y personality when I'm speaking French, because
d a man of few words." I can now appreciate her
wiry, bearded man in his late forties, Burney has the
energy of the Boy Scout he once was. He's also been, at various times,
a farm manager, commercial fisherman, tech writer for a biological
supply house, and naturalist for the state parks of North Carolina,
where he grew up. These days he throws his spinning-top vigor into
conducting science—in the lab, in the field, and in his head. "I've got
a million theories," he told me. "Trouble is, life is short, and you don't
get a chance to test them all."

In the cramped space of the car, a lot of Burney's vitality has gone
into conversation. Perhaps he's wired from lack of sleep, or perhaps
he's only trying to ensure I get the full story in the few days I have to
spend with him. But I'm exhausted from trying to keep up. Besides
Anjohibe and his work there, topics have ranged from his six years in
Kenya studying cheetahs to the triggering mechanisms behind *lavaka,*
the erosion gullies that pockmark Madagascar; from how the island
was first settled by people sailing boats rigged with hemp ("They
were built out of pot plants!") to how he has used satellite imagery to
find fossil sites ("Sounds like disco science, but, hey, whatever it takes
to find them"). His knowledge is encyclopedic. Is it wrong to want to
share it?

While Leva changes the tire, I climb a rocky hill to get a better
view of a fire I can hear crackling not far away on the other side of
the road. Even a little brush fire like that sounds ominous, with dry
sticks literally exploding in the fierce heat. Watching the fire, and
another nearby, I wonder how this entire parched land doesn't simply
go up in flames.

At certain times of the year, it does. Alison Jolly estimates that
one-third of the island burns every year, as herders torch dead savanna grass at the end of the dry season to encourage the young shoots of
new grass cherished by livestock. If you fly over Madagascar in
October, just before the spring rains, you can't see the ground in
many places because of the smoke from countless fires in the hills.
The smoke from these myriad blazes mixes together into one island-
wide pall, till the whole minicontinent seems to be aflame.

The fires before me are controlled, however, and the landscape
appears not much changed perhaps from the time when the megafauna called it home. From my perch high over the plain, I try to picture

them going about their lives in this sere land. Flocks (herds?) of elephant birds thundering across that stretch of treeless savanna. Giant tortoises dozing in the shade of those fire-resistant satra palms. A troop of ground-dwelling *Archeolemurs* bounding across that patch of hummocky earth, crying to one another. Pygmy hippos . . . Hell, there's no water. Where should I put them?

It's no use. The megafauna seem as distant to me as they did back in the Tsimbazaza Museum in Tana, where the mounted skeletons of two or three giant lemurs, a pygmy hippo, and an elephant bird standing at least three feet over my head (I'm six feet one) just seemed like so many bones. In my mind, I couldn't put flesh on them, just as I can't seem to place them in the landscape here. I can't see the elephant. Will this journey to Anjohibe Cave, chock-full as it is of megafaunal remains, help bring these alluring creatures alive?

2.

PERHAPS A HYPOTHETICAL scenario will help. Imagine you're the first person ever to step onto Madagascar. As far as archeologists can tell, you would have done so about two millennia ago, possibly in the southwest of the island. So let's say for the sake of argument that the year is A.D. 1, and you come ashore in the southwest near Raxworthy's camp at Lake Ranobe. The landscape looks much as it does today, a flat, semiarid land of octopus trees and other cactuslike plants of the spiny desert. As soon as you clear the oceanside dunes, you begin seeing creatures that frequent the area today, including mouse lemurs and chameleons and radiated tortoises.

The spiny desert, as you quickly discover, is painfully hot during the day, so you head for the stand of transitional forest next to Lake Ranobe. As you slip into the welcome shade of a massive tamarind tree, something suddenly bounds off into the shrubby understory. It's as big as a baboon and runs in the same forward-facing fashion. The monkey, or whatever it was, causes a commotion. Slightly alarmed, you hike out of the copse onto the grassy area that surrounds the lake. You're glad to be in the open air. But then you notice something that stops you in your tracks. A giant, ostrichlike bird, perhaps twice your height, breaks out of the woods a stone's throw away. It's not flying, but running on thick, earth-pounding legs. As you watch, not believing a bird could be so big, your eye is caught by something

large raising its head out of the nearby lake. No, there are several of
them, and they've got tusks. You turn to run, now thoroughly
spooked, and tumble over a boulder. Even as you frantically pick
yourself off the ground, you notice that the boulder has a head and
has turned to look at you with round watery eyes, its beaklike mouth
full of grass. . . .

Baboons? Giant ostriches? Living boulders? Well, in a manner of
speaking. Those creatures all lived in that part of Madagascar 2,000
years ago. But they don't today. These animals—the baboonlike
Archeolemur, the elephant bird, the pygmy hippo, and the giant tortoise—are all extinct. And they are not alone. Since the bones of
beasts unlike any alive today began to turn up in the early nineteenth
century, paleontologists have identified, as we've seen, more than two
dozen species of extinct Malagasy megafauna.

The lemurs took the biggest beating. Eight genera, comprising at
least fifteen species, vanished forever sometime in those two millennia. Every one of them, with the possible exception of the smallest of
the giant lemurs, was larger in size than any lemur living today. In a
diagram of the skulls of all extant and extinct genera, the living
lemurs are as dimes and pennies to the quarters, half-dollars, and silver dollars of their lost relatives.

From what paleontologists like Bill Jungers can discern from
their bone structures, two types of giant lemur existed—the ground
dwellers and the tree dwellers. Notable among the former are two
genera, *Archeolemur* and *Hadropithecus*. (Perhaps because giant lemurs
exist only as fossils, they have no common names.) Both of these
short-limbed, powerfully built lemurs lived at least part of the time on
the ground, filling niches that similar-sized baboons would have taken
had they ever made it to Madagascar. Paleontologists believe *Archeolemur*, which weighed somewhere between thirty and fifty-five pounds,
was a dietary generalist. Its teeth suggest it dined on foods that called
for the ripping power of its enlarged incisors, such as fruit with tough
rinds or hard-shelled seeds. *Hadropithecus* is thought to have been a
more specialized form of ground dweller than *Archeolemur*. It was even
more terrestrial, with dentition that implies it lived on grasses and
seeds.

The tree-dwelling lemurs comprise the other fifteen species of
extinct giant lemur. These include large-sized versions of the extant
ruffed lemur and aye aye. Most of the remainder belong to the "sloth
lemurs," so-named because of their slothlike build and habits (again,
inferred from bone structure). These include *Archeoindris*, the largest

lemur that ever lived; at an estimated 450 pounds, it was the size of an adult male gorilla. Although characterized as a sloth lemur because of its anatomy, *Archeoindris* probably spent most or all of its time on the ground, like the now-vanished giant ground sloth of the Americas. *Megaladapis* was a lemur playing koala. The size of a female gorilla, it had a highly elongated head, almost like that of an ungulate, and very long hands and feet, perfect for grasping tree trunks.

Another sloth lemur, *Babakotia*, was only identified in 1986. It was the first giant lemur to be named since 1909. First found in the Ankarana, a cave system in the far north, *Babakotia* later appeared in Anjohibe, where Dave Burney's 1992 team discovered a mandible with a single tooth. As in modern-day South American sloths, *Babakotia*'s forelimbs were much longer than its hindlimbs—perfect for reaching for the branches of leaves on which it fed. Its most distinctive feature was its long finger and toe bones. Each bone was curved, like a pirate hook, to give the animal a solid grip as it hung upside down by all fours, just as sloths do. The real trapeze artist of the sloth lemurs was *Paleopropithecus*. This species took the slothlike model to its extreme: arms almost twice as long as its legs, hooklike hands and feet, and limb joints clearly designed for maximum flexibility rather than stability. Though subfossils of this 90 to 120-pound lemur have been found around the island, the most complete skeleton—indeed, one of the best skeletons of a fossil primate ever found—turned up in Anjohibe Cave in 1983.

Sadly, not a single written eyewitness account of a giant lemur exists, though evidence from radiocarbon dating suggests that some of them persisted in remote parts of the island for a millennium or more after humans arrived. The closest we can come—painfully close—is through Etienne de Flacourt. In his 1661 *Histoire*, Flacourt describes an animal that local people told him lived in the hinterland. Scholars believe that this animal, if it existed, may have been a very late surviving *Megaladapis* or *Paleopropithecus*:

> The Tretretretre, or Tratratratra, is an animal the size of a two-year-old calf, with a round head and a human face; its forefeet are like those of a monkey and its hindfeet likewise. It has a frizzy coat, short tail, and ears resembling those of a man. . . . It has been seen near the pond of Lipomami in whose vicinity it has its haunts. It is a very solitary animal, greatly feared by the people of those parts, who flee from it as it flees from them.

This passage may be the nearest we'll ever come to a *written* account. But astonishingly, Dave Burney and Ramilisonina, the archeologist-ethnographer on our team, recently collected a *verbal* account. In a village near Belo-sur-Mer, a coastal town in the most remote part of western Madagascar, villagers told them that a large, ground-dwelling lemur still exists in the scorching, uninhabited dry forests just back from the coast. Thoroughly intrigued though loath to be branded as cryptozoologists, Burney and Ramilisonina have treated this story with both great interest and great caution, as we'll see.

Giant lemurs may not have been the only members of the megafauna to survive into Flacourt's time. The Frenchman, whose descriptions of natural and cultural history in the region around Fort Dauphin where he lived are considered scrupulously accurate by modern scholars, describes a bird that local Antandroy tribespeople called the Vouron-Patra. "[T]his is a large bird haunting the Ampastres," Flacourt wrote. "[I]ts eggs are like those of an ostrich, it is a kind of ostrich; the people of these regions cannot catch it, as it seeks out the most deserted places." As with the giant lemurs, it is deeply disquieting to think how recently the world lost Madagascar's elephant birds. Had they survived just a few more centuries, elephant birds might have gained official protection and today might enjoy a haven inside Malagasy reserves.

The "elephant bird" acquired its name from the Roc of the *Arabian Nights*. In "Sinbad the Sailor," one of the tales in this oral collection, which Arab storytellers began circulating around the tenth century, the Roc was a bird so enormous that it could lift an elephant in its talons. Marco Polo picked up on this legend as he passed through Arabia on his way back from China in 1294:

> According to the report of those who have seen them . . . they are just like eagles but of the most colossal size. . . . They are so huge and bulky that one of them can pounce on an elephant and carry it up to a great height in the air. Then it lets go, so that the elephant drops to earth and is smashed to a pulp. Whereupon the gryphon bird perches on the carcass and feeds at its ease. They add that they have a wingspan of thirty paces and their wing-feathers are twelve paces long and of a thickness proportionate to their length. . . . I should explain that the islanders call them *rukhs* and know them by no other name.*

*The rook in chess got its name from this mythical creature.

The "islanders" Polo refers to are none other than the Malagasy, for this account comes from his brief description of Madagascar. While his account is quaintly erroneous, as we've seen, his rukh—and the *Arabian Nights'* Roc—had a basis in fact. Arabian merchants first began trading along the northern coasts of Madagascar in the ninth or tenth centuries, and they may very well have seen elephant birds or at least their outsized eggs. Travelers even brought back giant "quills" said to be from the feathers of the Roc.*

It wasn't until the mid-nineteenth century that people outside of Madagascar realized a kernel of truth might lie in such tales. In 1851, the French naturalist Isidore Saint-Hilaire uncovered eggshells and bones of a giant bird on the west coast. Neither he nor anyone else had any idea what it was. In the coming decades, more bones and eggshell fragments turned up, but nothing set heads spinning like the whole eggs that Malagasy occasionally brought forth. The eggs' Brobdingnagian proportions astounded scientists and others in the West. Nature's largest eggs, elephant-bird eggs measure a foot long and nine inches in diameter. Mervyn Brown has noted that a single elephant bird egg had a capacity equivalent to 150 hen's eggs and could therefore produce an omelette for fifty people. Complete eggs occasionally still erode out of the desiccated hillsides of the southwest, where local middlemen are becoming more savvy about the astronomical prices the eggs can bring in the West.

In the nineteenth century, scientists bent over backward trying to identify the owner of these remains. One thought it was a diving bird related to the auks and penguins. Another, perhaps a bit too swayed by legends of the Roc, wrote a series of papers aiming to demonstrate that it was a giant bird of prey, whose closest living relative was the condor. Eventually scientists settled on the ratites. Today, paleontologists identify between six and a dozen species of Malagasy elephant bird, in two genera. The smaller *Mullerornis* species were about the size of an ostrich, while the largest of the *Aepyornis* species stood ten feet tall and weighed half a ton. These gigantic birds likely filled the browsing niche left empty on Madagascar by the absence of competing large herbivores such as giraffes.

The dominant grazers, on the other hand, were the giant land tortoises (and pygmy hippo). As large as Galápagos tortoises, Madagas-

---

*One later writer argued sensibly that the "quills" were likely the midribs of leaves from the raffia, a native Malagasy palm whose leaves can reach twenty or thirty feet in length and, as he wrote, "are not at all unlike an enormous quill stripped of the feathery portion."

car's two species reached prodigious size, with the carapace of the larger one stretching a full four feet across. They once ranged widely across the island, grazing savanna grasses and shrubs from the Highlands to the coasts. Like surviving giant tortoises elsewhere in the world, they probably grew, matured, and moved with a ponderous slowness, which made them vulnerable to hunters. Today, only three much smaller tortoises survive on Madagascar—the radiated and spider tortoises of the southwest and the plowshare tortoise of the west coast, which is itself on the verge of extinction. Such vulnerability doomed giant tortoises in other parts of the Indian Ocean as well. In 1759, according to the French geologist Joël Mahé, four small ships began transporting giant tortoises between Rodrigues and Mauritius, where they were sold for their fresh meat. Over the next eighteen months, Mahé says, more than 30,000 tortoises made the trip. Today, save for a population clinging to the uninhabited island of Aldabra about 300 miles north of Madagascar, giant tortoises have vanished in the Indian Ocean region.

The only other megafaunal reptile that may have gone extinct on Madagascar was a giant crocodile. I say "may" because most scholars suspect that the subfossil specimens recovered merely represent oversized examples of the Nile crocodile that lives on the island today. The only other carnivore thought to have vanished bears the same caveat. Paleontologists have dug up remains of a giant fossa, but excepting its great size, the giant fossa appears to be remarkably similar to the existing fossa, and at least one scholar has suggested that the two animals were conspecific, meaning of the same species.

There is no such doubt about the megafaunal animal known, for lack of a better label, as the Malagasy aardvark. This animal has no living relatives on Madagascar or anywhere else. To scientists, it goes by the mellifluous Latin name of *Plesiorycteropus madagascariensis*. The first bone of this creature—a partial skull—was uncovered in the last century at Belo-sur-Mer, where Burney will head after our visit to Anjohibe. Since then, it has turned up in the Highlands and in the southwest, though its bones remain a rare discovery. In Anjohibe in 1992, for instance, Burney's team found the first *Plesiorycteropus* remains unearthed in fifty years. The animal's anatomical structure suggests it was a digger like the aardvark, an African animal that lives primarily on termites. However, the mammalogist Ross MacPhee of the American Museum of Natural History recently argued in a 200-page monograph that *Plesiorycteropus* is so unusual that it belongs in its own mammalian order. Possibly as baffled as anyone else about its true nature—even after 200 pages—MacPhee designated the order *Bibymalagasia,* which means simply "Malagasy animal."

Lastly, there is the pygmy hippo, the only other nonprimate megafaunal mammal to go extinct. For some reason, of all the lost megafauna it's my personal favorite. Call it a fondness for oxymorons. Or maybe it's because, of all the vanished "big animals," the pygmy hippo is the one whose remains I actually have a chance of finding myself at Anjohibe. Also, I've been intrigued by stories Burney has been telling me in the car about a mysterious animal recently seen in the far west. In the same region where villagers told Burney and Ramilisonina of a large, ground-dwelling lemur, others have spoken of another creature they call the *kilopilopitsofy* ("floppy ears"), a creature that in its particulars sounds suspiciously like the pygmy hippo. Could this stocky member of the megafauna yet survive?

My interest also lies in the fact that the pygmy hippo, because of the ubiquity of its remains, has been largely ignored by researchers in the face of the other, rarer megafauna. "There's one person in the world who studies Malagasy hippos and that's Solweig Stuenes, a Scandinavian lady who has done a bunch of taxonomic work," Burney told me. "Frankly, nobody's interested in these hippos except us." The hippo has less exotic appeal, he says, because unlike the lemurs and the Malagasy aardvark, it's not that different from what exists in Africa. "But that's always been an interest of mine, not only looking at how Madagascar's different, but how it's similar to Africa," Burney said. "It's interesting when nature runs the same experiment more than once." Further, he considers the hippo a more useful paleoecological tool than the lemurs, for its commonness means that it better reflects the kinds of environments that existed in the past.

Like those of tortoises, the remains of hippos have turned up in both paleontological and archeological contexts. Seven of the nine megafaunal bones ever unearthed on Madagascar that show unequivocal signs of butchery came from hippos. Two of these cut bones were pulled out of a dried-up swamp a few miles from Lake Ranobe by Alfred Grandidier, who made the first description of the animal in the mid-nineteenth century based on about fifty individual hippos he unearthed there. In the decades following, scholars declared the existence of up to four species of pygmy hippo in Madagascar, though Stuenes, in a careful revision of the animal's taxonomy published in 1989, concludes that there were just two.

In the world as a whole, two hippopotamus species have survived to the present day—the African common hippo and the West African pygmy hippo. Though Madagascar's pygmies were closer in size to the pygmy of West Africa, anatomically they more closely resemble the African common hippo, which is their likely ancestor. They are much smaller, of course. An adult African common hippo stretches a full ten feet and stands over four feet at the shoulder—a formidable beast. By contrast, the largest specimens of the two Malagasy pygmies, which apparently did not differ appreciably in size, reached no longer than six and a half feet in length and stood two and a half feet tall.

Dwarfism is a common adaptation for large animals that colonize new islands, where a smaller size is advantageous in the face of a limited food supply and few or no large predators. But the Malagasy pygmies did not dwarf to the degree that pygmy hippos did on other islands, such as Cyprus. The reason, Stuenes feels, is that Madagascar's hippos probably faced two conflicting selection pressures. To better facilitate proliferation of the species, one such pressure called for early sexual maturity, which thus favored a smaller size; the other called for a larger size to better cope with the threat from crocodiles, likely the hippos' only enemy until humans came.

African hippos that made their way to small Mediterranean islands like Cyprus discovered only one niche available. But Madagascar appears to have been large enough to provide two niches. The niches match those of the two African species. The African common hippo is a mostly amphibious animal that comes out at night to feed on savanna grasses, while the West African pygmy hippo is a land animal that lives in the forest. Evidence suggests that one of the Madagascan species took the former's niche and the other the latter's. Bones of *Hippopotamus lemerlei,* for example, have turned up along the west coast in regions of slow-moving rivers, while those of its cousin *H. madagascariensis* have been found in the Highlands. Differences in the species' skulls and lower jaws are even more conclusive. *Hippopotamus lemerlei* had higher orbits, the bony sockets that hold the eyes, than *H. madagascariensis*, just as the African common hippo does compared to the West African pygmy. The higher the orbits, the better able an animal is to keep its head and most of its body submerged in water, as an aquatic hippo would have done. Further, the wear facets on the teeth of *H. lemerlei* agree with those found on the African common hippo, while those on *H. madagascariensis* resemble those on the West African pygmy hippo.

Alas, the same fate awaited both species of Malagasy hippo, as it did all members of the megafauna. What is it, paleontologists wonder, that could have so neatly killed off such a wide variety of animals?

The "big animals" included mammals, reptiles, and birds. Some were arboreal, some ground-dwelling. Some were grazers or browsers, others meat-eaters. Most were probably slow-moving, though if their extant analogues are any indication, the elephant birds and pygmy hippos could have moved at quite a clip in a panic. They lived in a variety of habitats, from semiarid bushland to dense tropical forest to Highland prairie. And they lived all over an island as vast as France and Switzerland combined. How could all these creatures over such a large area have suddenly vanished?

It is instructive to look at megafaunal extirpations elsewhere in the world. Australia and New Guinea lost their large-animal assemblages sometime between 30,000 and 15,000 years ago. North and South America lost theirs around 12,000 to 10,000 years ago. The most recent losses occurred on islands such as New Zealand, Mallorca, and Madagascar, which suffered extinctions of their biggest denizens 6,000 to 1,000 years ago, or possibly more recently on Madagascar. Eurasia had fewer megafaunal extinctions, yet as the paintings at Lascaux and other caves in France and Spain attest, a number of large prehistoric animals vanished, including the woolly mammoth, woolly rhino, cave bear, and giant deer. Curiously, the continent where humans have lived the longest, Africa, has apparently had the fewest large-animal extinctions, only seven over the past 100,000 years. (Compare that to more than ninety *genera* lost in the Americas, most if not all within the past 15,000 years.) Africa still has most of its Pleistocene megafauna: elephants, rhinos, hippos, giraffes, and so on.

Many researchers feel no mystery surrounds these massive extinction events. They all have certain features in common: All the big animals disappeared, often very rapidly, and were not replaced by other big animals. Again perhaps most significantly, every one of the extinctions occurred not long after man appeared on the scene. Paul Martin of the University of Arizona put together a world map showing temporal and geographic patterns of such extinctions that shows vividly how the extinctions clearly follow the stepwise pattern of human colonization of ever more remote regions, with such outlying islands as Madagascar and New Zealand being hit last. The hypothesis put forth as to why Africa and Eurasia were spared major extinctions is that the giant animals there had lived with humans for a far longer period and had learned to flee man long before he had developed the technological skills to wipe out huge numbers of prey animals.

In the 1960s, Martin and several colleagues proposed a theory to explain the extinctions in North America, which they suggested could apply equally well to extinctions in other parts of the world, including Madagascar. It has come to be known as the "blitzkrieg hypothesis." According to this hypothesis, the earliest Native Americans, soon after crossing the Bering Land Bridge to the Americas 12,000 years ago, quickly precipitated the crash by overhunting the animals.*

One criticism of Martin's theory is that Paleo-Indian hunters would have been too few in number to be technically capable of such a massive, continent-wide slaughter. But in the late 1960s the Russian scientist M. I. Budyko published a series of equations showing that, theoretically, such extinction events would not require a huge human population. Under constant hunting, a small initial population could multiply fast enough to quickly outpace the ability of large animals, with their slow rate of reproduction, to replenish themselves. In the mid-1970s, Martin and James Mosimann of the National Institutes of Health developed a computer model that further honed Martin's original "blitzkrieg" idea. Moving across the Americas in a "front," the earliest big-game hunters would have multiplied rapidly if they had an unlimited food supply at hand, suffered little from disease, and were efficiently organized. Having no natural fear of humans, mammoths and giant sloths and other megafauna would have fallen easy prey to atlatl-bearing hunters.

The blitzkrieg is not the only theory that scientists have proposed for the megafaunal extinctions on Madagascar. Putative causes have ranged from a proposition in the last century that major geological events, including stepped-up volcanism, wiped them out, to a recently revived theory that a "hypervirulent" disease brought by early humans raced through native species, which had no natural defenses against the pathogens. Climate change and the introduction of exotic species have both been implicated as well. Whatever it was, it was so all-encompassing that only one creature managed to pull through: the Nile crocodile.

---

*Recent findings indicating that humans were in the Americas far earlier, perhaps as early as 35,000 years ago, have not gained wide acceptance in the scientific community. And some scholars have suggested that, even if they were there, they might represent an earlier colonization event that never took hold in the New World as did the one at the end of the Pleistocene.

3.

AT THE AMPIJEROA Forestry Station, a dry-forest reserve where we stop for the night, I get a chance to look for this sole survivor of the megafauna. Ampijeroa ("am-*pee*-tsa-roo") lies almost 300 miles from Tana, and we arrive long after dark. With the headlights of the trucks floodlighting the forest, we pitch our tents in a scatter of crackly teak leaves, then rip into the foodstuffs for a long-anticipated dinner. Back in Tana, when one of us mentioned his hope that we'd find whole skeletons at Anjohibe, Burney said, in reference to ever-dwindling camp food his team would have to subsist on for weeks, "*we'll* be the whole skeletons." But the food is abundant now. Feeling my nerves unjangle as I sip native Betsileo wine, I stuff myself silly on canned corned beef, Laughing Cow cheese, stale French bread, tangarines, and chocolate.

After fourteen hours on the road, crammed into the trucks like telephone-booth-stuffers, we're eager to stretch our legs, and Burney offers to lead us along the road to a nearby lake to look for crocs. The road is empty and dark, and the stars are magnificent. Acting on instinct after so many weeks with Raxworthy, I immediately start picking out chameleons sleeping in roadside shrubs and trees with my headlamp.

On the way, Burney primes us with croc stories. At almost every lake where he has worked in Madagascar, he says, villagers have told him that somebody had recently been killed there by a crocodile. So common is this tragedy that when all hope is lost in life, the Malagasy say "Got into the clutches of the crocodile." While coring lakebed sediments on a raft, Burney himself has had some close encounters. His depth finder has symbols for fish of different sizes, he says, and every now and then the largest symbol will light up. "Sometimes you'll see this big 'fish' swim up right under the boat and stay there. It's kind of an uneasy feeling." He tells us about his doctoral advisor at Duke, who was doing research with a graduate student on a lake in Africa. A crocodile took a bite out of their inflatable boat, sinking it. They had to swim two miles to shore through crocodile-infested waters. The next morning, they thought of heading out to retrieve the outboard, which lay in about seven feet of water, but so many crocs lay on the beach that they thought better of it.

We reach the lake. Vast and ominous in the pitch black, it lies at the bottom of a broad man-made bank. Seeing no crocodiles on the shore, I make my way down the bank to see if I can spy any out in the water. As I near the lakeside, Burney suddenly yells from behind me,

"Peter, get back! I see eyeshine!"

I don't see it myself, but I don't need any convincing. I am up that embankment so fast my sneakers leave skidmarks. Burney's stories have worked me up, and I've read many tales of encounters with these dangerous beasts. One of the earliest accounts, written in the 1670s by the French traveler Sieur Du Bois, describes a trial by crocodile:

> [T]he natives also swear by the Crocodile, which they name *Voa*, with which the Rivers and Lakes of this Island are full; saying that they wish to be eaten by them, if they have done that of which they are accus'd; this done, they are oblig'd to pass thru a river, which they do. It also happens often that in passing through the water, they are taken and eaten by these Crocodiles or *Voa*. The spectators of this fine proof of truth say that such a one has done the thing of which he was accus'd, 'tis wherefore he has been eaten.

Malagasy have other curious beliefs concerning this reptile. Henry Grainge, an Englishman who traveled in 1875 down the Betsiboka, the river we're shadowing on the way to Anjohibe, reported that local people believe that crocs "live chiefly on stones, stealing cattle, pigs, and people merely as a relish to the harder fare" and that, "smitten with the charms of the pretty little divers and other waterbirds, they choose their mates from among them, and so crocodile's eggs are produced."

Europeans of the last century who journeyed along the Betsiboka had no such fanciful notions about what one of them dubbed "crawling saw-backed monsters." They knew the crocodile's natural history, and they rightly feared it more than any other wild animal in Madagascar. Their apprehension wafts from early accounts like a dank odor. "I never saw such a dreadful snapping apparatus in my life; a shark's mouth looks innocent by comparison," Grainge wrote after shooting a "small" one of eight feet, nine inches. Another traveler, grounded overnight on a sandbar that "one could almost cross on the backs of its crocodiles," sat all night on the forward deck, his Springfield across his knees, "ready to drive 150 grains of steel-jacketed lead between one of those encircling pairs of eyes which glowed at us so malevolently from the darkness." Their numbers could be astounding. After spying about forty crocodiles sleeping on a sand spit, another traveler on the Betsiboka began counting. In the first hour, he counted 105; in the next half hour, 102. He estimates that in four days he and his party saw no fewer than 1,600 crocodiles.

Later, from a safe vantage high up on the embankment, I see eyeshine myself: two amber ovals glowing in the light of my headlamp. The great survivor, which outlasted everything from *Majungasaurus*, a dinosaur whose bones began turning to fossil in these parts during the Cretaceous, to the last giant tortoise that succumbed perhaps only a few hundred years ago. I look back up the road. Burney and the others are a quarter mile away. In this quiet, solitary moment, I imagine myself tumbling down that steep bank and being knocked unconscious, to be consumed by the last of the megafauna. A Malagasy proverb enjoins you to remember "the man who went fishing and got fished by the crocodile." I recall the story of one nineteenth-century resident, who reported that a friend saw an ox crossing a river suddenly vanish, only to turn up a half hour later as an empty skin floating on the surface. "I was somewhat inclined to think he was drawing on *his* imagination or *my* credulity," the writer declared, "but he assured me it was a simple statement of fact." Having no wish to become a floating skin, I turn and head back to join the others.

In the morning, I rise early and sit in silence watching a plowshare tortoise in a pen here at Ampijeroa. It was exciting to encounter the only extant member of the megafauna, but now I can almost imagine I'm looking at an extinct member. In any event, it's the closest I'll probably ever come to seeing one of the extinct species alive. The plowshare is not only a smaller version of a Malagasy giant tortoise, but it's on the verge of extinction, which brings it closer to its lost cousins than, say, most of the living lemurs are to theirs. This motionless creature certainly conjures up more than those museum bones back in Tana.

Sadly, the plowshare is the world's rarest tortoise. Only 300 to 1,000 individuals are thought to remain, and the largest concentration of them lives here, in this pen at Ampijeroa. The pen looks like any outdoor zoo enclosure anywhere: a dirt-floored pen with low concrete walls. A prickly pear grows in the center, its green lobes topped by one or two pink blossoms that offer the only relief to the unrelenting brown of dirt, rocks, dead grass, and teak trees rising into the dry forest canopy overhead. An orangey, early-morning sunlight, able to penetrate the dense, though now largely leafless forest, only enhances the dun color scheme.

I can see the rounded humps of two adult plowshares, one along a tree trunk like a fallen telephone pole that divides the pen in half, the other against the far concrete wall. They huddle in their shells against the morning chill. Their two-foot-long domed shells, each bearing an intricate design of black-and-orange plates, resemble rounded foot-stools of inlaid marquetry in need of a good dusting. The three vanished species were of the same genus as the plowshare, *Geochelone*. Did they have the same reticulated pattern on their backs? If I blow up the plowshare before me in my mind until the length of its carapace is four feet long, would I be looking at *G. grandidieri?*

As if intrigued by the idea, the nearer plowshare, only a few steps from me over the surrounding wall, sticks its head out of its shell and looks in my direction. Its head is but a grape next to the halved cantaloupe of its shell. From this angle, I can't make out the protuberance of the lower shell that gives the plowshare its name. Shaped like a person's tongue frozen in the midst of trying to lick his nose, the plowshare is used as a jousting lance by the males. In his delightful book *The Aye-Aye and I,* the late naturalist Gerald Durrell, founder of the Jersey Wildlife Preservation Trust, the Channel Islands–based foundation that launched this plowshare conservation effort in 1985, best describes the comical skirmishes that take place between males competing for a female's attention:

> The two males, rotund as Tweedledum and Tweedledee dressed for battle, approach each other at what, for a tortoise, is a smart trot. The shells clash together and then the [plowshare] comes into use. Each male struggles to get this projection beneath his opponent and overturn him to win a victory in this bloodless duel. They stagger to and fro like scaly Sumo wrestlers, the dust kicked up into little clouds around them, while the subject of their adoration gazes at their passionate endeavors, showing about as much excitement and enthusiasm as a plum pudding. Finally, one or other of the suitors gets his weapon in the right position and skidding along and heaving madly he at last overturns his opponent. Then, he turns and lumbers over to gain his just reward from the female, while the vanquished tortoise, with much leg-waving and effort, rights himself and wanders dispiritedly away.

Sitting here looking at a creature that might be extinct before my children are old enough to see it, I find it hard to be as lighthearted as Durrell. On the tortoise across from me, the scraggly line between its upper and lower jaws makes the animal appear to be grimacing. And it has reason to be. Its wild cousins, which live only in the Baly Bay region of western Madagascar, now face imminent extinction. They cannot outrun the periodic fires set by Malagasy herders. They lose their eggs and soft-shelled young to wild pigs. And they fall into the hands of Malagasy who believe that keeping plowshares in chicken yards will stave off a cholera-like poultry infection.

Even the tortoise before me is not safe. On the night of May 6, 1996, thieves broke into this enclosure, which only a groggy night guard watches over, and stole seventy-five plowshares—two adult females and seventy-three hatchlings. That is half of the entire stock here. Since the station's resident dogs did not bark that night, authorities suspect an inside job. Certainly it was well planned. A month before the heist, a Dutch dealer sent out faxes to potential customers, offering baby plowshares "available soon" for $3,000 apiece. No one could save the giant tortoises that used to lumber about Madagascar, and perhaps no one will be able to save their smaller relative either.

4.

WALK INLAND FROM the coast anywhere on Madagascar, and two things will happen, one of which bears significantly on the extinction of the giant tortoises and their fellow megafauna.

First, you'll head up to a higher elevation. If you start from most places on the west coast, your way will be gradual, along a smoothly rising plain that in some areas may not even seem to be rising at all. If, on the other hand, you leave from almost anywhere along the east coast, you'll find your path rising steeply, in some places precipitously, over high, forest-covered ridges that leave you winded and dripping in sweat. From either side, however, indeed from any spot along the 2,500-mile-long coast, you will eventually reach the central Highlands, which rise from about 4,000 feet in the valleys to over 9,000 feet on some of the highest peaks.

The second thing that will happen—and this is what bears on the extinctions—is that you'll sooner or later leave forest behind and enter a sea of grass. In the northeast, where the largest unbroken tract of rain forest remains, you may walk for thirty or forty miles before the forest ends. But it will end. The writer David Quammen has aptly likened Madagascar to a giant bald head, with just a few tufts of hair—the relict forests—poking up around the edges.

When you fly over that vast, undulating prairie and take it in in its entirety, it often looks as if there is not a single tree down below. But if you narrow your focus to finite parts of the landscape, you will soon notice a bushy cluster of trees thatching an old erosion gully. In a gulch between grassy slopes, a splash of dark green standing out from the surrounding yellow green. A wispy strand of tall trees keeping close company with a winding stream. Every now and then, you pass over an isle of dense, unviolated forest incongruously holding its own in that endless sea of grass. *Where did that come from?* you wonder. Even more incongruous are its perfectly straight edges, as if that patch had been clipped by a pair of scissors. You realize that it must be a remnant of a forest that used to cover the land below, a token, probably sacred grove that people living there have chosen to preserve. And as you look at those wavy moors stretching to the hazy horizon out your window and out the windows on the other side of the aisle, you can't help but wonder: Was Madagascar once completely forested?

That's the conclusion, as I've noted, that a pair of French botanists came to earlier this century. Working independently, Henri Perrier de la Bâthie and Henri Humbert put forward the notion that because most parts of the interior appear to be warm and wet enough to support forest, and because there are remnant stands of woodland throughout the region, and because Malagasy today are quickly slashing and burning their way through existing patches, then the Highlands—that great bald pate—must once have been thick with forest. By their way of thinking, the great rain forests of the east once met the great dry deciduous forests of the west, with presumably some sort of transitional wet-dry forest in the middle. Further, the argument goes, Madagascar had existed in this forested form for tens of thousands of years, from some time in the distant past right up until the first Malagasy arrived and apparently chopped and burned it all down, leaving today as much as 90 percent covered in savanna. Finally, this original, closed forest knew little or no wildfire until humans came and set them—or so a corollary to this theory held.

This theory, which is chiefly associated with Humbert and Perrier de la Bâthie but which other scholars built upon over the past few decades, is relevant for two reasons. For one, the argument handily explains the subfossil extinctions. It holds that after Malagasy pastoralists destroyed the forest to make rangeland for their cattle, secondary forest and grassland quickly took over. The Malagasy kept on burning, to bring forth new growth for their livestock, and this caused the erosion gullies to start forming.* The loss of forest habitat, with all the attendant problems that go with that loss—including shrinking food supplies, range sizes, and mating opportunities, among others—sent at least two dozen species of megafauna spiraling downward into extinction. Other secondary causes, such as incidental hunting and the silting of wetlands by eroded soil, also played a part, the theory maintains.

The other reason that the Frenchmen's theory is pertinent is that it became the received wisdom. For one thing, the idea that dense forests once covered Madagascar had been around for some time, as had the related notion that the Malagasy were responsible for the loss of the megafauna. As early as 1868, Grandidier had declared that "it was under [man's] eyes, and probably as a result of his action, that all these species have disappeared of which the subfossil remains only now reveal to us their former existence." Humbert and Perrier de la Bâthie's theory seemed so sensible an explanation that, in the decades after it was formulated, practically everyone took the theory as gospel. Throughout the colonial period, scientists and nonscientists alike spun off any number of new theories, using the "continuous-forest" theory as a given.

In the 1970s, however, cracks began to appear as a result of stepped-up research in the country. For decades, really for most of the sixty-five years that Madagascar was a French colony, the only fieldwork conducted on the island was done by French scientists. Then, in 1970, ten years after independence, Madagascar held its first International Conference on the Preservation of Nature. It was a watershed for science and conservation in the country. In the years immediately following the meeting, dozens of foreign researchers began working in Madagascar. Within a few years, the newly established socialist regime of Didier Ratsiraka booted them all out of the country, but the ball had been set rolling. In the late 1970s and especially in the 1980s, research and conservation in Madagascar took off.

In 1974, three French researchers published a book in which they challenged the continuous-forest theory. They held that soils in certain parts of the interior, especially in the loftiest and driest areas, would have precluded much growth of forest. The increasingly dry conditions in the late Quaternary period (two million years ago to the present), which occurred in tropical areas worldwide, would have

---

*Ironically, those gullies, however they originated, sent silt down into valley bottoms to form the flat plains that enabled the Malagasy to grow paddy rice, the country's staple.

made conditions unfavorable for unbroken forest cover in those parts of Madagascar with a lengthy dry season. Finally, certain differences in composition between some of those hard-edged forest islands seem to suggest that they had been isolated for a long time, much longer than the two millennia humans are supposed to have lived on the island.

The next sizable crack appeared ten years later when the archeologist Robert Dewar published an oft-cited paper entitled "Extinctions in Madagascar: The Loss of the Subfossil Fauna." Dewar attacked the continuous-forest model from all angles. He started with the idea that early Malagasy had destroyed the uniform forest by burning it. He dismissed the notion that these forests knew little or no natural fire before humans arrived and therefore were highly susceptible. No natural ecosystem on Earth is wholly free of fire, he noted; why should Madagascar's forests be any different? And archeological evidence indicates that the human population, which was only about two million in 1900, was never very large. How could so few people deforest such a vast area?

Dewar's most convincing argument against a solid forest came from evidence collected at the fossil site of Ampasambazimba, a dried-up lake bed and marsh located in the Highlands west of the capital. Accidentally discovered by a prospecting crew in 1902, Ampasambazimba is the richest bed of subfossil bones ever unearthed in Madagascar. More species of extinct animal have turned up there than in any other site, including some, like *Archeoindris*, not found anywhere else. There were arboreal sloth lemurs and ground-dwelling sloth lemurs; tortoises, dwarf hippos, the giant fossa; five species of elephant bird. How, Dewar asked, could such a diverse array of animals live in a single forest? The lemurs alone, of which no fewer than twenty-six extant and extinct species came out of the site, must have filled impossibly narrow niches. It was much more likely, Dewar said, that a diversity of habitats existed around Ampasambazimba in times past. Interestingly, Perrier de la Bâthie's own collection of fossil plants from the site lent support to that idea.

Next, Dewar turned to the notion that destruction of this purported closed forest spelled doom for the subfossil species. He conceded that certain arboreal lemurs known only from now treeless sites on the central plateau, such as Ampasambazimba, may have fallen victim to the loss of their habitat. But what about the elephant birds, tortoises, and ground-dwelling lemurs? Lightly wooded grasslands would have suited them better than dense forest. Plus, paleontologists have uncovered remains of these animals across the island in different vegetation zones, including some that are preserved to this day in sizable patches. Finally, if giant lemurs cherished unbroken forest, why are there no giant lemurs in existing tracts of eastern rain forest?

While he was at it, Dewar went on to attack Martin's overkill hypothesis. Again, he asked, how could such a small population have killed off all the megafauna? Even if one invokes Budyko's argument for the infinite killing capacity of a population that begins with a small initial group, how come so little incriminating evidence exists? As we've seen, one can count on two hands the number of bones of extinct Malagasy species that show unmistakable cut marks. Finally, Dewar noted, archeological evidence hints that the first Malagasy were not hunters but rather traders, farmers, and fishermen. They probably hunted opportunistically, but, he concludes, "whatever brought the first Malagasy to the island, it was not game."

Dewar offered his own theory for what happened inside Madagascar. Traders and fishermen stayed along the coast, but pastoralists worked their way inland. They moved with their herds of cattle and possibly sheep and goats, ever searching for open patches of savanna in the mosaic of dense forest, open woodland, and savanna in which to graze their livestock. The herds provide the "missing link" in previous models, Dewar says. Introduced livestock came in direct competition with the elephant birds, giant tortoises, and terrestrial lemurs, which had coevolved in isolation for millions of years with the native flora. The megafauna could not compete with the newcomers, and they went extinct. One of the biggest contrasts between the savannas of Madagascar and those of East Africa, Dewar noted, is the former's total absence of medium- or large-sized native herbivores. In this regard, he said, the Malagasy grasslands are "truly eerie." Dewar concluded that the loss of the native herbivores proved devastating to the natural environment, precipitating the sterile grasslands we see today.*

---

*Or could the giant tortoises, pygmy hippos, and elephant birds have created the grasslands naturally, before people and their companion species arrived? Burney considers it possible. Since these large plant-eaters probably had no predators—"after all," he asks, "what do we know of that ever lived on Madagascar that was big enough to kill an elephant bird or hippo or giant tortoise?"—they once likely proliferated on the island. (The fact that these three animals represent the most common subfossils in Madagascar supports this.) Burney cites two examples of what can happen as a result of such species domination: the Galápagos, where early explorers recorded vast expanses of "tortoise turf" grazed down by the giant tortoises; and Murcheson Falls in Africa, where colonial authorities protected hippos and other animals from hunting. "In those areas without predation," Burney told me, referring to the latter area, "hippos bult up to such incredible numbers that they cleared the vegetation along the riverbanks, causing a huge amount of erosion and creating these grasslands, which are really herbivore-maintained grasslands."

So, after decades of the continuous-forest theory, Dewar had offered up a new theory to challenge. Three findings would lend it support, he said. First, evidence shows that initial settlement of the western and southern interiors was by cattle-raising pastoralists who hunted on the side. Second, as-yet nonexistent analyses of fossil pollen reveal that the region had "a primeval savanna/forest mosaic, followed by a conversion to grasslands without major climatic fluctuation. . . ." And third, someone shows convincingly that the extinctions, the spread of the herders, and the radical changes in the plant community all happened at the same time. It was the second challenge that Dave Burney would tackle, with myth-exploding results.

5.

BURNEY IS NOW LEADING us on an introductory tour of Anjohibe Cave, which we reached after two solid days of driving. The tour should help put the megafaunal extinctions in context, at least in this one remote corner of Madagascar.

"You see this hole here?" Burney points to a perfectly round core, the diameter of a plum, drilled horizontally straight back into a gigantic white speleothem, like the base of some massive primeval tree. "We drilled this two years ago, and it opened our eyes to the extreme age of the cave. This speleothem started growing 140,000 years ago and stopped growing some 105,000 years ago. My inference from the pollen samples we pulled from there is that this was not an entrance 140,000 years ago. It was closed—clean cave air, pure white limestone. But by 105,000 years ago, there was some little opening forming."

I look back at Entrance A just behind us. It's the main entrance to Anjohibe, and it's far from a little opening now. The entrance is about fifty feet wide and ten feet high. Did members of the extinct megafauna trundle through there long ago? I can see vines and roots overhanging the opening, and one or two trees just outside. Beyond them is a circle of cement with a flagpole in the middle. It is the only reminder that, years back, the French housed a garrison in the cave. ("The posting from hell," Alan Cooper noted.) Burney's tour will also help familiarize the five or six of us who have never been here with both the cave's complicated layout and some of the interior sites he and his teams have worked in the past.

"Since that time," Burney continues, his voice reverberating off the close-in walls of the cave, "there have been at least two major events, in which collapses have opened up more entrances. Now, this is a honeycombed area, with just a thin veneer of ceiling in most places. A lot of Pleistocene material may be buried in here. It will be interesting to remove a lot of this soft fill that's come in. It's mixed with breakdown, so it's a big rock-removal job. I'm going to try to stick somebody else with that task."

He rubs his hands together vigorously and smiles. He's in park-naturalist mode.

"Ready for a bit of cave darkness? Watch your step down at the far end of the room."

After so much buildup, including repeated references to a beautiful "sunken garden," I can hardly wait to explore this cave. Generally I despise tours; I'm too impatient. But since this cave is so vast—more than three miles of passages—and my cave experience is limited to a single spelunking trip in Austria back in college, I'm happy to submit to this one. The ceiling here is high enough to stand upright. The others have turned their headlamps on, but so much light pours in from Entrance A that, as we start walking, I leave mine off to conserve the batteries. *Wrong.* Suddenly, it seems, I step into pitch blackness. I'm literally blinded by the dark. I stop in midstep, instantly conscious of the possibility of tripping and spearing myself on the cave's sharp flowstone, some lying in jagged pieces and some all too firmly planted in the cavern floor. I reach up, throw the switch to my light, and wait a few seconds until my eyes adjust. The others are already out of sight up ahead. I quick-step to rejoin them.

The group has stopped beneath a broad drapery of white flowstone. It looks like four or five theater curtains sewn together end to end. Burney is gesturing beneath it at the dirt floor.

"At one standard deviation, it was between 7,000 and 7,100 years ago that all that mud came in, apparently after a big collapse," he says. "It's about three meters thick. Since then only a thin layer of red aeolian material—tiny, perfectly spherical particles—has been deposited, as if it was blown in. Below that, there are black, rolled bones, including very large bones that"—his voice drops conspiratorially—"could be *Megaladapis* or *Archeolemurs* or something like that. I would dearly love it if people want to dig close to this wall."

We continue on. This "room" is about the size of a small gymnasium. Four or five of me could stand on my shoulders and still not touch the ceiling. That 7,000-year-old silt is soft underfoot, and the cool cave air is now and then leavened with a surge of warmer air coursing between Entrance A and Entrance M ahead of us. I'm eager to see M, site of the much-ballyhooed sunken garden.

Soon we come to a rusted iron ladder. Burney tells us that during the colonial period the French developed the cave for tourism. They put in two ladders—one up, one down—to get over a house-sized block of stone separating this room from M. They also put in electric lights, powered by a generator, and charged people to come in. That explains the white electrical insulators I've seen high on some of the walls.

I climb the steps to the top and there, spread out below me, is the sunken garden. It is indeed eerily beautiful. Unlike Entrance A, which is in the side of a hill, the opening here is in the ceiling, and it is expansive enough that I can see several discrete cumulus clouds drifting by in an aquamarine sky. Direct sun spotlights the higher walls on one side of the gallery, while even the deepest, most hidden crevices are illuminated. My pupils are going the other way now, contracting in the bright light.

Dazzling green ferns, their hue enhanced by the absence of color for the past half hour, blanket the floor. A few cling precariously to nooks high in the cavern walls. Long, twisted roots of surface trees stretch sixty feet down to the floor, and other trees grow right in the cave soil itself, among the ferns. One is as thick as a telephone pole; others are more like overgrown shrubs. All of them are green, green, green. I try to imagine an *Archeolemur* rooting around in the ferns, or a *Babakotia* clinging to those long, dangling roots. It's possible that they were in this very room at some point, and that it looked largely as it does today.

Except for the light that floods it, M reminds me of how I pictured Gollum's haunt in *The Hobbit*. The walls can hardly be called that. They are more like giant church grottoes, vast hemispherical spaces carved over eons by water making its way farther and farther down into the earth. Somehow it's hard to get around the idea that the oldest parts of this cave, which plunges farther and farther down until it reaches an underground river, are actually the highest points, right up there at the top of M, for instance, far over my head. I'm used to thinking of archeological sites and geological strata, in which deeper normally means older. Except for a modest portion of the floor, on which centuries' worth of deposition has resulted in a smooth earthen base, there are no flat surfaces here: Everything is craggy, eroded, spiky, broken. The room looks like Bryce Canyon turned outside in. You have to watch your every move; one false step, and you could impale parts of your body.

M lies more or less at the center of Anjohibe and is thus a hub of sorts, though Anjohibe is nothing like a wheel. This is the place you're always trying to get back to, Burney says, for it's an easy stroll from here to Entrance A. It's also an ideal place to break off into remote parts of the cave. Here you can clamber up a small cliff to the east and then walk north into the Grande Avenue, the largest cavern in the complex, and from there to the north, east, and southeast branches of the cave. Or you can head back to A and then continue on either into the west branch or along the Central Gallery to meet up with the north end of the Grande Avenue. Finally, you can plunge due south out of M and down into the Salle R. de Joly—where Helen James, on Burney's expedition here in 1992, came upon that herd of subfossil hippos—and farther down the south branch to the underground Decary River, named for a French naturalist who visited the cave in the 1930s.

We choose the latter route and are soon heading out of the light past a large chunk of broken speleothem, like a piece of Corinthian column from some Roman temple. Cave darkness soon regains ascendancy, and this time I'm ready, my headlamp on in plenty of time. Even so, I slow to let my eyes adjust. As we approach the Salle R. de Joly—Burney wants to show us the hippo herd site—we are forced to clamber over sharp flowstone and ankle-scraping rimstone dams, which are places where minerals have precipitated out of a pool of water and built up a rim of rock. It's tougher going, though so far headroom has not been a problem. Stalactites and stalagmites of all sizes and shades of off-white hang from above or rise from below. Any light from skylights has long since disappeared; we're in pure cave darkness now.

In a few minutes, we arrive at the site. It's a much more enclosed space, a low alcove with the jagged ceiling arching down uncomfortably close to the tops of our heads. We crowd in. I can see a brown line staining the wall a few feet up—a pool of water once filled the bottom of this niche—and the floor of the alcove is all dug up. James is telling us how she and Grady discovered the site.

"Fred looked down and saw a little hole in the sediment. He glanced to either side, and there were these two rounded things, which he realized were the occipital condyles of a hippo skull. We were gazing right into the foramen magnum."

That is, the base of the skull. They came back later and began digging. Two square yards later, they had uncovered the well-preserved bones of at least eight hippos—five adults and three juveniles—including one that might have been unborn when it perished. Radio-carbon dating of collagen in the bones revealed that the hippos had died roughly 3,700 years ago—long before the advent of humans on the island. Since all the bones lay together, partially articulated and as tight as puzzle pieces, Burney's team concluded they must have died together.

"We think this is one herd," James says, her headlamp blinding each of us in turn as she swings her head around to take in her whole audience. "Which is extremely interesting, because it can tell you a lot about the social systems of these probably fairly terrestrial pygmy hippos." Along with anatomy, the discovery was another indication that Madagascar's hippos, while dwarfed, were probably more closely related to the common African hippo, which forms large, mixed-sex herds, than to the West African pygmy hippo, which travels alone or in pairs.

"Can I interrupt you for just one second?" Burney says to James. It's not a question. "If I can spin one short, fanciful scenario, imagine these hippos entered the cave at Entrance I, which is an easy walk in, and followed the skylights in the Grande Avenue. Maybe it was very dry, and they heard drips. When they got to M, they couldn't climb that ledge where the ladder is. So if they were going to continue instead of turning back, this was the only choice, and they would have soon found themselves in cave darkness."

He lowers his voice slightly, which I've come to realize signals an aside.

"Hippos have excellent night vision, but there's no animal that can see in cave darkness. Which, of course, is one reason why bats developed sonar."

He raises his voice again.

"So you can imagine them lost in here. I guess the question one has to ask is why they didn't mill around and eventually find their way out. Possibly they drowned in a flood."

"Be hard to drown a hippo," James says, laughing.

"Well, if they can't breathe, they drown, same as a whale," Burney responds, looking around as if seeking support from us. The good-natured debates between Burney and James have made for a lot of lively discussion in the past few days. "Anybody who has an end to the story . . ." he adds, letting his sentence finish itself.

Burney and his team are trying to write that end. Toussaint and

Gervais, Burney's Malagasy graduate students, will continue excavating this site tomorrow, in an effort to get the entire herd out. That should provide a good indication of the herd's size and demography. Egan, the other Burney graduate student on the trip, is eager to find teeth, for DNA might be preserved in the pulp, and DNA might reveal the pygmy hippo's evolutionary relationship to the living hippos. Teeth can also be useful for radiocarbon dating, and Burney wants a second date to verify the first one. For Burney himself, the most exciting find would be an intact sacrum, the fused portion of the lower backbone adjacent to the colon. ("It's like an ossified asshole, if you'll pardon the expression," he whispers to me behind a raised hand.) If any colon contents were preserved, they'd be in the sacrum. Pollen and plant fibers, for example, would offer long-sought clues to the pygmy hippo's diet, which is a mystery. Common hippos are primarily grazers, coming out of their water refuges at night to feed on short savanna grasses, while the West African forest-dwelling variety is both a grazer and a browser.

"Okay," Burney says, his voice bouncing off the low walls. "We should probably head back rather than indulging in the deeper parts of the cave. Everybody ready?"

No! I want to shout. Fired up by the thrill of exploration, I want to keep on going, right down to the underground river gurgling in the deepest recesses of the cave hundreds of feet below the Salle R. de Joly. Either that or start digging right here beneath our feet. I mean, only time separates us from *real hippos*, real grunting, stinky, big-nostriled hippos splashing around in this erstwhile pool.

But I can't speak for everybody, so I hold my tongue and we start back. I'll just have to savor this first close brush with my favorite of the "big animals."

6.

MY DESIRE TO FIND A HIPPO would have paled beside Burney's wish, when Bob Dewar's seminal extinction paper came out in 1984, to find fossil pollen whose analysis might help shed light on the continuous-forest enigma, not to mention that of the megafaunal extinctions. At the time, Burney was a second-year doctoral student in zoology at Duke University, undertaking just the sort of pollen work on Madagascar that Dewar said was then entirely missing. Burney's doctoral dissertation would be entitled *Late Quaternary Environmental Dynamics of Madagascar*, but that wouldn't appear until 1986. In 1984, he was busy trying to discern those "Late Quaternary environmental dynamics" at a site that must have seemed the logical first step: Ampasambazimba. Was Dewar right in proposing a mosaic of plant communities in this area?

Ampasambazimba is located due west of Tana, at the foot of an extinct volcanic massif known as Itasy. Lying at about 3,000 feet in the central Highlands, the site is a place where the surrounding grasslands dip down into a sort of shallow bowl. Here, a long time ago, a lake and accompanying marshland formed when a lava flow dammed up the Mazy River. Bones of the creatures eventually found there as subfossils washed down the river and collected in the now-dried lake between about 9,000 and 4,000 years ago.

Burney was aware that Perrier de la Bâthie had uncovered remains of plants that he said supported his notion for continuous forest in the area. And he was aware of Dewar's argument that, taken in the aggregate, the plant fossils Perrier de la Bâthie identified at Ampasambazimba hinted rather at a mosaic of plant formations, ranging from savanna to forest. But Burney took issue with the very idea of using Perrier de la Bâthie's plant macrofossils to make inferences about the paleoecology of an area.* For one thing, the man who actually collected the macrofossils in the 1920s, a French paleontologist named Charles Lamberton, failed to note the depth or context of his finds, so no one knows whether they lay together or at different levels. Neither Lamberton nor Perrier de la Bâthie appears to have had a sampling method, and Burney fears they were biased toward "the large and the distinctive." Finally, since macrofossils by their sheer size often rest where or near where they fell, they are really only good for making inferences about plants that grew on or near the spot where they were found, not over the entire region.

Burney knew that a better way to establish a past environment was to look at microfossils such as fossil pollen. Pollen disperses far and wide on the breeze and survives well in nonoxidizing environments such as lake and marsh sediments. Collect fossil pollen and match it to the plants that produced it, and voilà, you know what plants were growing in an area. Sounds easy, right? Well, it's not. Try this from one of Burney's papers on the subject: "Pollen zonation was

---

*Plant macrofossils are ones you can see with your naked eye, including stems and leaves, as compared to invisible microfossils such as pollen grains and diatoms.

discerned by Ward's Minimum Variance Hierarchical Cluster Analysis on a mainframe computer, a procedure in which the distance between two clusters is the sum of squares between the two clusters summed over all the variables. At each generation, the within-cluster sum of squares is minimized over all partitions obtainable by merging two clusters from the previous generation." Did you get that?

Fortunately, we laypeople don't need to know the techniques or technologies that are the palynologist's toolkit. We just want to hear the results. Burney sampled fossil pollen from the sediments at Ampasambazimba, while other members of his Cenozoic Research Group collected more bones of the dwarf hippos, elephant birds, and so on. After processing the samples and identifying them by referencing an extensive collection of pollen of tropical African plants in his advisor's laboratory back at Duke, Burney vindicated Dewar. Between 7,000 and 8,000 years ago, the region around Ampasambazimba appears to have had—and these are Burney's exact words—"a mosaic of woodlands, bushlands, and savanna. . . ." He published his results, which were the first dated pollen spectra from Madagascar, in 1985. It was the start of a long, distinguished career as the father of paleoecology in Madagascar, a career characterized by challenging long-standing hypotheses.

At the time, however, Burney advised caution about claiming any broader implications of his finding. After all, this was the first solid data on ancient environments in Madagascar ever collected, much less published, and it covered just one tiny corner of the Highlands. He had taken only the first step in a long uphill slog. He needed more sites to gain a better picture of Madagascar's past ecosystems. Before he could do that, however, he had to take what he called the palynologist's "calibration step." He had to validate his procedure for reconstructing past plant communities by doing the same work on present-day pollen from present-day plant communities.

So he did that at various sites around the island. Using Ward's Minimum Variance Hierarchical Cluster Analysis and other techniques, Burney not only demonstrated the validity of his procedure, but along the way he also stumbled upon a modern site that closely resembles what Ampasambazimba must have looked like 8,000 years ago. When you take in the woodlands, forest, and savanna surrounding Lake Mahery in northern Madagascar, he says, you take in Ampasambazimba when pygmy hippos wallowed in its waters, elephant birds grazed its upland slopes, and giant lemurs sampled leaves and fruit high in forest trees.

"For me," Burney says, "the big thrill is being on a site and trying, as Kurt Vonnegut describes in *The Sirens of Titan,* to see a place in a fourth dimension. How would this have looked a thousand years ago? Ten thousand years ago? A hundred thousand years ago?" He particularly enjoys visualizing the extinct megafauna going about their lives in the landscape, just what I failed to do when we had the flat tire. "It's the little boy in me that likes to imagine big prehistoric beasts," he says. "But it's also that those animals represent a major lost component of the ecosystem. In most parts of the world today, what you see is not a true reflection of what the place was like over the last several millennia. The whole idea of the Lost World as popularized by Spielberg's recent movie applies *everywhere.* It's all a Lost World."

7.

AT THE MOMENT, I FEEL as if I'm in a different sort of Lost World. My watch reads 2:23 A.M., and I'm wide awake, staring at the red Gore-Tex ceiling of my tent and wondering what the hell is going on outside. A few minutes ago I caught myself muttering the Doors lyric "weird scenes inside the gold mine," for weird things have indeed been happening.

Our tents here at Burney's camp outside Anjohibe range beneath a stand of bushy old mango trees. Dead leaves litter the bone-dry earth, along with cow patties in various stages of desiccation. A little while ago I woke to the sound of rain and of people moving around camp, closing windows of trucks, perhaps, or battening down the hatches of their tents. But when I unzipped my tent flap and went out to relieve myself, first checking my sneakers carefully for scorpions, I found a perfectly clear sky, the Milky Way cottony overhead. The dry, crackly mango leaves were as dry and crackly as ever, and there were no droplets on top of my tent. When I tried to hear "rain" back here inside the tent, I couldn't.

That's not the only strange thing. A few minutes later, I heard the urgent lowing of a zebu off in the palm savanna in which we're encamped. Was it giving birth? Dying in the jaws of a predator? As far as I know, nothing eats zebu except people and crocodiles, and this desertlike environment is no place for crocs. Afterward I heard the rumble of a zebu cart along the road. *At this hour?* I thought. Then I remembered there was a boxing match in the nearby village earlier in the evening; the cart driver was probably returning from that. But still, with everything else going on, it was a tad creepy to hear that rumble, with no accompanying voices, at such wee hours. Not long after the cart passed, someone in the camp suddenly sprinted away from his or her tent and threw up. What's going on here? Have we angered the Vazimba, whom Malagasy believe inhabited Madagascar long before they first set foot on the island and whose spirits are said to still dwell within Anjohibe?

This place is spooky enough as it is. We're in such a remote area of Madagascar that our encampment is apparently the second-largest settlement in the district. The mango trees are so broad, densely leaved, and close together that the sun barely penetrates our campsite during the day, and at night they take on a particularly lugubrious aspect. But they are downright comforting compared to the landscape that lies beyond them. A kind of scorched-earth policy is in force here. The ground is black with ash from cindered savanna grass that the locals must have torched very recently. Everywhere you look are browned plants and blackened soil. Half the plants are dead, and the other half are moribund. That's the way they look anyhow; maybe they'll bounce back. What with the fierce heat of this part of Madagascar and the annual fires, not much else grows here besides the satra palm and a few hardy grasses.

What's more, this is cave country. It's a place where if you're not mindful, you can drop 100 feet to a rocky death through a hole so small you never even saw it. ("We didn't lose anybody, that's a good sign," Cooper said after our first day.) Burney has repeatedly warned us about this since we got here: Be careful when you walk away from the camp, *especially* when you go out to take a pee in the middle of the night. Cavers have found human skeletons beneath some of the skylights that bring welcome illumination to many parts of Anjohibe Cave. Unseen karst holes are probably precisely how many of the megafauna in the cave wound up becoming personally extinct.

There are also nasty things that can get you out here. Last night a bombardier beetle spit hydrochloric acid at one of our party while she was putting up her tent. When Cooper went to her aid, he got his own dose. Burney has also warned us about a vicious vine that grows in this part of the country. "The most hated plant in Madagascar," he says of the *takilatra,* as the Malagasy call it. I came upon a sobering account in the *Antananarivo Annual and Madagascar Magazine* of what this otherwise nondescript climber, which also goes by the name *agy,* can do to you. It happened to an Englishman journeying by canoe down the Ikopa River on his way from Tana to Majunga in 1879:

> Walking under some trees and pushing aside the reeds and rushy grass, I was startled, in a moment, by a sudden tingling and pricking sensation over the back of my hands and fingers—a strange sensation, for never before had come the like to me, in Madagascar or elsewhere. I stopped at once in sudden surprise, for the pain was severe, and I had touched nothing, save perchance the long grass and rushes between the trees. But in another moment the pain increased, the tingling burning sensation seemed extending rapidly up my wrists; and as I bent my head down to look closely for the cause of the mischief, nothing was seen. But even as I lowered my head to look, pain, scalding pain, shot into my ears and neck, and growing worse too, every instant. Dazed and bewildered, I stood a few seconds in helplessness, for I could neither see nor guess at the cause of the terrible distress. Then with awakened instinct I drew softly back, away from the "uncannie" spot, and got back to my company with agony writ plain enough over every line of my face.
>
> The men started up when they saw me, some of them crying out, *"Efa voan' ny Agy hianao"* ("You have been smitten by the Agy tree"). Some of them led me to a seat, others rushed for water from the river, which they fetched plentifully, and two or three hurriedly brought sand and earth heaped in their hands. Then they chafed me with the sand and water that they might take out the *lay* (stinging hairs), they said, for that was the name by which they knew the cause of my sufferings. As they scrubbed me, I felt the pains abate; and after about a quarter of an hour's continuance of the operation, I was comparatively free from pain and was able to join in the rice-dinner which was soon ready.

The agy tree is really a vine, and the stinging hairs lie on the outside of the pod. If you touch the pod, they come out in your skin, but as this man discovered, even the wind blowing will dislodge the fine little spears, to shower down on unsuspecting Englishmen. Richard Baron, who once got a dose of his own, said the sensation reminded him of the sting of a nettle, though "ten times more virulent."

As unpleasant as it is, the agy is nothing compared to the so-called man-eating tree of Madagascar. We owe the initial European discovery of this carnivorous tree, a kind of elephantine Venus flytrap, to one Carl Liche, who reported on it during his visit to a Malagasy tribe known as the Mkodo. "The Mkodos," he wrote in the *South Australian Register* of 1881, "are a very primitive race, going entirely naked, having only faint vestiges of tribal relations, and no religion beyond that of the awful reverence which they pay to the sacred tree." Though Liche does not describe exactly where he found the man-eating tree, the area sounds remarkably similar to where I am right now—an area of caves and palms and infernal heat.

Liche relates how he and his companion, a man he identifies only as Hendrick, followed the Mkodos into an "impenetrable forest," where they came upon the dread tree. He was about to witness a human sacrifice so brutal that he later prayed, "May I never see such a sight again." The tree, he wrote, looked like an eight-foot-tall pineapple, with agave-like leaves drooping from its apex. Recessed in the top of the tree was a kind of natural receptacle, in which was found "a clear treacly liquid, honey sweet, and possessed of violent intoxicating and soporific properties." Long, green tendrils shot out from beneath the edge of this bowl, and above them, Liche wrote,

six white almost transparent palpi reared themselves towards the sky, twirling and twisting with a marvellous incessant motion, yet constantly reaching upwards. Thin as reeds and frail as quills, apparently, they were yet five or six feet tall, and were so constantly and vigorously in motion, with such a subtle, sinuous, silent throbbing against the air, that they made me want to shudder in spite of myself, with their suggestion of serpents flayed, yet dancing on their tails. . . .

My observations on this occasion were suddenly interrupted by the natives, who had been shrieking around the tree with their shrill voices, and chanting what Hendrick told me were propitiatory hymns to the great tree devil. With still wilder shrieks and chants they now surrounded one of the women, and urged her with the points of their javelins, until slowly, and with despairing face, she climbed up the stalk of the tree and stood on the summit of the cone, the palpi swirling all about her. *"Tsik! Tsik!"* ("Drink! Drink!") cried the men. Stooping, she drank of the viscid fluid in the cup, rising instantly again, with wild frenzy in her face and convulsive chords in her limbs. But she did not jump down, as she seemed to intend to do. Oh, no!

The atrocious cannibal tree that had been so inert and dead, came to sudden savage life. The slender delicate palpi, with the fury of starved serpents, quivered a moment over her head, then as if instinct with demoniac intelligence fastened upon her in sudden coils round and round her neck and arms; then while her awful screams and yet more awful laughter rose wildly to be instantly strangled down again into a gurgling moan, the tendrils one after another, like great green serpents, with brutal energy and infernal rapidity, rose, retracted themselves, and wrapped her about in fold after fold, ever tightening with cruel swiftness and savage tenacity of anacondas fastening upon their prey. It was the barbarity of the Laocoön without its beauty—this strange horrible murder.

And now the great leaves slowly rose and stiffly, like the arms of a derrick, erected themselves in the air, approached one another and closed about the dead and hampered victim with the silent force of a hydraulic press and the ruthless purpose of a thumb-screw. A moment more, and while I could see the bases of these great levers pressing more tightly towards each other from their interstices, there trickled down the stalk of the tree great streams of the viscid honey-like fluid mingled horribly with the blood and oozing viscera of the victim. At sight of this the savage hordes around me, yelling madly, bounded forward, crowded to the tree, clasped it, and with cups, leaves, hands, and tongues each one obtained enough of the liquid to send him mad and frantic. Then ensued a grotesque and indescribably hideous orgie, from which, even while its convulsive madness was turning rapidly into delirium and insensibility, Hendrick dragged me hurriedly away into the recesses of the forest, hiding me from the dangerous brutes.

To put your mind at ease, no one has ever again laid eyes on the man-eating tree. Or on the Mkodo tribe, for that matter. The afore-mentioned Liche purportedly sent the letter from which the narrative was pulled to one Dr. Omelius Fredlowski, a name so unlikely that it may give us some clue as to the veracity of the tale. Indeed, most deem the story a fabrication, and it does smack rather strongly of the handiwork of, say, Edgar Allan Poe.

Nevertheless, at this hour and in this place, alone in my tent, with phantom rain and animals screaming in agony and people throwing up and zebu carts passing mysteriously out on the road, I can't help but dwell on the bizarre.

And I haven't even mentioned the voices. That's right, voices. On and off throughout the night I've heard strange calls far out in the barren, burnt landscape. They sound like cries for help from a group of people. Are they lemurs? Ancestors? Or the Mkodo preparing another sacrifice at the long-lost man-eating tree of Madagascar? With no answer to hand, I bury my head under a balled-up sweatshirt and start counting sheep.

8.

BURNEY HAD EVEN GREATER mysteries to solve. Having gotten a sense of what the ecosystem near Ampasambazimba looked like, he then wanted to sample a site whose history extended right through the arrival of humans in the area. He discovered what he was looking for in Lake Kavitaha, a lava-barrier lake about eight miles southeast of Ampasambazimba.

Because this lake still exists, Burney turned to another of his tools, the piston-corer. The device sends a length of pipe down into the earth to extract a "core" of sediments; a piston suspended on a steel cable keeps the mud from dropping out. On Kavitaha, for example, Burney used a corer with a two-and-a-half-inch diameter to pull up a tube of muck about twelve feet long from the bottom of the lake. Inside that muck lay stratified layers of pollen, charcoal, and silt, which can offer clues to past vegetation. Burney calls his corer "a poor man's time machine," for it provides the most detailed picture of past environments that scientists can currently get.

Of course, that picture only appears after exhaustive analysis in the laboratory. In the field, it remains buried in wet, malodorous mud, the collection of which often leaves local Malagasy narrowing their eyes with suspicion. "The rumor went around, both in Africa and Madagascar, that I must be some kind of sorcerer, because I carry away trunks of plastic and metal tubes full of mud and come back the next year and buy a new Land Rover or something," Burney told me. "I collected mud from the bottom of lakes and deep inside caves and somehow turned that into money. In a way, I suppose that's a symbolic representation of what I do, though I turn it into information, which in this day and age is a valuable commodity."

The information from the Kavitaha core painted a picture of life in the region over the past 1,500 years—that is, before and after the advent of people in the region. At the time that inferences from arche-

ological evidence imply Malagasy arrived in the Highlands, Kavitaha's pollen spectra show a rise in pollen of grasses and weeds, a
decline in pollen of woody types, and a dramatic spike in microscopic
charcoal fragments—people were setting fires, presumably. By about
a thousand years ago, the pollen spectra reveal an environment not
too dissimilar to what it is today, namely, that undulating prairie
found throughout the interior of Madagascar.

At first glance, this reconstruction appeared to agree with the pre-
vailing notion of how people destroyed the forests of the central
plateau after their arrival. However, the canonical view was far more
extreme. If you recall, that theory called for an interior of closed forest, on which the various megafauna were dependent, and a corollary
to this view held that these forests and their wildlife disappeared after
early Malagasy "introduced" fire, which these forests had rarely or
never known before.* The earliest spectra from Kavitaha, however,
show a mixed-vegetation environment similar to that found around
Ampasambazimba 8,000 years ago and Lake Mahery today. Moreover, these early spectra showed that fires occurred naturally even
then, when no one was around to set them.

Burney had exploded another myth. But he didn't let it go to his
head. In his paper on Kavitaha, he pointed out that discussions of
environmental change in prehistoric Madagascar remained highly
speculative. For one thing, the Kavitaha findings simply did not go
back far enough in time. He needed a site that combined the time
lengths, essentially, of Ampasambazimba and Kavitaha, preferably
elsewhere in the Highlands.

He found it at Lake Tritrivakely. Tritrivakely is a crater lake in
the southern part of the Highlands near the town of Antsirabe. Surrounded by grasslands patched with terraced fields and rice paddies,
it stands much higher than Kavitaha, resting at an altitude of about
6,000 feet. In the dry season when Burney visited it, it was about
eleven acres in size and less than five feet deep. But it had tons of
muck. Standing precariously on a platform steadied by two rubber
rafts, Burney fired up the piston-corer and sent a series of four-foot-
long aluminum pipes into the flat bottom of the lake. The resulting
seventeen-foot-long core provided a record of vegetation changes at
the site over the past 11,000 years.

Now Burney was able to go back much further, right to the end of

---

*It is true that closed, humid forests normally are not very flammable, because the moist-
ness precludes the accumulation of thick, dry litter.

the Pleistocene—and many thousands of years before people stepped onto the stage. Pollen spectra showed a succession of different vegetation types over the millennia. At the end of the Pleistocene epoch 11,000 years ago, the site was dominated by heathlike shrubs and grasses, probably not unlike a similar floral community that now, in the warmer climate of the current Holocene epoch, thrives at higher elevations in Madagascar. Not closed forest, in other words. As the island warmed and dried from about 9,000 to 5,500 years ago, Tritrivakely's spectra show a shift to grasses and a few pollen grains of woody vegetation and herbs associated with dry, open environments. Even further from closed forest. At about 4,000 years ago, the spectra indicate a rise in pollen of trees and shrubs associated with forest edge, open woodland, and riverside environments, with grasses ranging between 25 and 38 percent. Getting woodier, but far from jungle. And things switch back when humans presumably entered the scene sometime in the first millennium after Christ. The spectra from that period record a shift back to a more grass-dominated environment. The modern spectrum from the site, which is forty-some miles west of the nearest large patch of rain forest, is similar to spectra from the postulated dry phase between 9,000 and 5,500 years ago.

The charcoal collected from Tritrivakely revealed a surprising fact. When he tested it, Burney found that charcoal concentrations in the lake's sediments were often *higher* in the late Pleistocene and early Holocene samples than in any time subsequent, including modern times. Natural fires were a happening thing way back when, and by the looks of it they were more intense even than today.* The burnt plant remains also allowed Burney to corroborate his findings from the 107 pollen types he teased from the sediments. Through microscopic analysis, he discovered that in most cases he could distinguish the source of the charcoal, whether it was derived from trees and shrubs, or from grasses, or from petroleum sources. A close inspection of the Tritrivakely core revealed that typically three-quarters of the charcoal in prehuman times came from grassy sources, as is the modern case at the same site, which features grassland pockmarked by terraced fields and rice paddies.

With the work at Ampasambazimba, Tritrivakely, and other Highland sites, Burney was beginning to get a handle on what was happening on the central plateau. He even got ever so slightly bolder

---

*A recent NASA study of satellite data, Burney told me, revealed that the latitudes in which Madagascar lies are among the most lightning-prone areas in the world.

in the conclusion of his next published paper, stating that "[e]very microfossil analysis published so far, as well as those in preparation, points to the same unavoidable impression"—namely, that it was time to put the closed-forest idea literally out to pasture.

9.

IT ALMOST FEELS AS IF I'm in a closed forest. I'm in the Grande Avenue, its great, dark ceiling arching over my head like a canopy, and I'm trying to imagine a herd of pygmy hippos sauntering through here, as they might have on their way to an early end in the Salle R. de Joly. I've left the others behind to take this solo *sangasanga* or walkabout. Burney's daughter Mara has a touch of "hotely belly,"* and so he's staying at the main entrance with her. On our orientation hike, he warned against walking through the cave alone, "especially if this is your first time." But I like nothing more than exploring on my own, especially a silent, secretive, thought-provoking place such as this.

So here I am, inside this enormous space. From where I'm standing, I can see the length of a football field in either direction. A cruise ship stuffed in here wouldn't touch the ceiling. Hard to believe that such spaces exist *underground*, yet this is merely the fifth-largest cave system in Madagascar.

It's inordinately quiet. Only the sound of my labored breathing and my boots crunching along the gravel-like breccia rend the fabric of silence. Until I stop and hold my breath, then nothing but the sound of dripping. Even though it's the dry season and the main areas of the cave are bone-dry, the little water coming down from the surface continues to chew away at the substrate, deep behind impassable draperies of organpipe speleothems. The quiet is oppressive—as in-your-face oppressive as complete cave darkness, like an enveloping blanket.

Yet comparatively speaking, the Grande Avenue is flooded with light. High above me I can see light-at-the-end-of-the-tunnel splotches of daylight coming through a series of small skylights ranged along the thousand-foot length of the gallery. These were the subject of

*Hotelys are the small restaurants found throughout Madagascar, and hotely belly is the island's version of Montezuma's revenge.

another of Burney's safety tips, this one about the danger of coming out of the cave through one of the twenty-eight entrances and making your way back overland to the main entrance at A. "If you fall through one of those, we'll find you eventually," he said as we did the sangasanga through here on our introductory walk yesterday morning. "You'll be a subfossil in the making."

The twilight in here might have been enough to draw the hippos in. The Grande Avenue is certainly wide enough, and in places the gravelly breccia gives way to downy soil—easy going for a hippo or human. From Entrance I, it would have been a long walk through the Avenue Nord into the Grande Avenue, over that steep but passable lip of stone at M, and down into the cave darkness of the Salle R. de Joly—maybe three-quarters of a mile or more. But it's physically possible. If they did come through here, did they walk purposefully, perhaps following the sound of running water? Or did they move hesitantly, grunting and swiveling on their uncloven hooves and bumping into one another? Were they scared? Did they come deliberately, or by chance, or even by force, with some predator at their heels? I can almost hear their snorting, feel their hooves pound the cave floor, smell their fetid breath.

I'm completely alone back here. Members of Burney's party are spread throughout Anjohibe Cave, but none is within earshot. The *crunch-crunch-crunch* of my boots on the loose breccia reverberates off the walls, even though they rear up many steps away on either side of me. There is a lot of debris on the floor, stuff that's fallen in from the roof. The light from my headlamp casts features at my feet in close relief—a broken stalactite, a cluster of tiny bones, the knobby curve of a rimstone dam. But it thins to nothing when I aim it at the faraway roof.

Some of the most important finds from Anjohibe came from the Grande Avenue. While wandering through here in 1983, Ross MacPhee, the mammalogist, came upon that nearly complete skeleton of *Paleopropithecus* just lying on the surface. Then, on Burney's expedition here in 1992, James and Grady looked into a crack in a fat speleothem that had fallen down onto a rimstone dam in the center of the room. They saw an *Archeolemur* femur lying on the surface, and some other bones. Digging deeper into that dam with rock hammers, they turned up one extinct species after another, including those first *Plesiorycteropus* remains found in more than a half century and the fragmentary mandible of a *Babakotia*, only the second site after the Ankarana known for this apparently geographically limited species.

They also found a number of extinct birds, including a cuckoo and two couas. (It's easy to forget that a number of "minifauna" also went extinct.)

Despite the pedigree of this part of the cave, I'm not looking very hard. I'm sure it has been pretty well picked clean over the years, starting with the early French colonials who came here. It's disheartening to think of what they might have taken out. They went for the big bones, and big bones—perhaps even full skeletons—might have been what they found. Even if they still exist somewhere in Madagascar or elsewhere, the lack of provenience—where they came from, what they were associated with—almost certainly does not. So they are lost to science. If only researchers like Burney and Jungers could have been the first to wander these passages.

Well, it's been picked clean of the unusual things anyway. I'm now in the northern end of the Grande Avenue, where a series of rooms about the size of real rooms in real houses have been worn out of the eastern wall. I'm watching where I'm going, because there are broken stalactites poking down dangerously close to my head. If I ran into one of those, I'd be a standing "subfossil in the making." I can feel the temperature change as air from some nearby entrance curls through here.

Ever since I left the main gallery, I have been coming upon fireplaces littered with potsherds and bones. Some of the bones are big—must be cow bones—and the same goes for the sherds. One fragment is the size of a serving dish and scrapes loudly when I pick it up off the stone floor. Malagasy probably broke these pots deliberately, which is a common ritual practice, Burney says. In 1992 in these same galleries, he discovered a ceramic incense burner and a multicolored glass-bead necklace that seemed to have been placed as an offering in a nook beneath a speleothem curtain.

It's hard to say how old these objects are. The beads are in a style that has been around for about a thousand years. ("They were encrusted with calcium carbonate, but I've seen beer cans encrusted with calcium carbonate," Burney told me.) Since the surrounding population is so light and probably always has been, I'm inclined to think that some of these remains are quite old. And earlier I found the tip of a whittled stick that had gone to subfossil, which means it had been there for hundreds if not thousands of years.

How long have people been coming here? Did they just make offerings or did they ever live here? (Burney found two postholes in this part of the cave.) What would they have thought to see how fascinated I am by their measly old fireplaces and smashed pots? Did they roast *Babakotia* over a spit, the smoke rising to the low-slung ceiling? Did they ever come upon living subfossils in their travels through here, say, a sloth lemur or a herd of pygmy hippos? As I head back through the Grande Avenue to the sunken garden, all I can do is wonder, a thing easy enough to do here.

10.

BURNEY, FOR HIS PART, wondered about sites in other parts of Madagascar. Would they corroborate what he'd found in the Highlands? He headed south to a subfossil locality almost as famous as Ampasambazimba. It was a site near Tuléar in the southwest where Grandidier had unearthed some of the first subfossils of the pygmy hippo. The Frenchman referred to the site as Ambolisatra ("place of the satra palms"), but when Burney got there he saw no satra palms and the locals were calling the place Andolonomby, so he does, too.

Andolonomby is a salty pond a stone's throw from Lake Ranobe, lying closer to the Mozambique Channel on the inland side of the coastal dunes. Andolonomby was ideal for Burney's next move, because it lay in a vegetation zone completely different from the grassy Highlands—namely, the spiny desert—and because it had a stratigraphically precise set of everything Burney needed for his analysis: fossil pollen, charcoal, plant macrofossils, and bones of extinct megafauna. Out again came the "poor man's time machine."

Andolonomby yielded a radiocarbon-dated record of changes to flora and, by inference, fauna over the past 5,000 years. By 5,000 years ago, a large, shallow freshwater lake formed on the site. At the time, a mosaic of western forest, arid woodlands, spiny bushlands, and grasslands surrounded the site. Hippos and other subfossil giants visited the lake then. Between 3,500 and 2,500 years ago, an apparent drop in precipitation caused the lake level to drop and surrounding vegetation to shift toward savanna-type plants. Megafauna became increasingly scarce there toward the end of this period. Around 2,000 years ago, previously low levels of charcoal abruptly increased and the savanna-type vegetation—palms and leguminous trees—gave way to arid bushland and grassland dominated by octopus trees and euphorbias.

Burney made two not mutually exclusive suggestions for what

occurred at Andolonomby. First, as the hippos and other grazers disappeared, grass and other dead-plant biomass may have accumulated, allowing fires to intensify in the increasingly dried-out environment. Second, these fires may have been human-set, which means that people arrived 500 years earlier than most archeologists say—that is, they appeared about 2,000 years ago.* Burney postulates that the fauna at Andolonomby may have been caught between a rock and a hard place—or several hard places. They may have been predated upon by humans and perhaps by introduced cats and dogs, of which they had no natural fear. They may have found their migration patterns, which may have helped them squeeze through deleterious climatic changes before humans arrived, interrupted. They may have faced competition from people for scarce sources of freshwater. Finally, and perhaps most devastatingly, they may have lost their forage to livestock, with a consequent proliferation of less palatable plants.

Burney's findings at Andolonomby led him to propose a cause for Madagascar's extinctions—and for environmental deterioration—that he still abides by as the most likely. He calls it the "recipe for disaster." The recipe is simply a host of different ingredients acting synergistically to precipitate the disaster. Not just hunting. Not just climate change. Not just humans slashing and burning forests. Not just introduced species. But a combination of all of them, acting at a time when most of the island was suffering a period of desiccation.† Such multiple, complex causes appear to have been in play as well during extinctions that followed human arrival on other oceanic islands. On Madagascar, evidence shows that the megafauna survived many shifts in vegetation types and fire regimes before humans came. The influx of our species may have been, he says, "the proverbial straw that broke the hippo's back."‡

Unexplained prehistoric extinctions, in Madagascar and elsewhere, should serve as cautionary tales for us today, Burney says. They remind us that we're living in the last remnants of a Lost World,

---

*Which is also the time, incidentally, when Burney found pollen of introduced plants in several lakes, and when he and Ross MacPhee carbon-dated collagen in human-modified bones of the extinct hippo from Andolonomby.

†The only putative cause Burney dismisses as untenable is his friend Ross MacPhee's "hypervirulent disease" hypothesis.

‡Despite the implication of their involvement, the Malagasy felt a kind of "cultural release," Burney says, upon learning of Dewar's and Burney's results, namely, that the French colonial theory that humans alone caused the extinctions is flawed. "They'd always been told by the French that they'd screwed everything up here," Burney says. "Madagascar was a wonderful paradise, and people ruined it."

where a sizable proportion of the most magnificent beasts that our ancestors beheld are now gone forever. They disappeared overnight as it were, most likely at the hands of people far less lethally equipped than we are today. Unlike the other massive extinction events in the world's history, the one begun by our earliest ancestors in Europe and continuing today is not allowing, on our time-scale at least, for replacement of the extinct animals with new forms—the very basis of evolution. We're depauperating an entire planet. And even though extinction is one of biology's fundamental concepts and one of paleontology's most well-studied subjects, we still know little about how extinctions occur, particularly massive ones.

Burney sees the late prehistoric extinctions not only as warnings but as learning opportunities. Investigating past extinctions can help in future conservation efforts. For instance, after studying the ecology of big mammals in his native South Africa and comparing them to the Pleistocene extinctions in North America, the mammalogist Norman Owen-Smith proposed the "Keystone Herbivore Hypothesis." Owen-Smith said that such overnight extinctions of seemingly whole animal communities is not surprising, because the fossil record shows that the first animals to go extinct are the largest herbivores, such as the mammoths and giant ground sloths in North America. These animals, as we know from watching extant large plant-eaters, play a crucial role in shaping their environment and providing niches for smaller creatures. Their loss can unleash a cascade of extinctions of their predators as well as of smaller herbivores and their predators. If Owen-Smith is right, Burney asks in an article in *American Scientist*, "what can we expect in Africa in the wake of the decline of the rhino and the elephant?"

Learning from catastrophic extinction is exactly what he is doing on Madagascar, where, as he says, the extinction spasm that began soon after humans arrived appears to be continuing today, working its way down through smaller species. His ongoing paleoecological research in Madagascar has three goals. Expand and diversify Madagascar's paleoecology during the Holocene—that is, the past 11,000 years. Extend paleoecological investigations further back in time, well into the Pleistocene. And improve the temporal resolution of events in and around the extinctions.

The first of these goals means filling in geographic gaps. Working independently, Toussaint and Gervais have added new sites from the Highlands. Their results confirm Burney's earlier findings. The second is being done, among other places, at Anjohibe, where uranium-series dating has brought dates back to 140,000 years ago. Under-standing the all-but-unknown Pleistocene in Madagascar will help Burney and others better understand past extinction patterns as well as present species distributions. It will be especially illuminating, he says, to see if previous interglacials experienced vegetation changes similar to those in the late Holocene. Burney hopes to accomplish the third goal using sediments from deep crater lakes, laminated cave for-mations, and the tree rings of Madagascar's enormous baobabs.* These should help clarify fire periodicity before and after human arrival as well as provide a more accurate chronology for Holocene climate and plant changes.

Alas, there's no smoking gun in this business, only smoking *guns*. Scholars will likely debate which of those guns participated in the dirty deed—and precisely how—for decades to come. Burney, for his part, is only getting started.

"I still have the sense of being a scientific tourist," he told me on the drive up here, referring to working in Madagascar. "They have a saying in Africa that an African expert is a person who has been there less than two weeks or more than thirty years. In other words, there are people who think they're experts and they're not, and there are people who've been there so long that they might actually know something."

He smiled.

"In the middle there's a group of people who feel great humility, because they've been there awhile, and they're still finding new things. I'm in that middle group. I feel really humbled by a place like this, because everywhere we turn we find new things, and they don't always fit our preconceived notions."

11.

"GET MARY!"

The call comes from behind an immense speleothem, so broad it's like part of a wall. It's Trevor Worthy, the ratite expert, and he's found something. We're deep in the cave, against one side where the ceiling arcs down to meet the floor.

---

*In modern tree-ring dating, researchers only remove a small-diameter core, leaving the tree itself unharmed.

I yell to Mary Egan, who is out of sight nearby, somewhere away from the edge in the main part of the cave. I can see momentary flashes of light from her headlamp on the cave walls. Hearing her response, I step between two speleothems like cathedral columns and onto a wet, asphaltlike surface. The roof is so low here that my helmet scrapes on the dry, white surface, even though I'm bent over. His back to me, Worthy crouches down near the point where the ceiling meets the floor, leaning over something in the fossilizing muck.

"It's the pelvis," he says. "Of a hippo."

I turn to my right, away from Worthy, and see something else sticking out of the flat floor.

"What's this?" I ask, even though I know.

"You've just found the jaw," he says, swinging around. A single tusk juts up from the lower mandible.

Egan, a dark-haired woman with big eyes, steps in behind us, shines her light on the bones, and then eases through a tiny opening at one end of the cramped, low-ceilinged room we're in. Charged up, she wants to see the extent of this find right away. As soon as she's through the gap, which lies between a giant speleothem and the low-slung wall, she exclaims.

"Here's the cranium!"

Worthy and I get on our hands and knees and crawl through the hole. There, facing us, is a pair of eye sockets. They're as round and big as silver dollars. Finding any bones is exciting, but there's something about looking into eyes, even if they're just gaping holes where eyes once were. Bones appeal to the mind, but eyes stir the soul.

Worthy sets to work. With his weathered features and tousled hair, he has the rugged, outdoorsy look of a park ranger. Lifting his rock hammer and chisel, he begins chipping away at the muck around the cranium. Sitting across from him, Egan does the same on the other side. She's careful not to get too close to the skull itself. There's the echoing thud of their hammers against the rocklike mud, followed by the scrape of gravelly junk out of the way. Bits of stone fly everywhere. Though the ceiling overhead is dry, the flowstone on the speleothem on one side and the low wall on the other is moist. It glistens in the reflected glare from our lights. Again, the air smells of charcoal.

Since the space where they're working is no bigger than an attic crawl space, I exit through the opening and sit on a dry rimstone dam on the high side of the other room, facing the jawbone that I "discovered." Worthy deserves full credit for this find, but I feel a slight thrill at the thought that maybe, just maybe, I'm the first person to lay eyes on that jaw since it was attached to a living, breathing animal. An animal now gone from the face of the earth. I remember only yesterday snickering privately at how charged up these guys get about a bit of bone. "It's a sweet specimen," Bill Jungers said while fingering a fragmentary mandible of an *Archeolemur*; a look approaching piousness on his big, expressive face. I'm snickering no more.

Like the pelvis and skull, the jawbone is half-sunk in the hard muck and is the color of the red, iron-rich soil of Madagascar (the muck itself is brown). Clearly the mud has gone to subfossil, just like the hippo that died here who knows how long ago. Above the bone, I can see water stains on the wall, showing various water levels over the millennia; the highest is three or four feet up. Did the hippo find a pool of water here and wallow in it until it died of starvation?

"This should come out pretty easily," Worthy says oddly, since he's been chopping at it for fifteen minutes already. The air down here is cool but humid, and through the opening I can see sweat stains on his back. He's breathing hard. When he shifts positions, his boots grinding on the breccia pebbles, there's a crack as his helmet hits rock.

"What we need is the spray-on fossil remover," says Travis Olson, Helen James's fourteen-year-old son, who has joined us. He sits behind me, looking on impassively. In between Worthy's thuds, which seem to shake the very walls, I can hear the incessant *drip, drip, drip* as Anjohibe continues to grow, inch by inch, stalactite by stalactite, as it has for hundreds of thousands of years. Egan has stopped, perhaps exhausted by the work, perhaps pondering the fate of this species, the focus of her doctoral studies under Dave Burney.

I, too, am thinking about the fate of this species, but probably not in the same way. I'm thinking about an animal called the "not-cow-cow" and another known as "floppy ears." Both are fascinating legends about creatures that just might be the pygmy hippo, and one of them is current.

In 1876, the German zoologist Josef-Peter Audebert, while collecting specimens on Madagascar for the Royal Museum of Natural History in Leiden, was shown an animal hide stamped with Arabic script. Its Malagasy owner said the skin had come from a secretive creature known as the *tsy-aomby-aomby*, or "not-cow-cow," that hunters had killed in the south of the island. This animal, he said, lived in rocky caves littered with the bones of humans and wild boars it devoured.

Audebert, who thought the hide resembled that of an African antelope, might have dismissed the man's story as the product of a lively imagination, except that everywhere he went in Madagascar he had heard tales of the tsy-aomby-aomby. Even the Betsimisaraka, who live at the opposite end of the island from where this specimen was supposedly procured, had a legend about the beast. Betsimisaraka in the northeastern coastal village of Mahavelona told a French folklorist in 1893 that the tsy-aomby-aomby had the body of a cow, but was hornless and had uncloven hooves. Fast-moving and savage, it consumed everything from insects to people, which, incidentally, it caught with a stunning blast of urine in the face. Moreover, Flacourt described an animal like the tsy-aomby-aomby that he knew as the *mangarsahoc* ("the beast whose ears hide its chin"). In his *Histoire*, Flacourt posits that the mangarsahoc, which he never personally saw, must be some kind of long-eared ass—a belief that Audebert held as well two centuries later.

Intrigued by these tales and hoping to procure a tsy-aomby-aomby for his employers, Audebert hired the skin's owner to take him to the spot where it had been killed. He hired porters and traveled by canoe up the crocodile-infested Manambato River, heading for the country of the Vohilakatra people. Near their land his bearers learned that the Vohilakatra had massacred members of the neighboring Sakavoay tribe, and they refused to go on. By the time Audebert assembled another team and reached the region, there was no sign of the tsy-aomby-aomby, living or dead. Profoundly disappointed, he retreated, writing later that "I had to admit finally that all of the accounts of the habits and existence of this beast were nothing but pure inventions."

But were they? Laurie Godfrey, an anthropologist at the University of Massachusetts at Amherst, who excavated the hippo herd site a year or two after Burney's team in 1992, believes the tsy-aomby-aomby was none other than the pygmy hippo. "Strip the cumulative fantasy from the description of the tsy-aomby-aomby," Godfrey wrote in *The Sciences* in 1986, "and one is left with a set of behavioral and anatomical characteristics too strikingly hippolike to be dismissed as coincidental."

The common African hippo fights viciously and has been known to kill its opponent with an errant swipe of its lower canines. Before such battles, the animal will try to intimidate its opponent by, among other acts, urinating and defecating at once and spraying the waste around with rapid gyrations of its tail. Common hippos, which leave the water at night to graze on the surrounding savanna, have been known to attack people who block their escape routes to water. Indeed, they cause more human fatalities than any other African species. As Jolly pointed out to me, "they don't eat people, but they do bite them in two." And they trample people as well, leaving bodies that often look as if they have been partially consumed.* West African pygmy hippos, though never seen eating meat, act similarly. With a little license for wild imaginations, the Betsimisaraka and other legends about the tsy-aomby-aomby fit quite closely with these known behaviors.

What of the uncloven hooves and the "ears that hide the chin"? Extant hippos have hooves, though with four toes rather than a single one (like an ass, for instance, hence Flacourt's belief). The uncloven hoof remains a mystery. Perhaps it had one, perhaps its purported observers were mistaken. But for the long ears, we can turn to none other than Burney and Ramilisonina. Here the story gets truly nail-biting, for these two have interviewed people in and around the village of Belo-sur-Mer who claim to have seen a hippolike creature as recently as 1976. Can it be that the not-cow-cow lives still?

Belo-sur-Mer is a fishing village on the southwest coast. It lies in the heart of one of the most remote regions of Madagascar, with fewer than ten people per square mile across an area of some 4,000 square miles. Most of the people who live there are fishermen, hunters, and woodcutters. The ecosystem grades from mangroves in estuaries along the coast to spiny desert and finally dry forest farther inland. It is hot, dry, and extremely isolated.

Burney has spent several seasons at Belo excavating what he calls an "integrated site," that is, one that preserves pollen, charcoal, human artifacts, and other evidence of past landscapes, and also has remains of the extinct megafauna dating from both before and after human arrival in the area. Pygmy hippos as well as giant lemurs, elephant birds, and giant tortoises died here and became subfossils over the past few thousand years. In the search for such sites, he and Ramilisonina often interview locals.

It all started one day in late July 1995, when a self-described sor-

---

*And perhaps they were: In 1995, Earthwatch research teams working in Zimbabwe with Joseph Dudley of the University of Alaska, Fairbanks, watched in amazement as a herd of hippos, which are known as strict herbivores, killed and partially ate an impala.

cerer named Armand Dabelatombo visited Burney's camp at Belo. Dabelatombo, who appeared to be in his sixties, remembered a French paleontological expedition to Belo before World War II that he had watched at work when he was a boy. He said that some of the animals that expedition had dug up, or something like them, may have been seen by people still living in the area. Burney was intrigued but dubious.

A few days later, Ramilisonina met a man named Constant in the nearby village of Antsira. Constant and his wife and son, with apparent excitement and without any prompting from the researchers, related an experience they had had with an animal they called the kilopilopitsofy ("ke-*loop*-ee-*loop*-ee-*tsoof*-ee"), which had woken them and their dogs with its loud grunting, breathing, and walking. When they had gone out to investigate, the "floppy ears" had fled into nearby mangroves and vanished underwater. The year, they noted, was 1976. Another informant said that though he had not seen it himself, others had. When shown a book with color plates of African and Malagasy animals, he chose the common hippo. He added, though, that it was not exactly right, because the animal he'd heard about was generally described as having large ears.

Burney and Ramilisonina were now thoroughly captivated. They decided to interview other people who claimed to be eyewitnesses. They were careful to implement safeguards: They interviewed only people whom they could prove had lived in the area for a long time; they talked to them separately, without eliciting specific answers and after testing their knowledge of local wildlife using color plates. Finally, Ramilisonina, who conducted the interviews in Malagasy and then translated, took separate notes and compared them later with Burney's.

On August 2, an elderly man named Jean Noelson Pascou, who at eighty-five was said to be the oldest man in the area, dropped by the camp at Belo. During the course of the conversation, Pascou revealed that he had seen the kilopilopitsofy several times, most recently in 1976—the same year given by Constant, who had encountered it in his village about four miles to the north. Pascou described the animal as cow-sized, hornless, mostly dark-skinned save for pinkish areas around the eyes and mouth. And it had large ears that flopped about. On the suspicion that his tale might have come over somehow from the African coast, which is directly across from Belo (albeit several hundred miles away), Burney and Ramilisonina showed him a picture of an elephant. "Oh, no," he chuckled, "that's an elephant." (He had seen one once on a farm in Majunga.) He also imitated the kilopilopitsofy's call, giving a series of deep, drawn-out grunts that Burney says sounded much like that of the common hippo.

The next day they spoke to Pascou's wife, Fatima Soariko. She related the following story:

In 1946, when I was a girl of ten, some of my neighbors saw the kilopilopitsofy one night when it came near our compound. I was sleeping, but somebody came to warn us it was near and my parents got up to go scare it away. While they were gone, an insect crawled into my ear. It was very painful, and I cried out. My parents and neighbors came running back, carrying spears and sticks, because they had seen the kilopilopitsofy and were afraid that I was screaming because it had attacked me.

Burney was hesitant to publish these findings, which he termed "professionally perilous," until Jolly jokingly said that if he didn't publish them, she would.* Perhaps the people of Belo had only embellished legends they'd heard from their grandparents. Perhaps they had fabricated the stories using details they'd picked up somehow from the historical literature or from that earlier French expedition. Perhaps a lone common hippo had somehow swum across from Africa — which, after all, is how scholars believe the Malagasy hippo originally made its appearance. Anything, but that the pygmy hippo was still alive in western Madagascar.

Yet many aspects of the story ring true, Burney says. Each villager's account was unsolicited, internally consistent, and consistent with those told independently by others. They also jived with historical and legendary accounts, save a few probably exaggerated particulars, such as the notion in old accounts that they devoured people. The Belo villagers even had their own name for it and could imitate its call.

Burney points out that one should not forget once-legendary animals that live in other parts of the world. "My wife and I have a collection of old, rare books by people like Baker, Livingstone, Stanley, all the early explorers of Africa," he told me on the drive from Tana. "It's a sobering realization that those books discuss animals that all the

*He eventually did, in a paper coauthored by Ramilisonina that was published in 1998 in the prestigious scientific journal *American Anthropologist* (Vol. 100, No. 4).

Africans talked about but that no Europeans believe existed, and they turned out to be the okapi, the gorilla, and so forth."

At the risk of being branded a cryptozoologist, Burney dreams of one day returning to the Belo area with several Masai trackers he knows from his days in Kenya.

"I would just turn 'em loose," he told me. *"Pata kiboko, mabwana.* Find a hippo, gentlemen."

"Take me along," I said.

He laughed.

"I might want to write that one up myself. No offense."

Remember what I said about eye sockets? Well, it's one thing to look into the eyes of a hoofed mammal like a hippo, and quite another to look into the eyes of a primate, our own kind. I'm staring right into the long-lost orbs of an *Archeolemur,* whose cantaloupe-sized skull lies half-buried in breccia a few inches from my face.

Worthy discovered it a few minutes ago. Eager to keep exploring, he had given up on the hippo skull; it would have to wait for another day. So, while I wandered around in the main portion of the cave, feeling claustrophobic enough where the hippo lies, Worthy, ardent spelunker that he is, wedged himself into ever-smaller spaces between speleothems. In one such space, akin to that behind one's refrigerator, he came upon this *Archeolemur.* Now I'm here, getting some face-on photographs of it in situ before Worthy goes at it with the rock hammer. There is barely enough room for one person in here, in a twisted fetal position.

Like something off a *National Geographic* cover, the lemur stares back from the impenetrable depths of extinction. Like the hippo bones that lie not twenty feet away through several curtains of moist minispeleothems, the skull is reddish in color and sunk in brown, rock-hard former mud. The rest of the animal's bones lie in a heap in the farthest corner of this tiny hole, where some small mammal, a rat perhaps, dragged them to its twiggy nest wedged in the high, dry space between two white stalagmites.

As Worthy and I trade places and he begins excavating the skull—he's not content to leave this one behind even for a day—I muse on the giant lemur's fate. Did it plunge to a rocky death through a skylight, its remains later hauled off in pieces to this spot by some industrious rodent? Might it have entered the cave out of curiosity, or to get some shade from the blistering heat, and become trapped? Or did early hunters spear it in the trees outside and carry it in here to roast over a fire?

I'm also pondering the extinction of this species. Like the hippo, *Archeolemur* has also come up in a manner of speaking in those interviews with villagers around Belo-sur-Mer. Besides the kilopilopitsofy, the villagers had talked about two other animals said to live in the area that caught Burney and Ramilisonina's attention. The first they called *bokyboky* and claimed had magical powers. Dabelatombo, the first man to visit Burney's camp, said this "terrible animal" killed vermin by sticking its tail down the hole of a rat or snake and killing it instantly by, as he said in French, *peter* (farting). Pascou, the elderly man who had imitated the kilopilopitsofy's call, mentioned that he had one as a pet in 1942. Upon further investigation, it turned out that bokyboky was the vernacular name for the narrow-striped mongoose, a living Malagasy carnivore whose range extends to Belo.

The other animal, however, could not be so easily explained away. Pascou brought it up first. Without any prompting, he mentioned a large, ground-dwelling lemur that lived in the forests to the east of Belo. It was called the *kidoky* ("ki-*dook*-ee"). Now, there are no large, ground-dwelling lemurs known to exist in Madagascar today. The kidoky was very rare, but Pascou had gotten a good look at one in 1952. It was about the size of, well, he pointed to his seven-year-old granddaughter who was standing nearby. It had a dark coat with white spots above and below the face. The lemur was like a sifaka — any of the typically long-tailed, silky-furred lemurs of the genus *Propithecus* — but with a face like a man, and when alarmed, it escaped not by fleeing through the canopy like a sifaka but by leaping along the ground. Pascou, as he had done with the kilopilopitsofy, even imitated its call, giving a long, single whoop, not unlike the short call of the indri. When shown a picture of the latter, he said that wasn't it; the kidoky has a rounder face, more like a sifaka.

A few days after speaking with Pascou, Burney and his Malagasy students, while working near a third village northeast of Belo, on the edge of the dry forest, asked a group of six villagers about the kilopilopitsofy and the kidoky. None had personal knowledge of the former. But when Burney mentioned the kidoky, they got all worked up. Several of the men, who all spent time working in the forest, said they had seen and heard this lemur in recent years. Their description of the creature and their imitation of its call as a long *whoo* essentially matched Pascou's. They said they had never observed it climbing a tree but only bounding along the ground in short leaps. When Burney imitated the comical sideways leap of the sifaka, one of them said, "No, that's a sifaka." The man then demonstrated the kidoky's gait—a forward gallop like a baboon's. These villagers, like Pascou, also maintained the animal is solitary, which the social sifaka is decidedly not.

Could it be that a second member of the megafauna yet exists in the remote, bandit-infested wilderness east of Belo? Burney and Ramilisonina estimate the weight, based on the size of Pascou's granddaughter, at around fifty-five pounds. The only other relatively recent description of a giant lemur is Flacourt's account of the Tretretretre, which he reported was "as large as a two-year-old calf." That would seem to be larger than the kidoky, and if, as Burney notes, the name "Tretretretre" is onomatopoetic, as is not unreasonable to suppose, then the call was probably more like a chatter than a whoop. Yet Flacourt went on to say, as you'll recall, that the Tretretretre "has a round head and a man's face. . . . It is a very solitary animal, which the people of the country hold in great fear and run away from, as it also does from them." Those details conform nicely with those from Belo.

Burney and Ramilisonina conclude that, if the kidoky is a giant lemur, it is most likely either *Archeolemur* or its close relative *Hadropithecus*. Those two genera, as we've seen, show numerous adaptations to a life on the ground and have been compared to baboons in their size, dentition, and inferred means of locomotion. Needless to say, the researchers hold the same reservations about the existence of the kidoky as they do of the kilopilopitsofy. Perhaps someday Burney's friends, the Masai trackers, will put the mystery to rest.

Or better yet, maybe a local guide will. Just before he left Belo at the end of the field season in 1995, Burney met a local man named André, who the other villagers claimed spent more time in the forest than anyone. If anyone would know about the kidoky, or the kilopilopitsofy, André would. Before Burney's team left, André agreed to guide them into the forest the next time they came.

I, for one, will eagerly await the results of that foray. In the meantime, I'll have to make do with memories of my own brushes with the near-mythical megafauna, including Worthy's *Archeolemur*, which he eventually succeeded in excavating like one would a tree, leaving it buried in a ball of substrate.

"Sweet, sweet!" Bill Jungers cried as a beaming Worthy showed it to him upon our emergence from the cave.

Burney, too, was excited. Later he told me, "What's really nice about Old Crusty is the way it's integrated into the flowstone. It's so artful, and it really has a prehistoric look about it. If I had been the one to find it, I would have had a philosophical dilemma about whether to take it or leave it there, as it's quite an important cave feature. But it was in a place where no one would have ever seen it, and probably somebody else would have found it eventually who may have been a bit less kind to it. This way we've got the whole thing and can make casts of it that can be shared all around. It'll make a very nice museum exhibit at the Musée d'Art et d'Archéologie in Tana."

After my experiences in the depths of Madagascar's Big Cave, that's one museum fossil that would come alive for me no problem.

The Most Beautiful
Enigma in the World

People constitute a great, broad mat.

—*Malagasy proverb*

1.

AS YOU DRIVE INTO Sambava, a vanilla town on Madagascar's north-east coast, a sign declares in English THE CITY OF SAMBAVA LOVES SILENCE. But you'd never know it judging from the music here at Le Cocotier (the Coconut Palm). At this family-run hotel just back from the beach, the owners play Malagasy pop music as if to silence the Indian Ocean waves smacking the beach fifty yards away. The bass notes rattle the thatch walls of the hotel's restaurant, where I'm trying to hear the archeologist Robert Dewar outline our plans for the coming week.

"There are a bunch of places where archeological sites were recorded around the turn of the century," he yells, leaning over a map spread out on the table. "The most important one is Bemanevika, which is just on the other side of this big river."

"Presumably there's a bridge across that," I yell back, seeing none on the map.

"No, the map says there are ferries. This one is marked as a motorized ferry, and that one as a kind of pirogue." Pirogues are the hand-hewn canoes of Madagascar. "These maps are from about 1974, so maybe there's a bridge there now. We'll see."

A momentary pause between songs on the stereo catches Dewar off-guard, and he shouts his next few words in the silent room before catching himself.

"We'll go into this village," he says, pointing to a town on the map, "and see if we can't find somebody who will work for us as a guide, because the sites are distributed right along in here, which the map marks as 'vanilla.' " We're in the heart of vanilla country, where most of the nation's crop of this valuable orchid is grown.

Our lunch arrives, and with it more music. My ears hurt, but the food distracts my attention. Fish, rice, and little boxwoodlike leaves for greens. Three Horses beer. And a delectable-looking salad that Darsot Léon Rasolofomampianina, a Malagasy archeologist who is helping Dewar on this survey, nips into immediately. With too many sordid memories of hotely belly, I won't go near it. Nor will Dewar.

"I don't eat salad until I've been in Madagascar for a week or two," he says, raising his voice again to beat the music. "My gut flora is usually not acclimatized."

Dewar looks like just the kind of man that intestinal bugs could do a real number on. At forty-seven, he's slightly built, gray-haired, and, when not competing with fifty watts, soft-spoken. There is a touch of the absent-minded professor about him. He often lets his sentences trail off as if he's lost his train of thought, and he usually walks with his eyes aimed straight down at the ground. He goes in for white button-down shirts and tennis sneakers with purple socks (as he's wearing now) and bow ties and light-colored sport jackets slightly too big for him (as I saw him wearing recently at a conference). But that image competes with his field persona. In his well-used blue jeans, two-day stubble, and sunglasses, Dewar looks the part of classic field scientist. In fact, on looks alone he could play the part in a film, because he bears a striking resemblance to Peter O'Toole.

Playing such a part would be justified, for Dewar is one of the foremost archeologists working in Madagascar today. An associate professor at the University of Connecticut, Storrs, he has worked across the island for fifteen years, trying to answer one basic question through archeological means: Who are the Malagasy? He has become a leading expert on their origins, though he's the first to admit that those origins remain naggingly obscure. There are strong Indonesian, African, and Arab elements in the culture, each of which tried to exert cultural and even political hegemony on the island in the centuries before Europeans arrived in the sixteenth century. Dewar's view is that the Malagasy people have a diverse set of ancestors from around the Indian Ocean, and to untangle their origins, much less understand who they are today, one must investigate that diverse set.

Which is what we're doing here on the northeast coast. Dewar is undertaking a preliminary survey of known archeological sites in the region. His objective is to identify one area, perhaps a single river valley, where he can launch an intensive archeological investigation of all sites from all time periods. The goal would be to paint a picture of life here over the centuries, from initial settlement to the modern day, that could be compared to similar information from other parts of the island, all to gain a better understanding of the rise of Malagasy civilization. Based on our findings on this trip, Dewar hopes to submit a grant proposal to the National Science Foundation with Dave Burney and archeologist Henry Wright of the University of Michigan to do this work.

The northeast is a good place to look for the ancestors of the modern Malagasy. If the archeology of Madagascar as a whole is, in some ways, still in its infancy, then the archeology of the northeast coast has not even been born yet. Excavation in the region has been limited to treasure-seeking French colonialists, who coerced prison labor to rip apart ancient tombs up and down the coast around the turn of the century. They sought, and found, rich grave goods, including gold coinage, soapstone carvings, and fine Chinese and Persian ceramics.

The only legitimate archeology that has ever taken place here was a survey of those French colonial sites undertaken in 1969 by French archeologist Pierre Verin and his assistant Ramilisonina (the same Ramilisonina with whom I worked at Anjohibe Cave). Verin is considered the father of Malagasy archeology. He founded both the Musée d'Art et d'Archéologie in Tana and the Center for Art and Archeology at the University of Madagascar. He also wrote the definitive history of Malagasy civilization—to the extent that it is known—based in large measure on fieldwork he did with Ramilisonina. We will be using Verin's texts and accompanying hand-drawn maps to locate known sites.

"This is a rich field," Dewar yells, turning to look incredulously at the restaurant staff. The song is skipping—a female singer incessantly gags on "he-he-he-he-he-he-he-he"—yet no one has budged. Don't they notice? Dewar shakes his head and goes on.

"I was up in the far north in 1986, and we found some absolutely spectacular things. We had a French colleague with us, and she kept saying, 'Don't get excited, this is virgin territory.' As far as she was concerned, just because we found something was not particularly to our credit, because no one else had looked for it."

Because the northeast is virgin turf, it is a good place to attempt to solve arguably the biggest archeological mystery surrounding Malagasy origins. The Indonesian element in Malagasy culture is the most predominant, with the language, rice culture, and belief in ancestors, among other island-wide cultural traits, having clear origins in that part of the world. Why, then, has no archeological trace of Indonesian culture ever turned up in Madagascar?

I ask Dewar about pottery. Archeologists have unearthed many kinds on the island, including Persian sgraffito, Chinese celadon, and red-slipped ware resembling similar pottery in Africa. Are there no ceramics that hint at Indonesian origins?

"No, none!" Dewar has a way of blurting out select words when he gets animated. "This is the trip."

He turns to Darsot.

"Have you ever seen anything that looks Indonesian, except ethnographically?"

"No," Darsot says and smiles. It is the first word he has spoken since we sat down to lunch.

Darsot is a tall, willowy man in his forties, with a birdlike head and kindly eyes. He told me he is one of only five trained Malagasy archeologists, and one of them is now living in Sweden while another is no longer working. Since I met him yesterday, I've sensed his frustration at how little he is able to do as an archeologist. He teaches at the university in Tana, but through lack of funding, he does little fieldwork on his own. "The government does not understand the need to look at the past when there are so many problems today," he told me. When I said it must be very frustrating for him, he said "Ohhh" and let his head fall to one side. You have no idea, he seemed to say. Darsot seems very melancholy, as if the weight of Madagascar's troubles, encapsulated for him in the field of archeology, rests on his shoulders.

Dewar goes on. "There are various speculative theories about the arrival of people from Indonesia that scholars have used, the most famous of which is that the Indonesians came to Africa, settled there, married African women, and moved to Madagascar. Which is all right, except that no one has ever found a trace of any Indonesians on the African coastline either." He laughs dryly. "It just puts the problem 600 miles away to the northwest."

If any sign of that kind of early contact survives, a reasonable place to look for it is here in the northeast. If the first Indonesians sailed directly across the Indian Ocean to Madagascar, they very well may have landed on the northeast coast, which, as the crow flies, is the closest bit of Madagascar to Indonesia, albeit over 3,500 miles away. And legends of the Merina, the most Indonesian of Malagasy tribes, hold that their ancestors arrived somewhere on the northern part of the east coast. It's also a good place to investigate the Arab element in Malagasy culture, for the Rasikajy civilization, an Arab trading culture that was one of the most significant pre-European civilizations on Madagascar, thrived along the northeast coast from the ninth to the sixteenth centuries.

"A former French professor of history here has written an article called 'La Plus Belle Énigme du Monde,'" Dewar says, almost whispering the words. The music has ended and for once the restaurant is quiet. "'The Most Beautiful Enigma in the World.' And it's about Malagasy origins. So we haven't managed to fix it yet." He smiles. "But that's the real romance about Madagascar. The story doesn't get fixed. It just gets more complicated."

I'm eager to hear more of that story, but we're interrupted by a car grinding to a halt a few arm-lengths away from us out the open window, chickens scattering before it. It's the car Dewar has hired for the week, and he heads out to arrange our first foray, to a village and accompanying archeological site known as Benavony. No matter. There should be ample time in the coming week for Dewar to share his thoughts on that "most beautiful enigma."

2.

THE ENIGMA HAS INTRIGUED Westerners ever since Diogo Dias, Madagascar's Columbus, first laid eyes on the island in 1500. In the first century after "discovery," arriving Europeans heard several languages, including Bantu and Malagasy, spoken along the coast, but they soon realized that a single tongue predominated. By 1613, the Jesuit missionary Luis Mariano was able to report, "In the whole of the interior of the island and along the rest of the coast [there is a language] spoken by the native people [which is] completely different from the *kaffir* language [Bantu], but which is very similar to Malay. This means that it is almost certain that the first inhabitants came from the ports of Malacca."

Mariano was not far off the mark, but confusion about the origins of the Malagasy persisted into the nineteenth century, when Europeans finally succeeded in penetrating the interior Highlands. There they found a tribe of almost pure Indonesian stock, the Merina, which only added to their confusion. How could one account for the various physical types found throughout the island, with lighter-skinned Asian features predominating in the Highlands, darker-skinned African features on the coast, and a mix in between?

One nineteenth-century author suggested a Polynesian/Melanesian mix. "May it not be that we have in Madagascar, as in the Malayan and Polynesian archipelagoes, *two* races represented—one, an olive or light-brown people closely connected with the inhabitants of Eastern Polynesia (from the Sandwich Islands thru the Samoan, Marquesan, Society, [Tuamotu], and other groups, down to New Zealand); and also a darker race, allied to the Melanesian tribes inhabiting Western Polynesia, from Fiji to New Guinea?" Another declared "the Malagasy agree closely with the Siamese." Still another, noting the similarity between the language spoken on Madagascar and that spoken on Nias, an island off Sumatra, claimed that Indonesian island as the homeland of the first Malagasy.

And how to account for the African element? As I've mentioned, many Malagasy today believe an aboriginal people known as the Vazimba lived on Madagascar before the first Malagasy scraped ashore in their outrigger canoes, though Western anthropologists say they have discovered no evidence to back up this claim. One nineteenth-century commentator even went so far as to suggest that the African races—indeed, the human race itself—arose on Madagascar. "He is inclined to suppose," wrote a skeptic, "that many of the African races went *from* Madagascar to the continent, and that Madagascar or some spot farther eastward and now submerged, was the seat of man's primitive civilization."

We may chuckle today at the naïveté of such notions, but do we know any more? If you parachuted into Madagascar with the charge of guessing the origins of the people you encountered there, your answer likely would be different depending on where you touched down. If you landed in the central Highlands, where an almond-colored people with almost straight black hair tend terraced rice paddies cut into steep mountainsides, you'd think the original homeland of the Malagasy must lie in Indonesia. If, however, you came down in the semiarid Mahafaly country of the southwest, where dark-skinned boys with crinkly hair drive enormous herds of hump-backed cattle through an acacia-stippled savanna, you might figure Africa as the source. And if you dropped down on the southeast coast around Manakara, where men wearing turbans and claiming a Meccan ancestry read books of Arabic script, you'd swear the Malagasy came from the Arabian region.

Even the foremost experts on the origins of the Malagasy, including Bob Dewar, can only hazard guesses as to where the proto-Malagasy originally came from. People from the Indonesian region obviously did come, but no one knows for certain when, why, or how. The same could be said for the first Africans to arrive. Many feel that pure Indonesians and pure Africans must have mixed somewhere along the line before coming to Madagascar—perhaps on the African coast, perhaps in the Comoros Islands—but that's conjecture, too. "We are also bound to ask," writes Pierre Verin, "how this mass of people in Madagascar of African origin could have assimilated the Indonesian language so completely, without having been dominated by the Indonesian Madagascans (as the Gauls were, for example, by the Romans)." And if the Indonesian element in Malagasy culture was so potent that its linguistic and cultural elements won the day, how is it that Islamic influences beginning about the tenth century so thoroughly penetrated Malagasy culture, possibly even playing a role in the formation of the state society? One leading archeologist noted that the formation of Malagasy culture is like a detective novel in which you know who did it but not how.

Though Malagasy origins are obscure, it is only by trying to bring them into the light of understanding that one can begin to grasp who the Malagasy people are today. Who are these singular people who worship the Christian God yet rely on their dead ancestors to give them guidance in everyday life? Who so highly cherish the home and family, yet spend more on tombs than on their houses? Who readily absorb Western economic and religious concepts yet believe in many areas that white people want to eat their hearts? Who had no writing 150 years ago yet possess an oral literature among the world's richest? How is it that there is such diversity in people, customs, and ways of life in Madagascar, while at the same time real unity of culture and language?

It's as if the Malagasy have thrown down a challenge to the world: Here we are, now figure us out.

3.

PRESENTLY I'M TRYING to figure out just one Malagasy, our driver Marcelin, who is racing at unconscionable speeds toward Benavony. He's driving perilously fast. Why? Is he figuring that the faster he drives, the quicker he'll get his daily wage of 130,000 Malagasy francs (about $26)?

"This guy likes to fly," Dewar says unnecessarily, and tries to look nonchalant. No chance: I can see his white knuckles gripping the armrest.

He and I are stuffed in the backseat of Marcelin's tiny, white Renault taxicab. Darsot, poor fellow, is in the "death seat" up front. These pillbox Renaults are as ubiquitous in Madagascar as a yellow cab in New York City, but the Malagasy are generally a head smaller than we vazaha. I feel like an NFL linebacker back here, with my left shoulder occasionally bumping Dewar's right, and my right shoulder smooshed up against the window frame. My leg muscles tighten every time Marcelin slams on the brakes and hits another pothole, with which this road, like all roads in Madagascar, is grievously pockmarked. Despite the growing pain in my right shoulder, I press closer toward the door, because in between Dewar's thigh and mine is the business end of a rifle Marcelin has laid down between the front seats. What sized pothole do we have to hit before it goes off?

I can see Marcelin's beady black eyes in the rearview mirror, concentrating hard on the road. With those dark eyes, partially hidden beneath a knitted brow, and the heavy beard he'd have if he didn't shave, he appears to have a goodly dose of that ancestral Arab blood in him. Perhaps his ancestors include members of the Arab Rasikajy civilization that we've come to study. He's well dressed in a starched-white shirt and slacks—not a laboring man—and he keeps his car spanking clean. The only thing to mar the snow-white exterior is a decal on the trunk of a lunging black jaguar, which is probably just how he pictures himself.

Despite his hell-for-leather driving, Marcelin is not above pausing when it suits him. Seconds after we dash past a dust-raising eighteen-wheeler, he stomps on the brakes and comes to a full stop. Dewar and I swing our heads around, wondering about the quality of that truck's brakes. But Marcelin casually steps out of the car, walks back down the road a few steps, picks up something off the road, and comes back. It's a rusted trunk latch.

We both sigh. This is Madagascar, where even roadside junk is worth stopping for.

At least there are few cars on this road, and Marcelin does grip the wheel with both hands; Madagascar is not a place to palm the wheel. The road is like most on the island: red-dirt, with the occasional patch straight and smooth, but most stretches like ski mogul fields. Then, out of nowhere, a paved section. This is worse than the dirt, because it usually features a raised leading edge, and it's so old and broken and potholed that it looks as if locals have mined it for its stone, which they probably have. "How is it," a Malagasy proverb asks, "that the well-used road of yesterday has become a dangerous precipice today?"

Rather than dwell on a nasty end by errant bullet or high-speed crash, I take in the scenery. This highway, extending about a hundred miles along the northeast coast between the towns of Antalaha in the south and Vohemar in the north, is known as the Vanilla Road. Madagascar produces the world's most prized vanilla, and much of it grows along this highway. On both sides of the road and for some twenty-five miles inland, right to the base of the Marojejy Mountains, which I can see rising like battlements to the west, stretch vast vanilla plantations. The only member of the 35,000-strong orchid family that produces an edible fruit, vanilla thrives in warm, moist climates with steady rainfall, and the Sambava area gets six and a half feet of rain a year. Vanilla also prefers half sun, half shade, so growers have left gigantic trees from the rain forest that once blanketed this region standing over the vines, lending the plantations the feel of a well-tended forest park.

Well tended is an understatement, as I can see clearly looking at these endless plantations out my window. Vanilla is the most labor-intensive crop in the world, with fully five years elapsing between planting the vine and bottling the aged extract. Growers must carefully tend each vine on its own specially planted tree, and they must pollinate each vanilla blossom *by hand* in order to produce the coveted bean. (*Vanilla* is Spanish for "little bean.") A native of Mexico, vanilla was introduced to Madagascar in the nineteenth century, but without the native bees and hummingbirds that normally pollinate the plant. So the Malagasy hand-pollinate them with a pointed bamboo stick that they use to transfer the male pollen to the female stigma. (A former Malagasy slave perfected the technique in 1841.) By the early 1970s, Madagascar supplied 70 percent of the world's vanilla, though United Nations figures show that competition from other producers and especially artificial vanilla had reduced that to 40 percent by the early 1990s.

We pull off the Vanilla Road onto a dirt track, which soon brings us to Benavony. The village consists of a cluster of weathered wood-and-thatch huts on stilts, ranged along either side of the pathlike road. Chickens flee before our car, while a lone, haughty goose with head held high stands its ground, staring us down. Snotty-nosed children appear to ogle, sidestepping square patches of rice and coffee beans set out to dry. One of them has a distended belly, but the others appear healthy. These people are Betsimisaraka (*"bets*-see-me-*sah-*rock"), the second-largest tribe on Madagascar, who reach from this region of the northeast coast to about midway down the eastern seaboard and inland to the escarpment that divides the moist east coast from the drier Highlands. Mostly rice and vanilla farmers, the Betsimisaraka are a rain-forest people known for burying their dead on the ground in secret locations deep in the forest, which they believe is inhabited by ghosts, fabulous creatures, and *kalanoro,* or "wild men of the woods."

Dewar, who speaks but a little Malagasy, lets Darsot explain our purpose to an elderly man in a dark, sleeveless shirt and beat-up fatigue hat. These preliminaries can take hours or even days—as Dewar says, "it's one of the things that drives most American researchers crazy"—but happily we are shortly on our way. The village president, off hunting in the forest, will not return until the evening, but two men, whom I watch slip away as Darsot speaks and return in spotless white shirts, agree to lead us to the site.

Large trees typical of rain forest spread a canopy overhead, shading the vanilla vines growing everywhere beneath, each vine draped over its own specially grown four-foot tree. The sun scorched the village, but in here the air is cool and sweet. It is a well-traveled forest, with narrow paths angling off in all directions. Even so, our guides carry machetes, as rural Malagasy inevitably do. We walk single file, Malagasy up front, vazaha behind.

Dewar walks hunched over, looking at the ground at his feet. Is he looking for sherds, or just lost in thought? I fear he will hit an overhanging branch, and sure enough he does, then steps back and looks at the offending branch as if to say, How *could* you? Every now and then, he pauses by some obscure feature in the landscape—a hole in the ground, say, or a mound of dirt—and, holding his chin in hand, mumbles "Hmm" or "Well, that's interesting." He doesn't let me in on the secret, so I spend my time dreaming of treasure.

Benavony is famous as the site where a French engineer dug up a terra-cotta vase full of jewels and gold coins in 1897. Perhaps it was a pirate's treasure, loot left behind by, say, the Englishman William Kidd or the American Thomas Tew, two of the most infamous buccaneers to frequent these waters. The coins, which were described in various publications after their discovery but which subsequently disappeared, hint at the Arab trade that existed along this coast from about the ninth century onward. THERE IS NO GOD BUT GOD / MUHAMMAD IS THE ONE SENT BY GOD. ALI IS THE FAVORITE OF GOD reads an inscription on one of the coins, most of which are imitations of gold dinars struck at Zebid in Yemen during the reign of the eighth Fatimid khalif, Al Mustansir Billah (1036–1094). The vase was buried at the earliest in the sixteenth century, however, since it contained a Spanish piastre of Philip II (1556–1598), along with a mysterious coin showing a horse, its legs and tail ending in trefoils, surmounted by a person sitting Buddha-style.

The treasure lured others to try their luck here. Typically, French colonial administrators around the turn of the century would hit the local jails, grab 100 men, and dig like hell for tombs and treasure, Dewar told me. In Vohemar, which represented the center of the Arabic Rasikajy civilization, they ransacked hundreds of tombs, though here at Benavony they dug but a single long trench. If anything of value turned up, it wasn't recorded. Then, in 1943, a Protestant minister from Antalaha excavated some of the Rasikajy tombs at Benavony—perhaps those very holes Dewar "Hmm"'d over. The cleric uncovered some jewels, including several rings of silver and copper, chains, a small bell, and some bronze mirrors as well as a fragment of Islamic pottery and a small Chinese blue-and-white bowl. Finally, Verin and Ramilisonina showed up in 1969 and dug a test pit near the fabled trench. They, too, found a piece of Islamic pottery, along with some local pottery and soapstone.

Based on the accumulated evidence, which he admitted was scant, Verin tentatively dated the site of Benavony to between the sixteenth and eighteenth centuries. He also suggested that the entire dune on which we are now walking, from the village of Benavony in the north to the mouth of the Lohoko River in the south—a distance of about a mile, with a width of a few hundred yards—is an archeological site. Dewar wants to investigate whether Verin is right, and to see if various sites within the overall site are contemporaneous with one another, which would mean this late-surviving satellite of the Rasikajy capital at Vohemar was a significant outpost. As Dewar put it, "We've got to find some sherds so we know what the hell is going on." Ultimately, his goal is to better date and understand this Arab phase of Malagasy civilization, which played such a vital role in the development of Malagasy culture and society as we know it.

Our guides slip off the trail where we hear the sound of chopping. A sinewy young man in a faded red Lacoste cap pulled low over his eyes is singlehandedly carving up one of those colossal rain-forest trees, which has toppled over. Darsot asks him a few questions, which he answers in a near whisper. The man then leads us back across the path and into a clearing in the forest.

"Don't take any pictures, okay?" Dewar says under his breath.

He thinks it's a brand-new tomb, which would be impolite to photograph, but it turns out to be a brand-new charcoal pit, smoking away. All across Madagascar, locals build these taxi-brousse-sized mounds of stacked wood covered with sand or dirt, which they burn in order to make charcoal for cooking. As we saw at PK-32, woodcutting for charcoal is one of the leading causes of deforestation on the island.

But the whisper-voiced man has not come to show us charcoal. He leans down beside the mound and pulls a potsherd the size of a dinner plate out of the sandy soil. Affixed with a handle, it's part of the lid of a large pot, and I can see right away it's made of chlorite-schist.

Chlorite-schist is a kind of gray soapstone, and it's one of the principal reasons the Arabs settled along this coast. They came to trade for copal resin, beef, rice, sea-turtle shell, mangrove wood, iron—and soapstone. Indeed, "Rasikajy" is the Malagasy name used by the inhabitants of Vohemar to label the people who were responsible for the region's early soapstone remains. During the heyday of the Rasikajy civilization in the fourteenth to sixteenth centuries, laborers worked some thirty chlorite-schist quarries in the hills just inland from this coast. Slaves hefted giant blocks of this easily carved stone down to the coast, where craftsmen shaped the rock into lidded vessels, flat-bottomed dishes, incense burners, lamps, spindle whorls, even well shafts.

While Dewar examines the soapstone lid, Darsot turns up a ceramic sherd sticking out of the heaped-up dirt of the charcoal mound. The sherds come fast after that, mostly red clay but some gray, all unglazed. Benavony's charcoal cutters, who have apparently been making braziers in this clearing for years, have dug up the soil over an area the size of a tennis court. All around us are those fifty-foot trees, with vanilla vines heavy with oblong green seed pods sheltered beneath them. But here there is just soil and ash and smoke.

Herpetologist Chris Raxworthy and friend at Lokobe Reserve

A leaf-tailed gecko guards its perch in a Malagasy rain forest

Malagasy twins
Angelin and Angeluc
Razafimanantsoa

Brookesias and
a small leaf-
tailed gecko
(middle) cupped in
Raxworthy's
hands

Surprised
at night,
a chameleon
in the
PK-32
proposed
reserve

Baobab in the spiny
desert at PK-32

Massif at Isalo National Park

Paleoecologist
Dave Burney
on the road to
Majunga

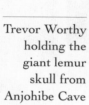

Trevor Worthy
holding the
giant lemur
skull from
Anjohibe Cave

Lower jaw of extinct pygmy hippo in situ inside Anjohibe Cave

The Sunken Garden in Anjohibe Cave

Archeologist Bob Dewar taking a GPS reading at the rockshelter

The king of Ambatovory in his home near Ranomafana

Primatologist Pat Wright in
Ranomafana National Park
Photograph by David Lowe

A tenrec, about which the
Malagasy say, "If you
don't see much, you
won't need much."

Betsileo
farmer
Pierre
Rakoto
before his
house at
Ranomafana

A Betsileo house in the Highlands

Verreaux's sifaka *(left)* and sportive lemur at Beza-Mahafaly Reserve

We sherd. Bemused, the Malagasy sherd, too, tapping the soil gently with the tips of their machetes, turning up anything hard. These are the cultural remains of their forebears, many centuries old, yet I'm not sure they make the connection. People along this coast claim they are descended from "the Arabs," as they call them, but at the same time "the Arabs" are not their ancestors. The same paradoxical situation seems to hold true across Madagascar: The ancestors are the recent dead, people you know.

While we sherd, Dewar sits cross-legged on the ground, taking notes and fixing our location using a handheld Global Positioning System unit. A shirtless Malagasy youth bends over him, no doubt wondering what the heck he is doing. We've been breathing smoke so long I can't remember what fresh air smells like. The sky is clouding over—rain?

"A rule of thumb in archeology: get a hundred sherds," Dewar says suddenly. It's the first thing he's said in an hour. "We've done that, so . . ."

Not finishing his thought, he gets up, and soon we're heading south through the vanilla forest. We've sherded in the center of the forested dune; now we'll sherd in other spots, to gain a picture of settlement here over space and time. Like other sites around Madagascar's coasts, this dune met the requirements for a suitable trading site: dry ground by a river mouth just back from the sea.

As we leave the forest at the southern end of the dune, we surprise a child carrying a fish on the end of a string. He can't be more than four years old. Far from his mommy, suddenly confronted by a group of grown-ups, including tall, bone-white vazaha, he bursts into tears and flees on stubby legs. Our guides break into warm laughter, as do other Malagasy far off in the rice paddies. For a minute, there is much amusement all around—grins, chortling, reassuring calls to the child.

Rice paddies stretch east for half a mile to another dune and then the sea. To the south I can see the former bed of the river, whose mouth now lies farther south. We sherd; Dewar takes notes and fixes our position; the shirtless Malagasy youth stands over him. Then it's back through the forest to the western edge of the dune. Here we find a small charcoal brazier, smoking away on the edge of a rice paddy so resplendently green it almost hurts the eyes. A single black zebu, tethered to a stake, grazes on the edge of the paddy, not remotely surprised to see us. We sherd; Dewar takes notes and fixes our position; the bare-backed youth hovers over him. No treasure, no fine porcelain, no golden tombs: such is true archeology.

Darsot has hired several locals to dig two yard-square test pits, one at the charcoal mound and another at the southern end of the dune. Test pits give archeologists a snapshot of a site, from the present day back to the earliest settlements and before. Once the whisper-voiced man and several others are safely under way, each with his own *angady*, the thin-headed spade of Madagascar, Dewar and I hike to the eastern part of the dune, the only spot we haven't yet sampled. We look for sites by the ocean, to no avail. "Well, that's interesting," Dewar says, then clams up for another lengthy silence. Lunch is a piece of French bread with sardines and a piece of cheese each; I'm looking forward to dinner.

"Archeologists do a lot of wandering around aimlessly," Dewar says finally, as if I didn't know. Trying to be patient, I think, *At least we've found the site.* Dewar told me of the time Verin was in a Betsileo village and asked in French if anyone knew of caves in the area. *"Oui, grotti!"* a villager piped up enthusiastically, offering to lead him there. Several hours and many miles later, they came upon a small church. The man led Verin around back and proudly pointed out the "grotto": a Virgin in a niche.

But we've been out here six hours already—might it be time to retreat?

Alas, there is some excitement back at the test pit by the charcoal mound. Darsot hands Dewar a small piece of glazed green ceramic that the angady-wielders unearthed at the bottom of the pit, a full yard down. Dewar hops into the square hole with a trowel and scans the sides, examining the stratigraphy. He actually looks fired up, for him. Is he thinking treasure? Suddenly he's talking—a lot.

"It's Chinese celadon," he says. "I'm not a Chinese porcelain expert, but it's probably somewhere between the eleventh and fourteenth centuries, possibly tenth century. It's interesting because it's down so deep."

He spins around in the pit.

"I'm just trying to figure out what we've got. It doesn't look like a tomb, and there is all kinds of mottling on the floor down here, and lots of charcoal. See the quantity of charcoal coming out of here?"

He scrapes, then scoops the remains into a blue bucket, which the Malagasy can't seem to empty fast enough for him.

"No evidence just yet of any kind of intrusive pit," he says, standing up. "The soil layers seem to be more or less stacked on top of one another, and they continue from one wall all the way to the other."

He climbs out and stands beside the pit, hands on hips.

"This is a deeper site than I imagined it to be—almost a meter deposit. It's probably in sand, which, you know, gets cracks. You just walk on it and you can throw something down fifteen centimeters."

He looks at me. He's *talking* to me.

"So the kind of stuff we're getting at the bottom—did you see those little stains of charcoal? They were kind of irregular, and when I scraped them, poof, they'd disappear. Probably means the occupation stopped some distance above."

I don't know which to be more enthused about, the find or Dewar's newfound tongue.

The site appears to be far older than Verin surmised in 1969 from his own test pit, Dewar says, and therefore it could shed light on the early years of the Rasikajy civilization. The diggers have also turned up moderately well preserved cow bones, a find arguably even more compelling for Dewar. In the moist tropical east of Madagascar, bones do not preserve well and are rarely found. But their rarity is not the only reason why this is an important discovery. A full three feet down, the bones may date to the eleventh or twelfth century, a time when some of the extinct megafauna may still have been roaming Madagascar's forests. Dewar is in the vanguard of experts trying to solve the mystery of who or what killed off the big beasts of Madagascar. With only that double handful of butchered bones, Dewar would like nothing more than to find evidence that the Rasikajy hunted these animals. Even if these bones aren't megafaunal, they can help clarify economic activities of the Rasikajy: What were they eating? Were the animals native or imported? and so on.

"We quadrupled what is known of this site just with what we collected today," Dewar says triumphantly as we hike back to Benavony after eight hours out. Despite my grumpiness about the day's tedium, we have made some interesting discoveries. And if his recent loquaciousness is any indication, perhaps Dewar will finally begin sharing his thoughts with me.

Or will he? As Marcelin speeds back to Sambava, two defeathered birds he shot rolling around in the trunk, Dewar suddenly says, "Oh, there was a story I wanted to tell you." Long pause. I wait, delighting in the prospect of something substantial, wondering whether I'll be able to hear him over the racing engine. "But I can't remember what it is."

4.

FOR ALL ITS PROMISE, Benavony, because of its late date, will likely not reveal much about the enigma. Some core element of the population, as we've seen, clearly came from the Indonesian region. The language is Malayo-Polynesian. The Malagasy belief in the spirits of their recently dead ancestors has Southeast Asian roots, as does the outrigger canoe, weaving and early iron-making technologies, and the rectangular house found island-wide (as compared to the round hut common in Africa). Many of the people in the Highlands look as if they just got off a flight from Jakarta, while the carefully tended rice terraces in the southern Highlands, stepping up the sheer mountainsides like stairways to heaven, come straight out of rural Java.

Rice is a cultural obsession of the Malagasy. "There are many plants," they say, "but it's the rice that's sweet." The Malagasy equivalent of our phrase "to have a meal" is "to eat rice," and all other food is called *laoka*, meaning an accompaniment. A heaped-up cup of rice grains is still a standard unit of measurement in Madagascar, as is, metaphorically, the rice cooking pot. The ethnographer John Mack once asked a Malagasy how far it was to a neighboring village; the man's answer was the number of pots of rice one could cook in the time it would take to walk there. Summing up her people's belief, an elderly Malagasy woman once told the Reverend S. E. Jorgensen, "Rice is Andriamanitra" (God).

As I've noted, not a single artifact in any early archeological site on Madagascar definitively points to Indonesia. Nor have any archeological sites in the Indonesian archipelago suggested even a tentative link with Madagascar, nothing that even remotely hints, for instance, at a mass migration away from the islands. With 13,700 islands, Indonesia is a big place, and archeologists could be forgiven for not putting their finger directly on the ancestral home of the Malagasy, particularly since archeology in that country is, like that in Madagascar, still getting off the ground. But on Madagascar? Why is there no hint, however tenuous, however circumstantial, of Indonesia?

The earliest archeological site so far identified is a rockshelter that Dewar and Malagasy colleague Solo Rakotovololona discovered in 1986 at the island's northern tip. Exploring in a limestone canyon above the Bay of Antsiranana, the archeologists stumbled upon twenty-four small caverns and rockshelters lining the base of the canyon. The deepest occupied levels of Lakaton'i Anja, one of three shelters they excavated, gave two radiocarbon dates calibrating to the fourth and eighth centuries A.D., respectively. Pottery turned up at Lakaton'i Anja points in the opposite direction, away from Indonesia. The ceramics resemble pottery of the same age found in the Comoros. Dewar believes the people who left their remains in the rockshelter's oldest levels were transients anyway, not settlers but hunters who used the shelter as a jumping-off point for forays into the forested interior.

But even the first continuously occupied sites on Madagascar show no sign of Indonesian origins. In fact, those on the northeastern island of Nosy Mangabe and on the mainland nearby show no clear links with the outside world at all; all artifacts appear to be indigenous products. Later sites, those dating from the tenth century onward, do show unequivocal evidence of a link with the Indian Ocean trade network. Pottery in these sites features not only red-slipped ware typical of the African coast, but Persian sgraffito ceramics and even celadon and blue-and-white porcelain from China. But as Dewar stressed back at the restaurant, even those sites do not contain a single piece of pottery from Indonesia.

Though many scholars feel the first Malagasy must have come to Madagascar by way of the African coast, having worked their way from Indonesia around the Indian Ocean along its north coast using ancient trade routes, no archeological trace of early Indonesians has ever showed up on the African coast either, as Dewar pointed out at the Coconut Palm. One compelling reason for the belief that Indonesians came that way is that Indonesian *crops* appeared in eastern Africa during this early period, foremost among them bananas. But archeologists have located neither Indonesian nor even Afro-Indonesian settlements along the African coast or on offshore islands such as Pemba, Zanzibar, or Mafia. Or, for that matter, in the Comoros.

"Frustratingly indecisive" is how Dewar sums up the archeological evidence for the advent of Indonesians in Madagascar.

Historical evidence is only slightly less sketchy. There are no oral or written records on either side of the Indian Ocean that shed light on the mystery. In Indonesia, the only historical texts written before the coming of the Portuguese in the sixteenth century appear on epigraphic stelae, carved stones that were mainly political in origin. "These were not in any sense records of legend, but rather the history of someone who put up a lot of money to make his political position clear," Dewar says. Indeed, much of what is known about pre-sixteenth-century Indonesia comes from accounts of merchants and other travelers visiting from other lands, principally China, and whose interests ran to trade, religion, and politics. No early accounts, whether on stelae or within traveler's tales, hint at any legends of a migration out of the region, either massive or modest, single or multiple.

The historical evidence in Madagascar is equally threadbare, though there is one titillating piece—that oral tradition among the Merina that holds that their ancestors came ashore somewhere along the northeast coast. "Today, the Malagasy publish comic books about their origins," says Dewar. "In the ones I've seen in Tana, the storyline is always one of a hero leading people from a homeland in Indonesia and settling in Madagascar."

This tradition was first recorded in the *Tantaran'ny Andriana* (History of the Nobility), a monumental history of the Merina that dates to the 1830s, when Merina elders responsible for maintaining the oral history of the royal family related it to a French missionary. The *Tantaran'ny Andriana* is extremely valuable to scholars of Malagasy history, both because it chronicles the history of the Merina kings back through many centuries, and because it is one of the first books ever published in Madagascar. The Malagasy people had no written language before European missionaries introduced one in the nineteenth century, so no early Malagasy texts exist that historians can turn to.

Interestingly, there are *Arabic* texts that are written in Malagasy using the Arabic script and that predate the advent of written Malagasy. Known as *Sorabe* or "great writings," they are the sacred texts of the Antaimoro, a Malagasy tribe with Arab ancestry who have lived on the southeast coast since the sixteenth century. Surviving examples of Sorabe, which scribes composed on bark paper, appear to date no earlier than the mid-nineteenth century, though they are probably copies of manuscripts written centuries before. Sorabe are an invaluable source of information about Arab religious and historical tradition in Madagascar, but they say nothing of pre-Arab times, much less of an Indonesian migration.

Ironically, it is to Arabs outside of Madagascar that we owe the only surviving records of possible early Indonesian journeys to Madagascar. These come from the writings of medieval Arab geographers, who chronicled what they knew of trade and polities around the Indian Ocean from the tenth through the thirteenth centuries. For a number of reasons, their works are not very accurate. First, the Arabs held to the system of geography propounded by the Alexandrian astronomer Ptolemy, who believed that the south of Africa joined the eastern part of the Indian Ocean. Moreover, they often simply repeated or elaborated upon earlier Arab authors, and they relied mostly on hearsay rather than personal observation, particularly when it came to those lands at the ends of the known world. Finally, different authors used one of two different names for Madagascar, Komr and Waqwaq. Nevertheless, these texts fascinate historians, for they provide the only clues to early Indonesian movements in and around Madagascar.

The earliest known Arab text with hints of Indonesians is also one of the most compelling. It comes from the *Book of Marvels of India* by one Bozorg Ibn Shamriyar, a Persian from Ramhormoz in what is today western Iran. It reports on a raid by the Waqwaq people on the island of Qanbaloh, which modern historians think could be either one of the Comoros Islands or possibly the island of Pemba along the African coast.

> Ibn Lakis has informed me that the people of Waqwaq have been seen doing extraordinary things. In the year 334 [that is, 945–946], they arrived in a thousand boats and fought with the utmost vigor. . . . When they had landed [at Qanbaloh], they were asked why they had come there and had not gone elsewhere. They replied that it was because they found there the products that suited their country and China, such things as ivory, turtle shell, panthers' skins and ambergris, and because they were looking for the Zeng [or Zanj, the name by which medieval Arab traders referred to the coast and inhabitants of Africa south of modern-day Somalia], because of the ease with which they bore slavery and because of their physical strength. They said that they had come from a distance which took a year to sail and that they had plundered islands six days' sailing from Qanbaloh. . . . If those people were speaking the truth, and if their report was exact, that would confirm what Ibn Lakis had said about the islands of Waqwaq, that is, that they were facing China.

This passage has several intriguing elements. First, take the unusual name of Waqwaq. While one scholar has suggested a possible connection with the South African Bushman click, most scholars believe the word has an Indonesian etymology. This seems plausible when you consider that, in Bahasa Indonesia, to pluralize a word you say it twice, as in *anak* (child) and *anak anak* (children). More importantly, Arab authors used Waqwaq, however confusedly, to refer to two lands, one on the coast of Africa and another in the eastern Indian Ocean. "The (islands of the) Waqwaq of China differ from (those of) the Waqwaq of the south in that the (islands of the) Waqwaq of the south produce gold of bad quality," wrote one Arab geographer of this period, implying this duality. It's possible that, because they relied on the misleading Ptolemaic system, the geographers were erroneously referring to one and the same land when they mentioned two Waqwaqs. But it seems just as likely that the two Waqwaqs referred to are Madagascar and Indonesia. Should we be surprised that two lands with the same people speaking the same language would have the same name?

The passage has other provocative details. The way the Waqwaq spoke of seeking African slaves "because of their physical strength" implies that they were a different race, a physically less robust people, which the Indonesians are compared to black Africans. The Waqwaq said they crossed "a distance which took a year to sail" (the Indian Ocean?) and had plundered islands "six days' sailing from Qanbaloh" (Madagascar, which has several small satellite islands of its own?). Finally, the passage declares that the "islands of Waqwaq . . . were facing China," which even today one could conceivably say of the Indonesian archipelago.

A passage from another text describes an actual migration from Southeast Asia to Komr, which may refer to both Madagascar and the Comoros, a name clearly derived from "Komr." This passage comes from Ibn Said, who lived in Baghdad in the thirteenth century:

> . . . the Komr people, who have given their name to the mountain [of their name], are the brothers of the Chinese. . . . This town [of Komoriyya, the capital of the Island of Komr] takes its name from the Komrs, who are descended from Amir, the son of Japheth. [The Komrs] lived with the Chinese in these eastern parts of the earth. Discord arose between them, the Chinese drove them out to the islands and they stayed there for a time. The title of their king was Kamrun. Then discord arose between them when they were in those islands of which we shall speak later. Then the people who were not part of the royal family went to that great island and their sultan lived in the town of Komoriyya.

This excerpt raises a number of questions. What was the Komr's homeland that they "lived with the Chinese"? To which islands did the Chinese initially drive the Komrs? Did they lie in Indonesia? Along the Southeast Asian coast? Could the Komrs be the Khmers, whose empire thrived at Angkor Wat in modern-day Cambodia from the ninth century onward? One can speculate wildly on reading such obscure texts as this one, yet the basic story beguiles historians: The Komrs migrated away from Chinese peoples in the east to a great island they named Komr.

One account even appears to chronicle the end of Indonesian migrations to the African region. It also hints at how widely they might have journeyed through this part of the Indian Ocean. The passage appears in the thirteenth-century writings of Ibn al Mujawir:

> The people of Al Komr invaded Aden [on the southern tip of the Arabian Peninsula]. . . . They sailed together in one monsoon. These peoples are now dead and their migrations have come to an end. From Aden to Mogadishu there is one monsoon, from Mogadishu to Kilwa [a town on the African coast across from the Comoros] a second monsoon, and from Kilwa to Al Komr a third. The people of Al Komr had combined these three monsoons into a single monsoon. A ship from Al Komr had come to Aden by this route in the year 626 of the hegira [1228]; sailing in the direction of Kilwa, it had arrived by error at Aden. . . . But the Barabars drove them out of Aden. Now there is no one who knows about the voyages of these peoples or who can tell us anything about the conditions in which they lived or what they did.

The finality of this passage is striking. It's as if the author were closing the book on speculations about the people who populated Madagascar, leaving the mystery of the Malagasy's Indonesian roots to tantalize and confuse later chroniclers, from the first European traders to land on the island in the sixteenth century to scholars like Bob Dewar today.

In the near-total absence of archeological and historical evidence that can help solve the "beautiful enigma," scholars have turned to the linguistic evidence—with astonishing results. The Malagasy language allows for such an analysis because, despite having engendered some eighteen different dialects on Madagascar, it remains a single language spoken by everyone from Vezo fishermen on the southwest coast to Merina politicians in Tana to Betsimisaraka ombiasy on the northeast coast. This is remarkable. That an island as large as Madagascar, with diverse ecosystems cut off from one another by forests, deserts, mountains, or rivers, should have but one language baffles linguists. Madagascar's neighbor, Africa, has 1,500 languages. The island of New Guinea, only a third larger than Madagascar, has 700 languages. Why does Madagascar have only one?

Ever since Luis Mariano first claimed in 1613 that the Malagasy tongue was "very similar to Malay [meaning] that it is almost certain that the first inhabitants came from the ports of Malacca," commentators have sought to pinpoint the origins of the language and, through it, the people of Madagascar. Some of these authors only took general aim. "If we take such a work as Mr. Alfred Wallace's *Malay Archipelago* and examine the vocabularies given at the end of the second volume," wrote the missionary and author James Sibree, "we shall find that in the list of 'Nine Words in Fifty-nine Languages' of that region, the Malagasy words are discovered in every one of the nine columns; and not only so, but in most of these nine examples words exactly like, or most closely identical with, the Malagasy are found in a great number of the fifty-nine languages."

Others went for the bull's-eye, such as this nineteenth-century writer, who imputed with singular confidence the birthplace of Malagasy to the Indonesian island of Nias:

> However anomalous may be thought the introduction in [Nias] of the language spoken in the great island of Madagascar . . . there are not wanting grounds to justify it, for not only are the small islands which lie off the western coast of Sumatra the least remote from Madagascar of any in the Archipelago of which they form the western boundary, but . . . the dialects spoken in them, and particularly that of Nias, bear a stronger affinity to the Madagash than those which prevail in the islands situated to the eastward of Sumatra—an affinity the indisputable existence of which is one of the most extraordinary facts in the history of language; when we take into consideration the immensity of the intervening ocean, combined with what we may presume to have been the state of navigation at the very remote period when the communication must have taken place, unnoticed as such a circumstance is by record or tradition.

Alas, modern linguistic research proves this author to be off the mark by about a thousand miles. The late Otto Christian Dahl, a Norwegian linguist, demonstrably showed that the modern language that most closely resembles Malagasy is Ma'anyan, a Malayo-Polynesian tongue spoken today in the Barito Valley of southeastern Borneo.

Dewar points out, however, that Malagasy and Ma'anyan may have begun diverging from each other before the ancestral Malagasy left Indonesia for Madagascar—that is, the people who would become the first Malagasy may have lived elsewhere in Indonesia, or even Southeast Asia, before beginning their migration or migrations west. But *malgachisants*, as specialists in Malagasy studies are known, generally agree that Dahl's findings, in the absence of hard physical evidence, provide the most compelling clue to the Malagasy's original homeland.

Moreover, within his broader finding, Dahl made an intriguing discovery that he believed might point to where the proto-Malagasy first landed in Madagascar. Dahl found that Malagasy words for the cardinal points of the compass resemble those of Ma'anyan, but only if those points are shifted by ninety degrees. Thus, while *timor* means east and *barat* west in Ma'anyan, the corresponding Malagasy words, *atsimo* and *avaratra,* mean south and north, respectively. One can explain the shift, Dahl wrote, by the fact that early sailors fixed the points of the compass by the winds. Indonesia's dry eastern trade wind meets its match in Madagascar's dry south wind, while the moist west wind of Indonesia corresponds to the north wind of Madagascar, which brings rains to the northwest coast. Based on this argument, Dahl felt that the Indonesians who first pioneered Madagascar must have landed on the northwest coast.

But the French scholar J. C. Hebert questioned this theory. If the Indonesian immigrants to Madagascar had attached more importance to the *qualities* of the wind—whether it brought rain, for instance, or whether it blew in the dry season—than to its direction, then those early pioneers would have used the word "avaratra," for example, to refer to the north from which the rains came. And in the north of the island, rains on the northwest coast come from the northwest, while those on the northeast coast blow in from the northeast. Hebert argued convincingly that Dahl's hypothesis about the initial landing is not defensible—that, in fact, a first landfall here on the northeast coast is just as likely.

5.

HEBERT'S ARGUMENT SUITS me just fine, as we're now hiking to a cliff that Dewar thinks may harbor rockshelters. Benavony holds clues to the Rasikajy, but rockshelters just might hold clues to the premiere Malagasy. The going's rough, however, and since Dewar's got a debilitating case of hotely belly, it's possible he may not feel quite as enthusiastic about this hike as I do.

We're fighting our way up a fifty-degree slope through thick undergrowth, the secondary forest that invades a landscape after primary forest is razed. Our guide is doing his best to machete a path through, but it's like trying to climb through a Brillo pad. Thick rainforest shrubs—would-be trees—fill every available space, and whipcord vines take whatever's left. It's impossible to put a foot right in here. You step, grab, stumble, grab something else, and pray.

"Fuckin-A!" Dewar cries as a low-hanging branch snags on his daypack, stopping him cold. As he tries to extricate himself, a tennis shoe slips out from under him on the muddy slope, and he goes down, tangled in viney vegetation. It's the first I've heard him curse, but I guess he has a pretty good excuse.

We're trying to get to the crest of a ridge that, from the road far below, appears to have overhangs. Forest of either the primary or secondary kind mantles most of the hills around here, including what I can see of the towering Marojejy Mountains to the west. But this ridge bears an exposed wall of granite perhaps 1,000 feet long by 75 high. It's just the kind of cliff to get the heart of a rockclimber pumping— and that of a rockshelter specialist like Dewar. Rockshelters are dry nooks beneath cliffs where people take refuge and, if you're an archeologist and get lucky, live for a time, leaving behind traces of their life in the accumulating sediment. The lowest levels of Lakaton'i Anja, Dewar's rockshelter in the far north, produced very modest amounts of indigenous pottery, for example. The pottery was nondescript, however, offering no clues as to who these people were or where they came from. "I am confident," Dewar wrote of the site, "that this deposit is in fact the product of a brief visit in the eighth century, but we will have to find more before we can characterize the visitors."

Hence our search today with Nestor, a guide we hired from the Betsimisaraka village nearest the cliff. Far from the Nestor of the *Iliad*, who incited warriors to battle with tales of his legendary exploits as king of Pylos, this one appears to have seen better days. Probably only in his thirties, he looks much older. His head hangs off his shoulders as if the muscles in his neck are barely serviceable anymore. To look at you, he swings his whole head around like a marionette to face you square on. And look at you he does, with droopy, bloodshot eyes that suggest a close affiliation with the local palm wine. His blue sport jacket is literally threadbare. The back consists of nothing but a thin rope of thread running down his spine.

But Nestor is earning his keep. Before we left the village, he upped his originally negotiated fee of 10,000 Malagasy francs to 40,000, or about $10, complaining that he'd have to do a lot of cutting. ("A fortune to him, two movie tickets to us," Dewar said, a bit ungraciously I thought.) Nestor was right: His machete is getting a lot of use. Without it, we'd never get through this stuff. Now if only we could be as sure-footed as Nestor, who never misses a step in his wide-splayed, tough-as-tires bare feet.

An hour and a half after starting from the village, we reach the cliff face. The view back to the west is stunning. The air smells moist with imminent rain, yet sunlight still fills the wide-spreading valley, with its backdrop of mountains and heavy clouds. Far below us, beyond the white trunks of a few rain-forest giants inexplicably left standing on the hill we've just ascended, lies a patchwork quilt of rice paddies, some stippled bright-green with hand-planted rice shoots, others filled with water and reflecting the sky like mirrors, still others lying fallow and straw-colored. Beyond them, up a small rise, sits the village of six or eight huts—home of Nestor. Then low mounds of secondary forest before a wide river sweeping inland to the Marojejy massif. These rugged mountains, the highest soaring more than 7,000 feet into the clouds, are trackless and all but unexplored. Who knows what secrets might lie within them?

Nestor has an idea. As we headed up here, he told us through Darsot that his people believe these hills are inhabited by the "wild people," who he said only eat fruit and crabs. As Dewar and Darsot begin exploring various niches for remains of the Kalanoro—or anyone else who may have lived here—I gaze out over the postcard scene and recollect what is known of the "wild men of the woods." "Kalanoro are by far the most mysterious, frightening, and bizarre (*hafahafa*) of the Malagasy spirits," writes the anthropologist Lesley Sharp. They are short, naked, and covered in hair, including great, flowing beards. They have long fingernails, eyes that glow like hot coals, and feet that point backward. Inhabiting the deepest recesses of the forest, they are rarely seen, though they are said to have a love of fire and sometimes appear at hunters' campfires.

An English missionary wrote a vivid description of one of these "wild men," who was captured in 1879 about eighty miles west of Maroantsetra, a coast town south of Antalaha:

We were informed by a trader from Mauritius, a Mr. Carmes, who saw him, that . . . [he] was caught by some Malagasy in the employ of a Manahar trader, while asleep on the branch of a tree, and when taken resisted violently, biting his captors severely; after a few days' confinement, however, he ceased to be aggressive. Mr. Carmes describes him as being a powerfully built man of about five feet nine inches in height, his face and body being thickly covered with long black hair; his mode of walking was peculiar, as he traveled very fast, with his head down, occasionally going on all-fours, his eyes (which resembled in expression those of an animal rather than of a human being) invariably being fixed on the ground. When caught he was perfectly nude, but wore clothes when provided with them. He could never be induced to eat flesh or any kind of cooked food, subsisting entirely on manioc and other roots; nor would he sleep in a recumbent position, but when resting preferred to squat on hands and feet on a stool in a corner of the house. After some weeks he commenced to learn a few words, and by means of these and signs it was understood that he had a father and two brothers in the forest where he was taken. These were found and surrounded by a search party one night, but being disturbed, easily eluded their pursuers, jumping from tree to tree like monkeys and running on all-fours. The captured man died five months after being taken.

Other unusual creatures inhabit the landscapes of Madagascar, according to the Malagasy. The *kinoly* are human beings with red eyes and long nails who are actually ghosts of the dead. Legend has it that a Malagasy who once met a kinoly asked it, "How is it your eyes are so red?" The kinoly replied, "God passed by them." The Malagasy then asked, "How is it your nails are so long?" and the creature said "That I may tear out your liver" and quickly eviscerated the man. There is the *songomby,* an ox that is fleet of foot and eats people. *Songa* means "having the upper lip turned upward, uncovered," and *omby* means ox, though the Malagasy use the word "songomby" figuratively to mean "lion-hearted." Another fearsome beast, the *lalomena,* also resembles an ox, but has very red horns and lives in the water. Then there is the *tokantongotra* ("single-foot"), a large white animal that has only two legs, one sticking out of its chest and the other jutting out behind its paps. Despite this odd physique, the animal is exceedingly swift, and like the songomby goes about at night and devours Malagasy. The *fanany* is a snakelike creature with seven heads, each with its own horn, while the *dona* is big and long and bellows like a bull.

Western skeptics have explanations for all these creatures. The fanany and dona, they say, simply embody the widespread Malagasy fear of snakes. As one writer has pointed out, the "single-foot" of the tokantongotra probably originally referred to the feet as not being cloven, which may be a lingering recollection in Malagasy tradition of the pygmy hippo with its possibly uncloven hooves. The water-dwelling ox lalomena is an even more likely candidate for retained memory of the hippo, which the songomby may be as well. As for the kinoly story, well, that smacks suspiciously of "Little Red Riding Hood," doesn't it? ("Oh, Grandmother, what great big nails you've got!" "The better to rip out your liver with, my child!") But the Malagasy believe in them, and who are vazaha to say they don't exist, particularly when vast expanses of Madagascar, including these very Marojejy Mountains, remain largely terra incognita?

Dewar's voice rings out from behind some rocks farther along the cliff.

"I just found a potsherd!"

I make my way over, managing to stay out of the light rain that has begun to fall by keeping under the edge of the cliff, which conveniently juts out here. I can see why the ancients might have cherished a rockshelter such as this—it can be the only dry place around. Dewar and Darsot are squatting beneath an overhang where the distance between dirt floor and stone ceiling is but four feet. I clamber up to help in the search, while Nestor hovers nearby, staring blankly with those bloodshot eyes.

The shelter is filled with sherds. In a niche at one end of the shelter, which is about twenty feet long by six feet wide, I find five or six plain brown sherds, while Darsot turns up a fragment the size of a bread plate. Soon we've filled several of Dewar's white nylon sacks.

If I begin to dream about this being a rockshelter of the first Malagasy, Dewar quickly quashes such thoughts. The sherds are anonymous-looking, he says, and are unlikely to supply any indication of chronology; they may even be recent. And there is no sign of a hearth or other human artifact, nor any animal remains beyond terrestrial snail shells, whose owners more than likely got there on their

own. Finally, the shelter contains no sediment that might hold material. The floor bears only lichen-covered breakdown that looks as
though it gets well rinsed in the rainy season.

"Well, it's not going to shake up science," Dewar says, snapping
shut his notebook. "The only curious feature is this area over here."

He shuffles over to the place where I uncovered the sherds,
where some blocks of stone form a kind of low wall.

"I'm guessing that that's a place where people pushed the rock
away so they could put down some vegetation and make a nice place
to sleep. Nestor says people sometimes sleep up here if they're working high on the slopes and are surprised by early evening."

When he hears his name, Nestor swings his head around as if on
springs and looks squarely at Dewar. Suddenly I begin to wonder
about Nestor and his story of the "wild people." If they lived up here,
would his fellow villagers risk sleeping in these lonely cliffs?

I really start to wonder about Nestor as we head back down. The
sun has come out and is rapidly drying everything up. I begin to enjoy
the hike. Then it strikes me: Where is all that vicious vegetation we
fought through on the way up? We're hiking freely through bean
fields and small forest stands *on a trail.* I glance at Nestor up ahead,
his shoulder blades jutting out where the erstwhile blue fabric of his
shredded jacket should have been. Our man Nestor took us for a ride,
he did. And the "wild people" who eat nothing but crabs and fruit? A
figment of his wine-soaked mind?

As Dewar pays him off back at the village, Nestor swivels his
head my way, levels a watery stare at me, and smiles. He knows I'm
on to him. But it doesn't matter; he got his 40,000 Malagasy francs.
We've been gone only four hours. That's 10,000 an hour, surely the
best money he's ever made. Be tellin' that story to the day he dies, I
suspect: "Climbed up, climbed down, 40,000."

6.

THE ROCKSHELTER, ALAS, sheds no light on the enigma, nor on three
mysteries that nag at those studying Malagasy origins: when, why, or
how they came.

Scholars still debate when the Malagasy first arrived. Recent estimated dates of initial colonization range from 1000 B.C. to A.D. 1000.
In his classic 1951 work *Malagasy and Ma'anyan: A Linguistic Compari-*

son, Otto Christian Dahl proposed A.D. 400 as an approximate date of departure of proto-Malagasy from Indonesia, based on the limited number of Sanskrit loanwords in Malagasy. Another scholar argues that these loanwords came via Old Malay or Old Java, rather than directly from an Indian language, and prefers a more recent date of first emigration, seventh century or later. Dahl's hypothesis seems to have won out with most malgachisants, however, who, based on his work and that of others who have used various linguistic methods to estimate the time when speakers of Malagasy first diverged from speakers of Ma'anyan, have reached a consensus estimate of 1,900 to 2,000 years ago.

The admittedly limited archeological evidence seems to corroborate that estimate. The first traces of human activity on Madagascar appear at two sites in the southwest, Ambolisatra and Lamboharana.* At these two sites around the turn of the century, Grandidier unearthed bones of the extinct hippo that clearly bear the marks of butchery with metal tools. There are no associated artifacts, however, which is why Lakaton'i Anja still holds pride of place as the island's oldest archeological site.

Another compelling piece of evidence for an arrival no earlier than 2,000 years ago is the fact that Madagascar apparently had no Stone Age. Archeologists have failed to identify a single site on the island bearing a stone-tool industry. This stands in marked contrast to prehistoric Africa and prehistoric Southeast Asia, and allows scholars to infer that no significant occupation of the island occurred before people had switched to iron tools exclusively. Though their understanding of when the Iron Age began in earnest in both coastal Africa and Southeast Asia is still incomplete, specialists believe that iron did not completely replace stone in either region before 200 B.C. at the earliest.

Thus, like linguists, archeologists generally agree that Madagascar remained uninhabited at least until the time of Christ. Burney, for one, deems this "one of the greatest ironies in the history of human migrations and colonizations." The irony is especially thick considering that modern humans apparently evolved less than a thousand miles away on the African savanna. It wasn't until long after humans had colonized the entire globe, including such inhospitable regions as the Arctic and the Mongolian steppe, that they finally reached Madagas-

---

*If you recall, Ambolisatra, close to Raxworthy's site at Ranobe, is the hypersaline basin also known as Andolonomby, where Dave Burney revealed, among other findings, a sharp spike in charcoal in the region around 2,000 years ago—possible evidence, he suggests, of a human presence on the island at that time.

car. Along with New Zealand, Madagascar was the last large land-mass on Earth colonized by people. It is astonishing to think that some of the most advanced civilizations that ever existed—the Assyrian and Mycenaean civilizations, the Egyptian dynasties, the Olmec of Central America, the Shang of China, the Indus civilization of India—all rose *and fell* before a human being likely ever set foot on Madagascar.

The late arrival of the first Malagasy bears on the unanswered question of *why* they came. Did a single oceangoing outrigger canoe from Indonesia, perhaps trading along the Indian coast, get blown in a storm to Madagascar? We know this is possible. In 1930, a boat of fishermen from the Laccadive Islands off India's southwest coast drifted all the way to Madagascar, coming ashore safely at Cape Est, on the northeast coast south of Antalaha. But scholars have long dismissed the notion of a single, accidental arrival that spawned the Malagasy people and culture. For one thing, the range of physical types seems too broad to account for a single coming. Many experts hold with the ideas of French historian Hubert Deschamps, author of the influential 1960 *History of Madagascar,* who believed multiple migrations took place, with the Merina appearing last, perhaps as late as the twelfth century.

If they came deliberately, why did they come? Was home, wherever that was, becoming so crowded that they were forced to seek more land? Did they want to expand their territory for political reasons? Were they fleeing persecution, as the Arab geographer Ibn Said intimated when he claimed that "discord" between the Komrs and the Chinese drove the Komrs to "that great island"? Did some religious leader convince his flock to seek a hoped-for promised land in the west?

Dewar's take on why the proto-Malagasy journeyed to Madagascar may explain when they arrived. The first Malagasy, he feels, came for purposes of trade. "As far as I'm concerned, Madagascar was settled by people who were looking for stuff to sell, not principally because they needed new territory," he says. "What other explanation makes sense, for folk to leave Indonesia behind and sail across an ocean? Must have been a profit motive."

Scholars know that a widespread trade network that included the Indonesians was in place along the entire north coast of the Indian Ocean 2,000 years ago. Here they can actually turn to a document produced roughly during the time in question. *The Periplus of the Erythraean Sea,* a document written by a Greek merchant living in

Alexandria in the first century A.D., describes thriving ports stretching from modern Kenya all the way to what is today Sri Lanka.* The anonymous author tells how ships sailed from the African coast to India, where they traded aromatic gums, tortoiseshell, ivory, and slaves for cotton cloth, sugar, grain, oil, and ghee. "It's clear that he knew of, though he never visited, ports on the southeast coast of India," says Dewar, "and that he knew that some of the materials offered in Sri Lanka, including spices, were coming from farther east."

Dewar's hypothesis of a trade motive also helps to provide possible answers to the question of *how* the first Malagasy came. Carvings on the eighth-century Borobudur Temple on Java hint at what kind of vessel they likely used: They depict a two-masted outrigger canoe with sails. But questions remain. First, did the proto-Malagasy make a single trip or many? Dewar holds with the latter. "Otherwise you have this terrible problem," he says, "because they *brought all this stuff* with them. There is Indonesian iron-making technology in Madagascar and all these Indonesian crops over in Africa. I suppose if you hold with the romantic vision of loading up the boats for the trip to wherever, they might have put in a little of everything—a Noah's ark of Indonesian culture. But it seems much more likely that they came over, spotted good land, went home, and came back the next time with sprouts of sugarcane, taro sets, bags of seed rice, and set about building Malagasy culture."†

Second, whether they came once or in waves, what route did they take? Did they come directly across the Indian Ocean, or did they make their way along its northern coast, trading as they went? The direct route is possible, as trade winds and currents and the lack of tropical cyclones in the region favor sea travel from east to west during the Southern Hemisphere summer. But as the scholar Gerald Donque has written, deep-seated problems attend the notion of a direct route, which

> represents a voyage of about 6,000 kilometers [3,720 miles] through a desert sea without a single port of call. There is no

---

*The Erythraean, or Eritrean, Sea included the Persian Gulf and the northwestern part of the Indian Ocean.

†"Much more important, they brought *women*," Alison Jolly emphasizes. "For people to look wholly Indonesian today, which upper-class Merina do, there must have been a sizable deliberate colonization by families intending to settle, going on information brought by traders of a place worth settling—much as Europeans went to America. Also, of course, strict rules about whom to marry."

island along this sea route which would enable successive reconnaissances to be made increasingly farther towards the west or where the migrating groups of people could take in fresh supplies. The only island in the west lies in the Mascarene archipelago [Mauritius, Réunion, and Rodrigues]. Is it reasonable to believe that these islands, providing the only natural harbors on the route, would have remained uninhabited before the coming of the Europeans, while the great land farther to the west had ceased to be uninhabited a long time before this?

Scholars, including Dewar, favor a coastal route for the first Malagasy. They lived and traded successively along the Southeast Asian coast to Sri Lanka and India, across to Arabia, and down to the coast of Africa. As Dewar told me on that first day at the Coconut Palm, "several scholars have proposed that folk made it around to the East African coast, *found* Madagascar, found it as a fairly unoccupied place where they could set up a kind of forward-position trade zone where they didn't have to compete with folk on the African coast, and then made trips back and forth across the Indian Ocean. I'm perfectly willing to believe they did that. But as for a direct route . . ." He laughed. "Maybe it's just that I'm not sufficiently romantic to imagine that they loaded up the boats and set sail into the setting sun."

Partly because of the lack of unequivocal answers, and partly because of the nature of what little evidence there is, Dewar prefers, as he has written, "to consider both the initial settlement of Madagascar and the origins of Malagasy culture, in all of its diversity, as *processes*, and not as events. . . ." These processes can only be understood in the context of Indian Ocean history, he adds, which provides "a further confirmation of the need, if indeed another is wanted, for further research into the still poorly known prehistory of the Indian Ocean."

7.

THE INDIAN OCEAN stretches out before me now in all its incomprehensible vastness. It's after dark, and I'm sitting on the beach in front of my hotel, trying to imagine what it would be like to travel 3,700 miles in a small boat. Would it be possible today, much less two millennia ago? I look at the horizon. Thataway lies Indonesia. Thataway, perhaps, came an outrigger canoe or three 2,000 years ago with the first Malagasy.

I wonder if they would have had a night like this one. A warm breeze is blowing in from the ocean, which beyond the breakers appears as placid as a lake. The moon is so bright it pierces the veil of diaphanous clouds with ease and, when it breaks free into open sky, leaves an afterimage on the eye. Beneath the moon, far out to sea, a flat disc of light rests on the surface, like a distant city at night. The waves are silver envelopes, the stars like stabs of harmless needles. How wondrous the island would have appeared to them cast in such a light, with the Marojejy range rearing up regally just back from the coast.

Today we say it's a small world, and indeed, even the thousands of miles that separate me right now from Indonesia can seem negligible under certain circumstances. "A most curious thing happened today," wrote a resident of Tamatave, a town halfway down Madagascar's east coast, on August 27, 1883. "Four or five times in succession the sea receded, leaving the coast dry for many yards, while the reef stood some feet out of the water, and the water from the top thereof came rushing down like a river. Boats, lighters, etc., were left high and dry on the sand. We suppose there must have been an earthquake somewhere." In fact, it was the eruption of Krakatoa in Indonesia, one of the most powerful volcanic blasts in recorded history. In the months following the eruption, quantities of pumice washed up all along the east coast of Madagascar. Musing on Malagasy origins, another observer commenting on the pumice wrote that "clearly . . . there is a strong set of wind and current in this direction. So in prehistoric times it is possible that stray canoes may have been carried by the same forces to the same destination."

Stray boats have in modern times, as the remarkable story of the *Emulous* attests. In 1835, this English brig lay in the harbor of Port Louis, Mauritius, 600 miles east of Madagascar, waiting for cargo. Its captain decided to take a small boat out for a bit of fishing. He took provisions for the day and a young seaman to work the boat. He also brought along half a dozen casks called "breakers," which sailors of the day filled with salt water to serve as ballast. Finding the casks full of freshwater instead, the young seaman gave orders to have them emptied into the ship's tank and filled with seawater. But the captain was eager to be off and had them lowered into the boat as is. "Was that chance?" asked the nineteenth-century chronicler of this story. "No, the hand of Providence was visible there." As the two pushed away from the ship, the captain, hearing his dog whining piteously to be taken along, had the seaman turn back to fetch the animal. This, too, would prove providential.

They were out all day but had no luck. Determined not to return empty-handed, the captain fished until dark, then anchored under the lee of a lonely rock called the Gunner's Quoin, preparing to return to the ship in the morning. The night was fine and warm. They enjoyed the rest of their provisions sitting in the boat under a tent they made of the mainsail, then fell asleep. In the middle of the night, an uneasy motion of the boat awakened them. They found that a gale was blowing, and they were quickly drifting away from the island. They close-reefed the sail and tried to work up under the land, but they were too late. The waves rose higher and higher, threatening to swamp their fifteen-foot craft. So they tied their mainsail to the boat's painter and threw it overboard to serve as a drogue, or sea anchor, to keep the boat's head to the sea. Then they anxiously waited for dawn.

Toward morning the wind abated, but just as they were about to haul in the mainsail, the painter broke and they lost it. Beating up to Mauritius was now out of the question; there was nothing for it but to try for the island of Réunion, 100 miles to the southwest. Having lost their best sail, however, they could not keep the boat close enough to the wind, and they passed the island far to leeward. Their situation now became alarming. Drifting westward, their only hope lay in reaching Madagascar, some 500 miles away. They had no compass, so they set their course using the sun and stars. By luck, they had fresh water, but no food. They held out as long as they could, then killed the dog and carefully rationed its meat. Ten days after passing Réunion, they came in sight of the southern tip of Madagascar, but the wind soon failed, and they feared they would perish before they ever reached the island. "But they were not forsaken," hailed our chronicler. "He who notes the fall of a sparrow had them in His keeping."

Eventually they managed to get ashore, where three Malagasy women, "semisavages though they were," gave them food and led them to their village. After hearing them out, the chief supplied them with an escort of eight men and a canoe. They made their way by stages up the coast to Tamatave, which they reached in eleven days. There they boarded a bullock schooner trading between Madagascar and Mauritius, and, to the astonishment of their shipmates, returned to the *Emulous* after an absence of thirty-two days.

Providence may well have had a hand in their safe crossing, but 600 miles is a lot different than 3,700 miles. Even with the trade winds on their side, could the proto-Malagasy have crossed the entire Indian Ocean in their handmade outrigger canoes? To find out, a

Briton named Bob Hobman decided to build a replica of the kind of boat the first Malagasy might have used and, in the manner of Thor Heyerdahl, try to sail it from Java to Madagascar, making no landfalls, using no modern navigation aids, and subsisting solely on foods the ancient Malagasy might have eaten. The sixty-foot double outrigger canoe was built entirely of wood and bamboo, with palm-weave sails and rattan bindings instead of nails; it had no motor, radio, or sextant. On June 3, 1985, the *Sarimanok,* as the vessel was christened, set sail from Java.

"They had an unending, horrible voyage," Dewar told me. "There were problems with the boat, more or less continuous high seas, strong winds, and frequent storms. All the time they're filming this damn thing, filming the boat falling to pieces and so forth." After one stop on Cocos (Keeling) Island to let off a sick crew member (and bring on some tinned food), Hobman's crew, against all odds, managed to go the distance to Madagascar in forty-nine days. But by then they had lost their ability to steer the craft, and they drifted past the northern tip of the island and into the Mozambique Channel.

"On the boat they had this sealed, watertight container with a button," Dewar told me. "If they pushed the button, it would turn on a radio beacon that would identify where they were and would send out a distress signal."

"Just like the original Malagasy might have had," I said.

"Exactly. Well, they finally gave up and pushed the button."

A French coast guard ship came out from the Comoros and towed them back to the island of Mayotte, where they were promptly saddled with a hefty bill for the rescue. The crew then hired a local boat to tow the ailing craft to Madagascar, where, on September 5, the *Sarimanok* finally came to rest on Nosy Be, on the beach by the Holiday Inn.

"About a year later, a group of these people came back to try to raise money in Madagascar—which strikes one as a somewhat humorous effort—to refurbish the *Sarimanok* and memorialize it," Dewar said. "One of them gave a lecture in Diégo Suarez while I was in town. He delivered it in English, with simultaneous translation, to a crowd of about sixty, at least half of whom were under the age of twelve. I think they left disappointed in terms of finding anyone to take care of the *Sarimanok*."

But Jean-Aimé Rakotoarisoa, a leading Malagasy archeologist and a close friend of Dewar's, had a different take on what the *Sarimanok* voyagers had accomplished, Dewar told me. "They had done marvelous work, Jean-Aimé felt, solving problems that we archeologists had not been able to solve before. We now know that the first place settled in Madagascar was the Holiday Inn in Nosy Be, and we know that Americans must have settled the island first, because there we have proof: They built the Holiday Inn."

The hotel behind me on this beach is nothing like a Holiday Inn, unfortunately. The Carrefour ("Crossroads") is a dark, dingy, largely empty hotel that has no right to charge the comparatively exorbitant rate of sixteen dollars a night. Dewar and Darsot are paying a fraction of that at the Coconut Palm, but it's full, as is every other half-decent hotel in Sambava.

At first I was glad to be staying apart from the others. After a long day cooped up in Marcelin's Renault, I felt I could use a break from them, as I'm sure they could from me. But the longer I stay here, the more I wish I had a room at the Coconut Palm. At that first lunch, Dewar divulged most of the information I've received so far; perhaps he opens up more in off-hours. Getting the story out of him in the field is like getting change from a Malagasy taxi driver. I have no idea if our survey is going well or not, what he thinks of the sites we're visiting, or where he plans to go from here. He tells me nada.

Is he playing off Darsot's formidable silence? Is he loath to appear to be hogging the stage before his Malagasy counterpart? Does he want to withhold judgment about our work until the end? Is he pissed off about something? Does he dislike me? I have no idea, but it's beginning to frustrate the hell out of me. How can I tell his story if he doesn't give it to me?

Gazing down the beach, I watch four dogs parry and thrust at the surf. The tide is out, and dark, forbidding rocks rise out of the water behind them, as if climbing ashore. The dogs canter in my direction, confident and unaware, until they catch sight of me, a tall, still figure high on the sandbank beneath a cluster of coconut palms. They stop, stand stock-still for a moment, then turn around and trot, then run, whence they came. Two silhouetted figures, the first people I've seen tonight, sidle out onto the beach a half-mile down. They sit for a minute or two on a washed-up snag, then are gone, just like the dogs.

As I head back to my stinky, airless room, Dewar's words on that first day come back to me: "The real romance about Madagascar is the story doesn't get fixed. It just gets more complicated." Somehow the story doesn't seem particularly romantic just now.

8.

THE MYSTERY OF THE first Malagasy does indeed get more complicated, though, infinitely so, when one factors in the very strong African element in Madagascar's history and culture. Exactly what role did Africans play in the advent of Malagasy society?

In his monumental *Physical, Natural, and Political History of Madagascar,* published in 1892, Grandidier gave pride of place to Indonesia as the ancestral homeland of the Malagasy. The diversity of physical types and especially of skin color in Madagascar, Grandidier believed, was due to the immigration of light-skinned Indonesians and darker Melanesians, with a later, limited influx of Africans. But later scholars have pooh-poohed this theory. As Verin has written, adequately summing up the current thinking about Grandidier's views, "it should be stated emphatically that the ancestors of the Madagascans were of Indonesian and African origin and that the predominantly Indonesian character of the language . . . does not give us the right to deny the important part played by Africa in populating the country."

If your first sight of Madagascar is Tana, you might be inclined to agree with Grandidier. Most people in the capital look distinctly Indonesian in stature, skin color, and hair texture. Yet come down out of the central Highlands, which comprise perhaps a fifth of the island's land area, and you find that a more African physical type predominates, with people becoming generally taller, darker, and more curly-haired the closer you get to the coasts. This is a generalization, for great diversity in physical types can be found throughout Madagascar, whether in the "Asian" Highlands or along the "African" coasts. Such diversity can manifest itself even within a single village. But it's clear from even a cursory glance at the people of Madagascar that the African gene is far more widespread than Grandidier would have had you believe. And results from limited genetic studies indicate that gene frequencies reveal major contributions from both Indonesia and East Africa.*

African cultural contributions run as deep as its genetic contributions. By far the most significant is the cattle culture. Together, rice and zebu form the pillars of the Malagasy economy and a kind of cultural obsession as well. Again, the farther you move away from the

---

*No Melanesian gene has turned up, though a team of South African geneticists recently claimed to have identified a Polynesian gene in Madagascar.

capital toward the coasts, the more a preoccupation with rice (brought by Asians) gives way to one with zebu (brought by Africans). In the great plain of Imahamasina ("having power to make sacred") just west of Tana, for instance, rice paddies seem to go on forever, while farmers use the few cattle seen mainly to tramp down their fields prior to sowing. But in the semiarid savannas of the south, herds of cattle many thousands strong roam the parched earth, where the only rice paddies are narrow, struggling patches wedged into the deepest clefts between scorched hills.

Again, this is a generalization. Throughout Madagascar, the zebu, like rice, plays a profound, even defining role in the lives of Malagasy. As the nineteenth-century missionary John Alden Houlder wrote comprehensively if rather grandiloquently, "Savage and free on the far distant wilds and in the inaccessible depths of the forest, herded by thousands on the nearer and well-known feeding grounds, on sale at every market, [fed] to bursting in the village cattle-pen, drummed into the town at the annual festival at the head of a proud and joyful troop of rustics, killed, feasted on by every one, [the zebu] is closely bound up with the very life of the people."

Indeed, the Malagasy make as much use of the zebu as American Indians once did of the bison. In his 1915 book *A Naturalist in Madagascar,* James Sibree records an old Malagasy invocation given at circumcisions, which details how elders divvied up a sacrificial zebu: "The ox's horns go to the spoon-maker; its molar teeth to the mat-maker (for smoothing out the *zozoro* peel [zozoro is a kind of reed]); its ears are for making medicine for nettle-rash; its hump for making ointment; its rump to the sovereign; its feet to the oil-maker; its spleen to the old man; its liver to the old woman; its lungs to the son-in-law; its intestines to those who brought the ropes; its neck to him who brought the axe; its haunch to the crier; its tail to the weaver; its suet to the soap-maker; its skin to the drummer; its head to the speech-maker; its eyes to be made into beads (used in the divination), and its hoofs to the gun-maker."

Zebu are bound up with the Malagasy not only physically but metaphysically. As in East Africa, cattle in Madagascar are at once expressions of wealth, emblems of might and majesty, and symbols invested with a deep spiritual significance. This has been the case for centuries. In former times, Malagasy kings were saluted as *Ombilahy* (Bulls), an acknowledgment of both their power and that of zebu, which are the island's largest land animals. Today, the net worth of entire villages, particularly among southern tribes, may be wrapped up in cattle. "I have never known a man commit suicide if he lost his family, only if he has lost his zebu," writes Madagascar's only Antandroy novelist.

Tribes often keep cattle solely for sacrificial purposes. Funerals of rich or important individuals may feature the killing of well over a hundred animals, and Malagasy regularly sacrifice zebu at ceremonies to seek blessings of the ancestors or to thank them for blessings received. One such sacrifice even preceded the maiden flight of Air Mad's first jumbo jet. "Cattle are thus converted from mere beasts into a channel of communication with the ancestors," writes John Mack in his incisive book *Island of the Ancestors*. This connection between cattle and the razana is embodied in the special mark that a zebu owner, using a pattern or formula characteristic of his clan of kinsmen, cuts into each of his cattle; this mark is known as *sofindrazana* ("the ancestral ear"). Like the razana, "the zebu," say the Malagasy, "will lick bare stone and die in the earth of the place he loves."

The zebu-ancestor bond shows up most strongly in Sakalava customs, in which African cultural contributions are perhaps most pronounced. If the Merina are the most Indonesian and the Antaimoro the most Arab of Malagasy tribes, the Sakalava are the most African. Customs showing clear African origins abound in their rituals. These include tromba, funerary rites (especially for royal funerals), and the cult of *dady*, in which Sakalava memorialize past kings by preserving their fingernails, hair, and bones. Other African contributions are island-wide, such as the dugout canoe, the calabash-resonated cordophone, and, of course, the national preoccupation with cattle.

Most scholars feel that since zebu are deeply rooted in the culture of all tribes, settlers must have brought them to Madagascar at a very early date. Indeed, this notion, and the assumption that the cattle culture came by way of Africa, is evident in the Malagasy language. Though at least 93 percent of the basic vocabulary is Malayo-Polynesian in origin, the words for ox (*omby*), sheep (*ondry*), chicken (*akoho*), and guinea fowl (*akanga*) are of Bantu origin, as are most of the words relating to animal husbandry. These Bantu terms exist in all eighteen Malagasy dialects, further strengthening the argument for an early arrival from Africa. Noting this, Verin posited that the role Africans played in populating the island "must go back to the very origins of the Madagascan civilization."

Most Malagasy believe, as we've seen, that Madagascar had an aboriginal race known as the Vazimba, who looked different from the

Malagasy and were most likely Africans. As evidence, they point to simple stone tombs scattered across the Highlands, to which no tribe today lays claim; those tombs belong to the unknown Vazimba, the people say.* The Malagasy also point to the western tribespeople who call themselves Vazimba. Merina history holds that when the Merina first penetrated the Highlands in the twelfth or thirteenth century, they discovered the Vazimba already living there. The Merina intermarried with the Vazimba but eventually forced them out of the Highlands. The Vazimba fled west, where they live today.

Dave Burney, who has worked among the Vazimba near his site at Belo-sur-Mer, says "their hair is almost peppercorn hair like that of the Hottentots, and they have a whole different skin coloring. They look like Bantu Africans. And it turns out there's a tribe directly across from there on the Mozambique coast called the Zimba. Go figure, as they say."

Archeologists have gone and figured about the Vazimba. They claim that no archeological evidence exists that anyone, Vazimba or otherwise, inhabited the island before the proto-Malagasy arrived. So-called Vazimba tombs are merely Malagasy tombs so old that any knowledge of their occupants, much less their builders, has long since passed out of the collective memory of the Malagasy people. Scholars do believe, however, that the Vazimba were likely part of the earliest African migrations to the island, though, as with the Indonesians, both whence and when they came remain unknown. As Verin has written, "We know that the Vazimbas have a cultural affiliation to the Indonesians from the earliest times in Madagascar, but they constitute an enigma which has so far proved insoluble."

In the final analysis, most historians agree that Asians and Africans must have mixed very early on. Again, scholars have uncovered possible clues to this intermixing in the writings of medieval Arab geographers. The prolific writer Masudi, who lived in the tenth century, wrote:

> The end of their voyage and of the tribe of the Azd on the sea of Zanj is the island of Qanbalu, to which we have already referred, and the land of Sofala and Waqwaq, situated on the borders of Zangebar and at the end of this arm of sea . . . Just as the China Sea ends at the land of Sila [Japan] . . . so too are the limits of

---

*The famous Highland fossil site of Ampasambazimba literally means "at the tomb of the Vazimba."

the sea of Zanj to be found at the land of Sofala and Waqwaq, countries which produce a great deal of gold and other marvels. The climate there is warm and the ground is fertile. The Zanj have built their capital there. Afterwards, they elected a king whom they called Waklumi. This was the name given to all their sovereigns at all times.

If you recall, Zanj was the Arab name for the African coastal region south of modern-day Somalia, and thus the "sea of Zanj" implies the westernmost arm of the Indian Ocean. Qanbalu is believed to refer either to an island off the coast of Africa, such as Pemba or Mafia, or to one of the Comoros Islands. Sofala corresponds to the modern town of Beira on the coast of Mozambique, and Waqwaq is the etymologically mysterious name for Madagascar. Scholars have posited that Waklumi is a Bantu name. "Is it therefore possible," asks Verin, "that this passage by Masudi reflects an encounter between the Bantu and the Indonesians and a consequent intermixing?"

Another excerpt is even more thought-provoking. This one comes from the twelfth-century writings of the Arab geographer Idrisi:

> The Zanj have no ships in which they can sail. But vessels from Oman and other countries come to their coast on their way to the Zabaj Islands [or Zabej, that is, Sumatra] . . . These foreigners sell their goods and buy the products of the land. The inhabitants of the Zabaj Islands also go to the Zanj in large and small ships and make use of their goods in trade, in view of the fact that they understand each others' languages.

How could people of Sumatra understand the language of people on the east coast of Africa? By speaking the same Malayo-Polynesian language, of course—namely, the tongue that became Malagasy. How early the proto-Malagasy began traveling along the African coast and intermixing with Bantus there, and presumably on Madagascar, is unknown. But this passage suggests they were doing so at least until the twelfth century.

The consensus among scholars that Indonesians and Africans have intermarried for well over a millennium, perhaps since the very initial colonization of Madagascar, introduces yet another Malagasy mystery: How did the Indonesian language and culture win out over the African? What was, Mack wonders, "the process by which Malagasy culture achieved so comprehensive an overlay of Asian features when over half of its population, on the basis of physical and other evidence, are of African background . . ."? Like the Vazimba, this enigma has thus far proved insoluble.

9.

BY THE SAME TOKEN, how is it that the Arabs, who came much later, as much as a thousand years after the first Indonesian or Afro-Indonesian immigrants, and whose genetic contribution to Malagasy blood is limited, had such a deep and lasting impact on Malagasy culture?

The Arab migrations to Madagascar, which began in the ninth century with Arabico-Swahili traders sailing over from the East African coast, lasted until the arrival of the Antaimoro, a group of Arabs claiming ancestry from Mecca who settled on the southeast coast around Manakara at the close of the fifteenth century. The Antaimoro created the Sorabe ("Great Writings," from Arabic *surat* and Malagasy *be*), which included historical literature, such as genealogy and folkore. But they are perhaps best known for their mastery of the magic arts.

The Antaimoro emphasis on divination and magic, along with the fact that the Sorabe constituted the first writing that the Malagasy had ever seen, enabled Antaimoro scribes and priest-handlers to wield enormous influence on surrounding tribes. Flacourt reports that the Antaimoro ombiasy, who practiced the arts of divination, astrology, or healing, held the Antanosy tribe around Fort Dauphin under their sway:

> These ombiasses are very feared, not just by the people, who regard them as sorcerers, but by their leaders [Antanosy chiefs], who employed them against the French. . . . [The ombiasses] sent to the French fort baskets full of papers with writing, eggs laid on Fridays covered with written symbols and script, unbaked clay pots with inscriptions both inside and out, small coffins, dugout canoes, paddles . . . all covered with inscribed signs, scissors, and tongs.

This influence spread far and wide in Madagascar. An Antaimoro chief named Rambo was the ancestor of the Zafi-Rambo, who founded the Malagasy tribe known as the Tanala. Certain traditions of the Betsileo hold that that tribe's ancestors came from the east, which some scholars suggest hints at an Antaimoro or Zafi-Rambo origin. The anthropologist Aidan Southall claims an "insistent" link between these Arabs of the east coast and the ancestors of the ruling lines that developed in the southwest and west of the island. The Antaimoro ombiasy even left their mark on the Merina. Andrianampoinimerina, the great Merina king, relied on Antaimoro priests as advisors in astrology and other magic arts. Before the Latin script was introduced in 1823, Merina officials conducted some of the affairs of state in Arabic script, and Radama I even considered using it exclusively to write Malagasy.

Some scholars have proposed that the Antaimoro ombiasy, with their powers of divination, prophecy, literacy, and even sorcery, played key if unintentional roles in enabling the formation of individual states, including the great Merina state. "These powers certainly provided a focus for the growth of centralized power and influenced some of the cultural forms of symbolic belief and action in which it was expressed," writes Southall, "although in every case it had to be fundamentally based on critical local changes. . . ." Those changes included economic developments such as the increased emphasis on paddy-rice farming by the Merina and Betsileo, for instance, or the varying combinations of cattle-herding and dry-land farming among southern tribes. Mervyn Brown has gone so far as to suggest that, with better leadership, the Antaimoro tribe might have played a more dominating role in the island's history, comparable to that of the Merina or Sakalava.

So what was it that made the Antaimoro chiefs and their "mysterious ritual potency" (as Southall has phrased it) so influential? In essence, it comes down to the concept of *vintana,* or destiny. This idea, the most significant Arab contribution to Malagasy culture, is simply that everyone, by virtue of the date and time of his or her birth, inherits a particular destiny. It may be a lucky one (*tsara*) or an unlucky one (*ratsy*). Either way, it is one's for life and must be taken into consideration when planning all major events, such as getting married or building a tomb. Though the laws of vintana are eternal and unchangeable, an ombiasy in some cases may be able to reverse a bad destiny through a complex process of divination.

Vintana is woven into the very fabric of life throughout the island. Along with the concept of the ancestors, that of vintana forms the bedrock of the Malagasy cosmology. Just as most Malagasy would do nothing of importance without first consulting their forebears, so they would do nothing willingly that went against vintana. To get an idea of how thoroughly vintana has implanted itself in Malagasy thought, it is worth examining the concept in some detail, while acknowledging that we as Westerners have little hope of plumbing its truly profound depths. Like the fady to which it is closely affiliated, vintana differs from region to region, even village to village, so the portrait I give here is a composite.

Vintana is, above all, temporal in nature. It is no mystery why the names of the months and the days of the week in Malagasy are of Arabic origin (along with terms used in astrology, divination, and arithmetic). For the diviner's purposes, the year is divided into twelve lunar months, with each month beginning with a new moon and ending with its wane. The names of the months, beginning with the equivalent of our January, are Alahamady, Adaoro, Adizaoza, Asorotany, Alahasaty, Asombola, Adimizana, Alakarabo, Alakaosy, Adijady, Adalo, and Alohotsy. The days of the week are Alahady (Sunday), Alatsinainy (Monday), Talata (Tuesday), Alarobia (Wednesday), Alakamisy (Thursday), Zoma (Friday), and Asabotsy (Saturday). The Arabic origin of these is clear: Zoma, for instance, is Dschuma', or "congregation day," the sacred day for Muslims, while the root of Asabotsy is simply the Hebrew Sabbath, slightly altered in its transmission through the Arabic.

Each of the months and days of the week has its own destiny, and each of those destinies is mind-numbingly complex. Take Sunday, for instance. Being "the oldest one of the days," Sunday is more powerful than all the others. Since the word for Sunday, Alahady, is similar to the Malagasy word *malady*, which means to hear well or apprehend quickly, things done on a Sunday will be done expeditiously and well. At the same time, Sunday's power makes it potentially hazardous, and there are certain dos and don'ts for the day. The morning is good, and the day in general is a good one for sacrifices. But the afternoon can be evil and dangerous, and it is taboo to work or to bury someone on a Sunday. If a burial is unavoidable on that day, it must take place when the sun is "dead," that is, after sunset. Otherwise it may result in the death of young people, who have little resistance to the day's powerful forces.

Each day has its own color, and Sunday's is white. This decidedly does not suggest purity, for the Malagasy concept of white is opposite to the Western one. "To white one's neighbor" (*mamotsifotsy namana*) is to slander; "to spit white" (*manao fotsy rora*) is what one does after losing in court or begging unsuccessfully. An unreliable person is one who "says white words" (*olona fotsy teny*). Malagasy call a Westerner vazaha, or foreigner, for to call him a "white man" would be to insult him in the worst way. The color's pejorative connotations—whose source may lie in a pervasive Malagasy fear of albinos—come into play on Sundays, when it is taboo to consume the meat of white animals or white-colored foods such as milk, indeed to come into contact with anything white. Paradoxically, when a baby is born on this day, diviners sacrifice a white fowl and rub it on the newborn to counteract the bad forces of the vintana.

Other factors further complicate successful divination. For one thing, each month has twenty-eight destinies that correspond to the twenty-eight days of the lunar month, and, as if to confuse matters, are named after the months themselves. Each day also has twenty-eight different destinies within it, beginning with the first destiny of the day at sunrise. Known as Alahamady's mouth, this time is an auspicious one both to be born and to practice divination, which is why the Malagasy sometimes call diviners *mpanandro,* or "makers of the day." And then there are joint destinies, when a child is born on the birthday of another member of the family. This can be hazardous, because the child's destiny has collided with that of a sibling or parent. Unless the diviner can diffuse the vintana, those with joint destinies must be separated, that is, the child must be brought up by someone else. Finally, when the month and the day have the same vintana, the destiny's power is doubled and therefore more perilous. "If the birthday is at Asorotany's beginning [meaning one of the two days of any given month that have the name Asorotany, which corresponds to our April] or at Alakaosy's beginning [Alakaosy equaling our September], and one is in one of these two months already, the child will be a very great danger and threat to parents, relatives, and all the people in the village," writes Jørgen Ruud in *Taboo.* "It is hard to get a diviner who is willing to change such a destiny."

A sign of how deeply the Arab concept of vintana entered the Malagasy mind-set and how closely the Malagasy adhered to its tenets is the fact that in olden times, babies born on unlucky days were put to death. They were placed face-down in a shallow dish of lukewarm water, left to be trampled to death before a cattle pen, or placed alive in an anthill. Olive Murray Chapman, an Englishwoman who traveled through Madagascar in 1939, met an Antandroy pastor whose face was horribly scarred from having been stepped on by zebu as an infant. (A Norwegian missionary rescued him and brought him up.) Later, in Sakalava country, Chapman learned of a childless couple who, by following an expectant young couple into the forest on a Thursday, an unlucky day, had plucked a baby boy from an anthill. "The top of the heap of earth had been knocked off, the baby laid on the remainder, and the top lightly replaced," Chapman wrote, "but so quickly was the child removed by his rescuers that he was mercifully unharmed." The new parents named him *Trabonjy*, meaning "Overtaken by salvation."

Europeans were naturally horrified by this custom. In his 1822 *History of the Island of Madagascar*, the vituperative Englishman Samuel Copland prefaced his comments about the practice by declaring that "the most horrid and execrable feature in their ritual remains to be related, the observance of which can only be accounted for on the general principle that man, when left to the dictates of natural reason, is liable to fall into the most dreadful errors." In the mid-nineteenth century, English missionaries succeeded in abolishing the practice in regions where they preached, though how long it persisted in remote areas is not known.

Vintana has not only temporal but spatial intricacies. To help keep track of them, the people of Madagascar use the domestic house as a metaphor for Malagasy cosmology (see diagram on page 234). Susan Kus, a social anthropologist who has worked among the Betsileo, has written regarding this metaphorical use of the house that "in societies where literacy is unknown or of limited service in daily activities, it is concrete and contextualized symbols and metaphors rather than abstract and decontextualized logical manipulations that serve as the tools for assigning and apprehending order and meaning in one's life." Stating it more simply, one early missionary wrote, "There is almost as much symbolism in a Betsileo house as in an Egyptian pyramid." The same can be said for the houses of all Malagasy, which are invariably rectangular in shape. There is a good reason why the round hut of Africa is missing in Madagascar. As John Mack has written, "The Malagasy conception of destiny with its roots in Islamic cosmology requires rectangles to make it 'thinkable.'"

The Malagasy house serves as a kind of vintana calendar, in which one can read and interpret the destiny of every hour of every day. The twelve lunar months are mapped on the four walls, beginning with Alahamady in the northeast corner and working around in a clockwise fashion. The months in the four corners are known as "destiny mothers," while the other eight arranged around the four walls are known as "destiny children." In order to accommodate the twenty-eight days of the month, destiny mothers are assigned three days each and destiny children two. As we've seen, each month as well as each of the days associated with each month have their own individual character. Diviners take this into account when determining vintana.

There are even more complications. First, there is the dynamic force known as *rohontany* that continually circulates around the house, changing its nature every day. Circulating clockwise around the house, it flows against the movement of the sun, which in the Southern Hemisphere rises from the east to the north before setting in the west. To go against rohontany is perilous. Thus, it is taboo to arrive at a house from the north and then walk along its western side. "The diviners watch carefully over this rule," Ruud says, "and they expect others to do the same."

The Malagasy are also mindful of opposing and agreeing destinies, again using the house as a model. Opposing destinies will be seen by drawing a line from the position of each month on the walls through the central post of the house to the corresponding month on the other side. In other words, Alahamady in the northeast corner opposes Adimizana in the southwest corner, like boxers in the ring. If lovers are found to have opposing destinies—say, she was born in Alahamady and he in Adimizana—their parents will refuse to allow them to marry. If, on the other hand, they have agreeing destinies—determined by drawing lines that pass closest to the central post without touching it; thus Alahamady agrees with both Asombola and Alakarabo—then the marriage will be seen as auspicious and pursued forthwith. Malagasy arrange all momentous events in their lives around these positive and negative vintana. In other words, if a farmer has Alahamady as his vintana, he will not plant his fields, start building a house, or marry off one of his children on an Adimizana day, which is opposed to his own destiny. Rather, he'll aim for either Asombola or Alakarabo, which are in agreement with his vintana.

The final element of Malagasy cosmology as visualized through the house is orientation, which is also an Arab contribution. In general, Malagasy orient their houses to the four cardinal points. The door is in the west wall, for west is the direction of ordinary life. East, associated with the dead, is the direction of the sacred and of prayer. If there is a door in the east wall, it is only opened in order to take a corpse out of the house. North is the direction of honor; this is where the father of the household sleeps. South is the direction of the impure; this is where the chickens sleep. Where directions overlap is

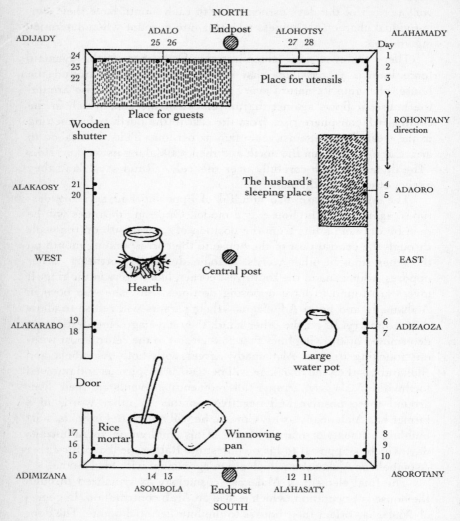

Plan of a typical Malagasy house, with the metaphorical elements used to determine vintana (destiny) mapped around the exterior.

important. Thus, the northeast corner, which combines the sacred and the honorable, is the most revered, the *zoro firarazana* or "corner of prayers." This corner is reserved for the ancestors, and as such, proper etiquette upon entering a Malagasy house is to step deferentially to the southwest.

As with Indonesian and African contributions to Malagasy culture, vintana, the practice of divination, and other Arab contributions have been fully absorbed into the Malagasy melting pot. Very few vestiges of Islam proper remain in eastern Madagascar, where people of Arab extraction settled until the end of the fifteenth century.* In fact, tribes of Arab descent along the east coast, including the Antaimoro, have little knowledge of some of the central features of the Muslim religion, including prayer, alms, the mosque, the name of Allah, or the pilgrimage to Mecca.

10.

NOR PERHAPS DID the Rasikajy, the Arabic people whose archeological remains we've come to study, nor certainly their descendants here along the Vanilla Road, down which we're now racing once again, on our way to the Rasikajy site known as Bemanevika. Marcelin is gripping the wheel like the lapels of a miscreant, staring at the road ahead with an intensity almost painful to watch through the rearview mirror.

Again, I concentrate on the scenery: Rice paddies so green they seem electrified. Discrete plantations of pineapple, banana, papaya, and raffia, an endemic palm used for construction. Higher up, hillsides of secondary forest, with towering bamboo stalks like rocket flares, and more remnant stands of rain forest. To the west, the Marojejy Mountains, scraping a sky hung with clouds like cotton balls.

Every few minutes we pass through another Betsimisaraka village. The houses are invariably small, rectangular affairs with roofs of thatch, though a few have tin roofs. The monotonous dun gives way to sudden splashes of color: a neon-green lamba, colorful squares of laundry laid on the road to dry, a giant flame tree with brilliant orange blossoms. Everything screams "indigenous," but then: a

---

*More recently, practicing Muslims from Pakistan and the Comoros have settled in the northwest of the island.

Chicago Bulls T-shirt. Malagasy turn to look into our car even while carrying heavy loads or peeing by the side of the road. They are healthy, often smiling, always curious. I see few old people, but there are lots of children. I'm afraid we'll hit one of them speeding through these hamlets, for nothing slows Marcelin except zebu and potholes.

My right shoulder feels like an open wound that gets punched every few seconds. Dewar and I are crowded into the backseat as usual, but now there's another person between us—Alison Richard, his wife. She's a powerful presence, as one of the leading lemur researchers and conservationists in Madagascar, not to mention the provost of Yale University, in charge of a budget reaching into the hundreds of millions of dollars. Happily, though, she's not a large woman; in fact, she borders on the petite. Nevertheless, she's taking up room that wasn't there to begin with, and every time Marcelin slams into another pothole, the door jamb buries itself a bit deeper into my deltoid muscle. At least he put the rifle in the trunk.

Dewar doesn't seem to notice the discomfort. He's got his face buried in Verin's notes, reading about Bemanevika. Dewar considers Bemanevika the most important site in this part of the northeast coast. Verin and Ramilisonina spent only a single day here in 1969, but it was long enough to determine that Bemanevika, at roughly twenty-five acres, is one of the largest archeological sites in the region. It is also the earliest site on record for the northeast coast, from Vohemar in the north to the Masoala Peninsula in the south. Verin dates it to the eleventh century, based on a single piece of Islamic sgraffito pottery, a yellow-and-green sherd with a red glaze that he found in the deepest level of one of the five test pits he dug. He also revealed a heap of iron slag, indicating that Bemanevika was an iron-making center, and a well shaft three feet in diameter carved out of chlorite-schist. ("The place is holy, and coins are left there," Verin wrote of the well.)

Bemanevika is also interesting for its links to Mahilaka, Madagascar's first city. Discovered by French colonials on the west coast across from Nosy Be, Mahilaka at its height in the eleventh to fourteenth centuries had an estimated 5,000 to 10,000 people. At a time when most villages on the coast were simple affairs with no stone architecture (villages did not yet exist in the Highlands), Mahilaka featured a stone wall enclosing some 170 acres of stone houses, workshop areas, a walled inner precinct, and at least one mosque. Near Eastern sgraffito and Far Eastern celadon pottery, which also turn up at contemporary sites along the African coast, suggest that Mahilaka was an entrepôt, where Malagasy exchanged island goods such as chlorite-schist, tortoiseshell, gold, and perhaps gum copal, iron, and wood for ceramics, glass vessels, beads, and possibly cloth. In their test pits at Bemanevika, Verin and Ramilisonina unearthed pottery just like that found on the other side of the island at Mahilaka, including, besides the sgraffito, imported celadon and thick-rimmed local ware.

Next to the time of earliest settlement, this period of Madagascar's past, roughly the eleventh to fourteenth centuries, most fascinates Dewar. For it is a time when an enormous expansion of activity took place throughout Madagascar. Save for a few gaps, the entire coastline was now settled, and most of the hallmarks of Malagasy traditional society were visible, including rice farming, livestock herding, and marine fishing. More importantly, toward the end of this period, people finally began to settle the Highlands.* What Dewar would like to know is: To what extent did this commercial activity along the coasts sponsor or motivate penetration of the interior? The link is not obvious. While earthenware ceramics of all coastal occupations are sufficiently similar to suggest a common tradition, the earliest pottery of the Highlands is distinct enough in form and decoration that its origins remain unclear.†

An hour after leaving Sambava, we arrive at Bemanevika village, for which the site is named. The village is substantial, with rows of thatch-roofed houses two or three deep on either side of a dusty main street. The sky has begun to cloud over. There's a cool breeze, and the air temperature is just on the edge of demanding more than a T-shirt. Darsot leads the way into the fokontany office, to which visitors must present themselves in the larger villages and towns. The office has a desk, a folding metal chair, and a faded poster of a young, uniformed Kim Il Sung standing amid a group of smiling lieutenants brandishing submachine guns—a leftover no doubt from the socialist regime of Didier Ratsiraka, who had close ties with the North Korean leader.

The mayor of Bemanevika arrives. He is a pleasant-looking man with a round face and thin mustache. Though barefoot and wearing an old panama-style hat, he bears the look of authority, enhanced by a black pen clipped into the breast pocket of his striped, button-down shirt.

The mayor delivers a blow. He knows of the site with the stone

---

*Of literally tens of thousands of archeological sites discovered so far in the Highlands, not one dates to before the thirteenth century.

†Yet another mystery of Malagasy history: Who first settled the mountainous center of Madagascar?

well and will take us to see it, but a coconut plantation stands there now, he says. It went up in 1978, nine years after Verin's visit.

"From the sounds of it, the site may be completely destroyed," Dewar says as we head out to the car, his disappointment palpable. "Since that is one of the three main sites in our grant proposal . . ." There is no need to finish his thought.

The mayor eases into the front seat with Darsot and we're off. Soon after leaving the village, we enter the plantation. As the Renault bounces along a little-used road between towering palms, I can feel coconut husks hitting the thin floor of the car beneath my feet. Many of the palms are bent double, a west-facing elbow joint halfway up their trunks attesting to the ravages of a cyclone that swept through here a decade or two ago.

"Landsat could pick up this plantation," Richard says. "Easily." She's trying to make conversation, but her husband remains silent.

He stays mum as an elderly local guide, whose new woven hat contrasts sharply with the shirt that literally hangs in tatters off his back, leads us to the site. We wade through the shallow end of a narrow lagoon and scramble onto a piece of raised ground several football fields long by one wide. Like Benavony, the site's advantages to early Arab traders is obvious. It sits on the north bank of a large river, the Bemarivo, a half mile from the ocean, which washes onto a protective dune backed up by that lagoon we just crossed. To the west, between the site and the village of Bemanevika, lies a broad marsh filled with watery rice paddies. We're standing on the only piece of dry ground in the area.

As we feared, coconut palms now totally cover the site. The guide shows us the spot where Verin discovered the chlorite-schist well in 1969, but there's no sign of it. The holy well was bulldozed, along with the rest of the site, to make way for what Richard deems "the moral equivalent of a supermarket." We do find the iron slag pile, hidden beneath a stand of bushy trees festooned with dozens of spider webs, but Dewar gives it the most cursory glance before wandering off to stare at the ground. Only later, as we tip back water coconuts the guide has kindly cut open for us with his machete, does Dewar finally speak.

"Well, short of setting out with a campaign of trying to figure out what is lost, there's no point being here. As a zone of research, it has essentially been wiped out. It's a dot on the map, and frankly it will probably stay that way forever."

He falls silent, and this time I can't fault him.

We decide to hike back to Bemanevika rather than wait for

Marcelin, who's gone off hunting. To get to the village, we have to cross those marshy rice paddies. The Rasikajy supposedly built an embankment across the wetland in order to transport heavy goods such as the pieces of that chlorite-schist well, but there is no sign of it now. Instead, we have to negotiate submerged tree trunks thrown down into the muck to serve as bridges. Each step features a careful exploration with the toe, a weak attempt to gain a foothold, and a throw-caution-to-the-winds step forward. Richard told me her husband wears sneakers even in three feet of snow; well, he's paying the price now, poor soul. In his smooth-soled tennis sneakers, Dewar cannot keep out of the soup. Time after time he plunges up to his waist in the custardlike mud.

These "bridges" have always challenged vazaha. "One of these bridges, consisting of the round trunk of a young tree, was so slippery that, when in the middle of it, I slipped off, hurt my shin, and fell on my back into the thick mud below," wrote an English missionary in the last century. "This happened at a rather unfortunate time, for I had just put on a clean suit of white clothes before entering Fenoarivo." Even Malagasy could not always hold their own. Sibree recalled crossing "deepish streams by bridges of a single round pole, a foot or two *under* water, a ticklish proceeding, which all our luggage bearers did not accomplish successfully." A Malagasy proverb holds that "it's wrong to say there's a bridge, and yet be afraid to cross it." Of this saying, one of hundreds he collected, John Alden Houlder wrote, "These rickety structures beggar description. We must leave them alone, and allow the proverb to suggest more than we can write. To ourselves it calls up scenes of hope and fear, of terror and dismay, and anon of ludicrous disasters that, alas! have given rise to well nigh uncontrollable laughter and endless mirth."

Indeed, the locals appear much amused at our travails. A middle-aged woman bearing a gigantic load on her head sashays past on one of these submerged trunks as if walking on a plank bridge. Then, as if to rub it in, she keeps turning *as she is walking* so as to not miss a moment of the spectacle. The mayor himself can't hide his enjoyment, even as he and the guide search for a firmer path across matted marsh weeds or offer Richard an arm for support. Darsot, although a city dweller, seems to be managing just fine, and my hiking boots are serving me well. But in the hour and a half it takes us to reach Bemane-vika—Marcelin is there; three birds—our unfortunate leader goes in again and again. Not a good day for him, all things considered.

11.

THE BETSIMISARAKA mayor, the Rasikajy descendant Marcelin, our Merina companion Darsot—all are part of the great melting pot that is Malagasy culture. The pot, as we've seen, blends Indonesian, African, and Arab ingredients spiced with French, English, and South Indian influences, resulting in a creolized culture not unlike that found in the Caribbean. Madagascar, the folklorist Lee Haring has observed, "has combined diverse imported elements to create something discontinuous and new, which could not have been predicted from its origins and which is no mere assemblage of foreign influences." Such creolization has taken place in its own way within different regions of the island. Marcelin's people here on the northeast coast, for example, syncretized diverse elements of immigrant cultures in a way different from, say, Darsot's people in the Highlands or James Bond's people in the deep south.

The real paradox of Malagasy culture is that despite its inherent diversity, brought about by those continual borrowings from other lands over the millennia, an undeniable unity exists. This paradox is reflected in the plasticity of Malagasy tribes. As Haring has written, "The ethnic map of Madagascar is a variable one, existing in people's minds." Tribal names themselves are suspect. French colonials imposed the name Mahafaly ("having the ability to taboo"), which tribal members apparently only use to label themselves for outsiders. "Sakalava," too, has disputed derivations and is also thought to have come from the outside.

The Malagasy themselves are loose about ethnic labels. Dewar once visited a Sakalava fishing village in the far north with Ramilisonina. "Ramili asked the village president what ethnic group he belonged to," Dewar told me. "The president said, 'Well, my mother was Merina and my father was Betsileo, but since I find myself in the land of the Sakalava, I call myself Sakalava.'" As Aidan Southall has written, "The relative meaninglessness of the current ethnic labels and divisions applied to the peoples of Madagascar . . . reflects their basic unity."

One of the best explanations for the essence of Malagasy culture comes from a wife-and-husband team of researchers who themselves represent this cultural amalgamation. The American Susan Kus and her Malagasy husband, Victor Raharijoana, two of the leading social anthropologists working in Madagascar today, have written that while "Malagasy culture may be . . . a hybrid culture whose elements show a rich borrowing from elsewhere and from different times, it is nevertheless a culture whose integrity has been forged in place through local practice and interpretation."

Written Malagasy symbolizes this unique hybridization perhaps more than any other aspect of Malagasy culture. Spelling combines Malagasy, English, French, and Arabic elements. When Welsh missionaries adapted the Latin alphabet to the Malagasy language in 1823, Radama I, trying to appease the two foreign powers then maneuvering to assume economic and religious control of Madagascar, decreed that the consonants would have their English values and the vowels, with two exceptions, their French ones. The two exceptions are the terminal *y*, which came from English,* and the letter *o*. When missionaries put Malagasy to the roman alphabet, they borrowed the romanized spellings *ao* for the "o" sound (as in "low") and *o* for the "oo" sound (as in "loot") from Arabic.

The melding of languages continues today. Malagasy politicians are wont to season their Malagasy with pinches of French, as in *"Ny extrémistes no atahorako,"* "It's the extremists I'm afraid of." In bourgeois Merina society, which mandates bilingualism, children learn to switch between Malagasy and French in a single breath, as this amusing rhyme attests:

> *Un, deux, trois: c'est gai!*
> *Et quatre, et cinq, et six: c'est gai!*
> *O Randria Maola [crazy teacher]*
> *Maola Firantsa [crazy France]*
> *Aza mandehandeha [don't move]*
> *Pas permis*
> *Pi pan do*
> *La re mi re mi re do*

The Malagasy language and its offspring, the verbal arts, offer the richest and most abundant clues to the nature of the people of Madagascar, because they most distinguish the Malagasy from any other people.

Vazaha have struggled to express the nature of Madagascar's mother tongue. Their descriptions have ranged from "a very soft and musically sounding language [that is] the 'Italian of the Southern

---

*Rumor has it the *y* was chosen from letters then available in the British-built print shop in Tana.

Hemisphere,'" to "a fine, rackity-clackity, ringing language which sounds not unlike someone carelessly emptying a barrel of glass marbles down a stone staircase." Unstressed syllables tend to evaporate, as in "Merina," which sounds like "mairn." One guidebook advises visitors who attempt to speak Malagasy to "swallow as many syllables as you can and drop the last one." For despite the valiant efforts of the Welsh missionaries who reduced the language to a written form in the early nineteenth century, Malagasy as she is written bears little resemblance to Malagasy as she is spoken. For example, Antananarivo is pronounced "Tananarive," which is the way French colonials spelled it, or simply "Tana."

Pronunciation, like the language itself, is a product of creolization. The vowels, as we've seen, are pronounced essentially as in French or Italian, with the exception of *o*, which is pronounced "oo," as in *veloma*, "ve-*loom*," meaning goodbye. The combination *ao* matches the sound of the French or Italian *o*. (Malagasy vowels are never pronounced as diphthongs, as in *toy*.) Consonants have their English values, though *h* is only slightly aspirated, *r* is a trilled sound, and *j* equals our "dz" sound, as in *adze*. Lexicographers describe the clusters *tr* and *dr* as, respectively, "voiceless and voiced blade aveolar affricates," which essentially means they are combination sounds you make using the tip of your tongue just behind your upper front teeth.

There are other things one should know before trying to speak Malagasy. First, the language offers no morphological indications of plurality. Thus, *jiolahy* can mean "bandit" or "bandits," depending on the context. Second, grammatical gender does not exist in Malagasy. *Izy*, for instance, can mean he, she, it, or they. Finally, the letters *c*, *q*, *u*, *w*, and *x* are altogether missing. When the Malagasy borrow words from other languages that use these missing consonants, they replace them as follows: *k* or *s* for *c*, *k* for *q*, and *ks* for *x*. Thus, "pencil" becomes *pensilihazo*.

Borrowed words are common in Malagasy. As one might expect, most of them are English or French. Thus, after attending a *kilasy* at your *kolejy*, you may dump your *boky* and have a lunch of *lasopy* and *salady*, followed by a glass of *lalikera* and a *sigara*, then ride your *bisikilety* to a *konserta* put on by an *ofisialy governemanta komity*. In the last century, some Malagasy were so taken with English missionaries that they took their names, placing the honorific *Ra* before such surnames as Johnson (*Rajaonisaonina*) and Wilson (*Raoilisaonin*). They even assumed first names, from Miss Daisy (*Radezy*) right on up to Victoria herself (*Ravikitoara*). The Malagasy appropriated whatever words—

or even partial words—that tickled their fancy. Hearing of the German politician Prince Otto Eduard Leopold von Bismarck's cunning exploits, the Malagasy made a root of the first syllable of his surname, creating the verb *mbizy*. *"Bizio izy!"* they'd say, or "Bis(marck) him!" Referring to this particular borrowing, one missionary, a proud speaker of proper English, wrote, "I most certainly wash my hands of any alliance with the good folks in making 'roots' of this kind."

The Malagasy are particularly adept at using roots to create whole forests of words. The English cleric Herbert Standing once drew up a list of words created from the root *vely* or "struck" (which itself strikes one as a curious choice for a missionary):

vely	struck
mively	to beat (a drum, etc.)
fively	an instrument used for striking
mamely	to strike or abuse
mpamely	the striker or abuser
mifamely	to strike each other
mampamely	to cause to strike
mampifamely	to cause to strike each other
velezina	being struck
amelezana	cause, time, place, etc., of striking
famelezana	ordinary instrument, time, place, etc., for striking
ampifamelezina	made to strike each other
hampifamelezina	will be made to strike each other
hampifamelivelezinareo	will be made by you to keep on striking or abusing each other

As you can see, Malagasy words can wind up being a mouthful, even if they're just a bite (*ambilombazana*) or even a crumb (*sombintsombiny*). A stroll (*fitsangantsanganana*) to the town hall (*tranompokonolona*) at midday (*antoandrobenanahary*) with a member of parliament (*solombavambahoaka*) might entice (*fanambatambazana*) or even dazzle (*mampipendrampendrana*) you, though it might leave you at a loss for words (*miambakavaka*).* Clearly, translating some English words into a single Malagasy term has proved challenging, and the

---

*Proper names can get particularly unwieldy. When he assumed the Merina throne in 1787, for instance, Rambosalama took the name Andrianampoinimerinandriantsimitoviaminandriampanjaka, which history has thankfully shortened to Andrianampoinimerina.

reverse is also true. The Malagasy term *miala manдry*, for example, consists of *miala* ("leave, go away") and *manдry* ("lie down, go to sleep"), yet used together they mean "to spend the night away from home and yet be back in the early morning as if never having been away."

Some of the longest words in Malagasy are onomatopoetic, and many of them blast our snap, crackle, and pop to smithereens:

hehy	laughing
mifofofofo	to blow hard
mipitripitrika	to sputter as grease in a frying pan
дoboдobohana	to be made to thump like a drum
mibitsibitsika	to whisper
misaosao	to make a rustling noise as of silk
mibobobobo	to make a bubbling noise

The words for taste are equally creative. Some resemble our own; the Malagasy know, for instance, when something's mouth-watering (*mampitelin-дrora*, "making the mouth water") or finger-licking good (*mampilela-tanana*, "causing to lick the hands"). But what, pray tell, do the following tastes taste like?

tsy hani-mpandova	not to be eaten by an heir
mampiteny ny moana	making the dumb to speak
mitsatoka amy ny tranon' aina	piercing into the house of life

Literal translations of many terms offer illuminating clues to Malagasy character and customs. Older people are often called out of respect *mainty molaly*, "the black soot." This term derives from the Merina practice in former times of deliberately not clearing away the strings of soot that hung from the rafters of their chimneyless wooden houses. The more soot you had, the longer your family had lived in the dwelling and thus the more established and hallowed it was.* The word for hypocrisy, *fihatsarambelatsihy*, means "the becoming good by spreading a mat," a reference to the time-honored Merina practice of laying down a good mat when visitors arrive at the house, to hide the

---

*Foreigners were often caught off-guard by the sheer quantity of soot. Joseph Mullens, who in 1874 became the first Englishman to make the overland crossing from Tana to Majunga, wrote of one evening meal that "everything would have gone well, if the long strings of venerable soot hanging from the roof had not diverted themselves by dropping contributions into our rice and tea."

old and worn ones the owners use every day. The term for the marriage of two closely related people, *lova-tsi-mifindra*, means
"inheritance-not-removing," a sign of how deeply the Malagasy cherish kinship bonds (not to mention wealth).

Some Malagasy terms literally tell it like it is. The word for taking
an examination means to "fight at writing," while the verb "to bar-
gain" translates as "a fought-out sale." The word for paying, *mandoa*,
is the same as that for vomiting. Since nothing ever cost over a million, the word for that number, *tapitrisa*, means "the finishing of
counting." The finishing of freedom—that is, to become a slave—was
to be *very* or "lost."

Other words are literally delightful. The word for sun, *masoandro*,
means "the eye of day," while that for glory, *voninahitra*, translates as
"flowers of the grass." When an earthquake strikes, "the whales are
turning over." A turkey is *vorontsiloza*, which means "the not terrible
bird." Did the first Malagasy to encounter this creature, with its gob-
bling call and formidable crest and wattles, flee in terror, only to later
discover its harmlessness? Or did they believe, before they actually
tried it, that such a homely bird must taste terrible? It was not unusual in former times for names of things to be changed on occasion. The
word for the silkworm moth was *soherina* until Queen Rasoherina
took that name when she assumed the Merina throne in 1862. Since
the people could not continue calling a lowly insect by the same name
as that for their exalted sovereign, they gave it a new one, *zana-dandy*,
which means "offspring of the silk."

The Malagasy have particularly inventive terms for the time of
day, which reveal the degree to which the largely agricultural Malagasy remain tied to the land. When "the eye of day" sets, it is dusk or
maizimbava vilany, which means "darken the mouth of the cooking
pot," while the wee hours of the morning are known as *misafo helika
ny kary*, "when the wild cat washes itself." The following comes from a
translated list Sibree made of terms that rural Malagasy use for
approximate times of the day:

12:00 A.M.	Halving of night
2:00 A.M.	Frog croaking
3:00 A.M.	Cock-crowing
4:00 A.M.	Morning also night
5:00 A.M.	Crow croaking
5:15 A.M.	Glimmer of day
5:30 A.M.	Diligent people awake

6:00 A.M.	Sunrise
6:15 A.M.	Dew-falls; also: Cattle go out (to pasture)
6:30 A.M.	Leaves are dry (from dew)
6:45 A.M.	The day chills the mouth [winter months only]
8:00 A.M.	Advance of the day
9:00 A.M.	Over the purlin [a horizontal roof beam that supports the rafters]
NOON	Over the ridge of the roof
12:30 P.M.	Day taking hold of the threshold
1:00 P.M.	Slipping of the day
1:30 to 2:00 P.M.	Decline of the day
3:00 P.M.	At the place of tying the calf
4:00 P.M.	At the sheep or poultry pen
4:30 P.M.	The cow newly calved comes home
5:00 P.M.	Sun touching (i.e., the western wall)
5:30 P.M.	Cattle come home
5:45 P.M.	Sunset flush
6:00 P.M.	Sunset (literally, "Sun dead")
6:15 P.M.	Fowls come in
6:30 P.M.	Dusk, twilight
6:45 P.M.	Edge of rice-cooking pan obscure
7:00 P.M.	People begin to cook rice
8:00 P.M.	People eat rice
8:30 P.M.	Finished eating
9:00 P.M.	People go to sleep
9:30 P.M.	Everyone in bed
10:00 P.M.	Gunfire

If the language itself is expressive, what the Malagasy make of it is remarkable for its range, depth, and sheer inventiveness. Though largely unknown to the outside world, the verbal arts of Madagascar, including oratory, fables, riddles, proverbs, and poetry, arguably rank among the richest the world's peoples have produced and deserve far wider recognition. For our purposes, they get to the heart of who the Malagasy people are and even give clues to their origins. As Sibree implored of his fellow missionaries in the *Antananarivo Annual and Madagascar Magazine*'s inaugural issue in 1875, "It is most desirable that any Traditions, Legends, Fables, or Folk-Lore that may be met with should be preserved, as throwing light on the origin of the different tribes."

The depth of Malagasy verbal arts grows out of the depth of the

Malagasy character. In *Hainteny,* his insightful book on the indigenous poetry of Madagascar, the ethnographer Leonard Fox defines two phrases that encapsulate the Merina attitude toward existence: *mamy ny aina,* "life is sweet," and *tsihibelambana ny olona,* "people constitute a great, broad mat." (One can safely argue that all Malagasy tribes adhere to this way of thinking to one degree or another, which serves as yet another element that unifies them.) The first phrase embodies the Buddhistic belief in enjoying life and what it has to offer without complaint and in harmony with nature, which is considered a priori as sacred. Sacred, too, are the ancestors, who are included along with the living in the metaphorical notion of humanity as a large, woven mat. The devotion of all Malagasy to their forebears forms the bedrock of the Malagasy worldview. As the French historian Hubert Deschamps put it, "A society and a morality inherited from the ancestors and modeled on a stable, just, durable world constitute the foundations of the wisdom of the ancient Malagasy."

This wisdom is expressed in Malagasy behavior, which places enormous stock in being modest, courteous, and respectful of one's elders. While the common farewell, for example, is *Veloma,* which literally means "May you live," to a superior one should say *Trarantitra,* "May you live long." When addressing the dead, as Malagasy do during bone-turnings and on many other occasions, one of the most polite turns of phrase is *Tompoko, haniko rofy,* "Sir, I eat (your) illness," according to Jørgen Ruud. "It is as if one simulates that the deceased is still alive, and the relatives ask that they may take upon themselves his illness in order that he may not die," Ruud writes. It is indelicate, indeed fady, to say outright that someone is dead. Instead, Ruud notes, "one must say that he is seriously ill (*marary mafy*), and everybody will grasp the meaning." Other acceptable terms denoting the recently deceased include *naratra* (wounded), *folaka* (broken), *very* (lost), and *Tsy hita intsony* ("He/she who never will be seen again").

Malagasy will go to extraordinary lengths to avoid seeming impolite. Sibree recalled the time his wife brought out a good-sized plum cake, cut a slice or two, and then held out the plate to a Malagasy guest. To his astonishment, the man quietly took the entire cake and started to eat it. "But finding himself, after a little time, rather embarrassed by its quantity, and that it was a good deal more than he could then comfortably manage, he gradually stowed it away in his pockets, remarking that his children would like it," Sibree wrote. "We altered our way of handing cake to native friends from that date." Malagasy themselves have to be on their guard against other Malagasy, who out of politeness must say yes to all questions. The Malagasy naturalist Guy Ramanantsoa, who was trying to arrange boat transport for himself and Alison Jolly during a visit to Nosy Be, told her what happened when he asked residents when he should expect high tide. "The villagers said . . . that high tide is two o'clock in the afternoon. I misheard and said, 'Ten o'clock?' They said, 'Yes, ten o'clock . . .' then added in a whisper, 'Two o'clock.' "

Such courteous behavior holds true even within large crowds. Jolly tells a delightful story of a political demonstration she witnessed in the capital during a tense time in 1991: "I . . . found a crowd of tens of thousands of cheerful people, mostly wearing pink sisal sunhats, listening to speeches for about three hours in the main street before they all went home to lunch." Of course, violent uprisings have occurred in Madagascar. But many familiar with the country might agree with a writer from the *Madagascar Times,* an English-language newspaper printed in Tana in the last century, who wrote after attending a royal *kabary,* a publicly held proclamation by the reigning monarch, that "any foreigner, be he of whatever nationality he may, after having mingled in a Malagasy crowd of some hundred thousand men, would be convinced that the people of this country are not difficult to live amongst."

The ancestors' wisdom is also expressed, naturally, in the verbal arts. Since there was no written word in Madagascar before the first half of the nineteenth century, save for the Sorabe of the southeast coast, all folklore was oral. That is not to say Madagascar had no literature before it adopted the Latin alphabet. As Lee Haring has put it, the Malagasy people "wrote their own culture before the letter." Literacy helped mold a concept of Malagasy culture as a unified whole, but unity was already there, if hidden — an Alexandrian library of Malagasy thought amassed over two millennia and stocked with riddles, oratory, proverbs, and poetry.

These four genres fall under the heading of *fitenin-ðrazana* or "sayings of the ancestors." Fitenin-drazana are characterized by two-sided interaction. This interaction can be overt as in riddling, in which the riddler seeks a response, or covert, as in *hainteny,* the traditional poetry of Madagascar, which often implies two protagonists waging a kind of verbal battle. Such binary opposition grows out of age-old village debates known as *mampiaðy karajia,* whose literal meaning is "fighting like a drunk." Village men conducted these verbal sparring contests to parody the solemn colloquies of the elders and to help their listeners escape for a moment from the grind of daily life. An anonymous Malagasy once defined mampiady karajia as "to talk and dispute for the mere purpose of wasting one's time."

Another way to waste one's time is by riddling. Often practiced by children, riddling is a good way for young people to learn the art of dialogue, which is so cherished in Madagascar. It teaches them the elements of speech as well as those of poetry, with the notion of a precedent and sequent. Riddles typically begin with the question "What then is this?" So what then are these?

> *If you try to lift it you can't, but it is very readily moved.*
> *God's little lake in which you cannot swim.*
> *At night they come without being fetched, and by day they are lost*
> *without being stolen.*
> *The dead cry out to sell the living.*

The answers are a shadow, the eye, the stars, and a drum beaten before an ox going to market. Madagascar even has a unique form of double riddle known as *safidy* (choice). One must identify two precedent riddles and then make the proper choice between them. "Which do you prefer, little eyes in the rocks or big eyes in the grass?" asks one safidy. Correct answers are first, "wildcat" and "ox," and second, "ox," for one would always prefer the domestic cow over the wild animal.

In olden days, riddling often went hand in hand with storytelling. "Often have I sat by the village or campfire . . . while the men, and women also, if any were present, squatted around enjoying the early hours of the evening, telling tales (fables) and putting conundrums," wrote a missionary in 1889. Often fables were little more than elaborate riddles. Why does the kite screech and the fowl scratch on the ground? Well, hold the Malagasy, the fowl lost a needle it borrowed from the kite, so it continually scratches the ground searching for it. The kite, meanwhile, screeches *"Filokohokoho"* (*filo*, needle, and *akoho*, fowl) to remind the fowl that it wants its needle, even as it steals the fowl's chicks as compensation.

Just as riddles can help teach the young the art of dialogue, fables can also impart valuable lessons. The tale of the wild hog and the chameleon, paraphrased here, is a Malagasy version of the tortoise and the hare that is shot through with moral lessons—learn from the past, plan for the future, listen and learn, brains are better than brawn, heed the ancestors:

One day a wild hog and a chameleon met in the road.

"How is it that you chameleons walk so slowly and awkwardly?" asked the hog.

"It is because we are meditating on the past and thinking of the future," replied the chameleon.

"But if you should be startled, can you run away?"

"Oh, yes," said the chameleon.

"Come, then, let's run a race to that rock down the road there."

"Very well," said the chameleon, "but first let me climb this tree and look along the road, to make sure nothing will hinder us."

"Good idea," said the hog.

Just before it said "Go!" the chameleon jumped on the back of the hog's neck. Not sensing this, the boar rushed off. Just as it reached the rock, the chameleon leaped off and stood on the rock looking relaxed, even as the hog fought for breath.

Astonished, the wild hog decided that he would follow the advice of the chameleon and that all his descendants should walk more slowly and gently. This wild boar, it is said, was the father of all domestic pigs.

The most highly esteemed storyteller in Madagascar is the *mpikabary* ("man of words"), the artistic speaker. The term comes from the word *kabary*, which has several meanings. Official kabary, introduced by Andrianampoinimerina in the late eighteenth century, consisted of huge public gatherings during which the monarch made proclamations to his subjects. Today, kabary refers to any ceremonial speech, such as a funeral oration, as well as to the formal debate held between the heads of families to establish the soundness and conditions of a proposed marriage. Finally, kabary also refers to the formal, stylized speech of adults, which comes into play at important events such as marriages, bone-turnings, and circumcisions, and which often features the deft use of riddles, proverbs, and traditional poetry. (Informal speech, that of greetings, gossip, and other everyday talk, is known as *resaka*.)

Registers, or the styles of speech that are appropriate to a given situation, play a significant role in the verbal arts of Madagascar. Kabary is associated with men, resaka with women, but men know the women's register and can drop into it if they choose, even during an official kabary (to parody the formal register, for instance). They may even use a third register that allows them to drop still farther down the social scale, as Haring has noted: "When driving their cattle (addressees of low status, after all), men make their indirect rejoinder to sixty years of colonial imperatives by barking orders to the cows in French."* Malagasy have other registers to choose from as well. Depending on the circumstances, they may speak "official" Malagasy, or a ritual language associated with the spirit world, or the "speaking in tongues" known from rites of possession.

Not surprisingly, formal oratory draws heavily on the other verbal arts. Speakers often lace their speeches with proverbs, riddles, and poetry, giving them the ideological strength of fable, as in this snippet of oratory recorded in the last century entitled "Every thing has its place":

> The whitebird [cattle egret] does not leave the oxen, the sandpiper does not forsake the ford, the hawk does not depart from the tree, the valley is the dwelling of the mosquito, the mountain is the home of the mist, the water holes are the lair of the crocodile. And the sovereign is the depository [literally "resting-place"] of the law, and the people are the depository of good sense.

Or this, taken from a piece of oratory titled "Desolate one forsaken by friends":

> I am but a straggling piece of peel from the shoots of the plantain-tree; and now I am left spent and desolate and having nothing, and hated by father's family, and cast off by mother's relations, and considered by them but a stone on which things are dried in the sun, and, when the day becomes cloudy, kicked away.

The Malagasy style of speech, particularly the more formal *kabary*, can leave Westerners, well, speechless. "[T]he Malagasy are ready and fluent speakers, although they certainly have also the power of saying a great deal without conveying any clear idea of what they mean," Sibree wrote. "It is often most difficult—in fact, next to

---

*Ironically, unemployed city kids in Tana today challenge Malagasy authority in a similar way by speaking a quasi-secret language called *zoam,* in which syllables and stresses are reversed.

impossible—to get a simple answer, Yes or No, or to make them come to the point; there is such a cloud of words, complimentary phrases, and vague nothings, that one often wonders what it really is they are driving at."

This indirect style extends to listening as well. Paul Siegel, an American technical adviser for the World Wildlife Fund in Tana, told me that, since Americans tend to be very direct, they can be real bulls in a china shop. "You have to realize that the person who is listening to you is not hearing what it is you're saying unless you're saying it the right way," Siegel said. "You may feel that you're being absolutely straightforward, but because you are being straightforward, they can't hear you, because they aren't *listening* straightforward.

"By the same token," he added, "you may find that what you consider to be very indirect is actually entering into people's ears as quite direct. You have to learn not only to hear between the lines but to *speak* between the lines. It's an exercise in cultural awareness I haven't run into in any other country I've ever lived in. It takes a lot of time and patience to learn, but once you get used to it, it's really fun. And then you can pick out the people who just aren't doing it right. You think, 'Oh, God, I wish he hadn't said it that way.'"

12.

I WISH DEWAR WOULD say something, *anything*, in any way at all. We're deep within a deciduous-looking forest about forty miles north of Bemanevika, at a site called Angolovato ("at the stone well"), and we've just found the well. I wish he'd say that this is what Bemanevika's holy well must have looked like, or that this is not what it would have looked like, or even that it's a nice day. But as usual, he has kept his lips sealed all morning. I haven't heard even a "Well, that's interesting" out of him.

The well is sunk into the moist earth. Looking down into it, I can see two pipelike sections set one atop the other; perhaps there are more beneath the soil that fills the bottom some six feet down. Darsot measures a third section that lies on its side out on the ground. It is about four feet long by three feet in diameter, with a rim some two and a half inches thick. It looks like a fluted column, with smooth grooves running down its outer surface.

All in all these pieces resemble cement piping you'd see at any construction site. Which is just what the two or three French colonials who visited the site between the 1890s and the 1940s thought the well was made of. Even Grandidier described it as "concrete packed together" with grooves "made by ropes." But what these early observers failed to notice is that this well was carved out of chlorite-schist. I kneel down to look inside the third section and have a sudden urge to crawl through it; it would easily accommodate me. Who were the people who would take the effort to carve these sections out of *solid rock*?

"Must have been a large settlement around here somewhere, for someone to go to all that work," Richard says, as if reading my thoughts.

Her husband, busy taking a satellite reading of our position, doesn't respond.

But Richard is right. What little is known about this area along the Mahanara River suggests that it was once a significant satellite of the Rasikajy civilization, which was centered at Vohemar just forty-five miles north of here. The first time the outside world learned of the site was in 1898. That year, the *Journal of Madagascar* reported that "M. Jully has revealed, on the River Mahanara, to the north of [Sambava], some ruins which must have been Arab buildings made of lime. The importance of these buildings would seem to indicate that a fairly large town once existed on this site."

Try as they might, succeeding visitors have been unable to locate any large town. But they have identified three archeological sites along a two-mile stretch at the mouth of the Mahanara that hint at a larger settlement. This well is one of them. Another is an island in the river mouth called Nosy Lava (Long Island). Jully's excavations there in 1898 revealed "a wall made of stones roughcast with mortar" and "fragments of pottery, including one of decorated and glazed sandstone, indicating an Arab origin." A year later, Guillaume Grandidier made a number of excavations there. According to his father, Alfred, who reported on the excavations in his monumental *Ethnography,* a multivolume work on Madagascar, Guillaume

exposed a surrounding wall which was very thick and more than 60 meters [200 feet] long and a well made of cement. The curb-stone of the latter has been marked with grooves made by ropes. Various fragments of pottery (water-coolers especially), a large glass flask, some little discs pierced in the middle, some pieces of rusty iron (knives, nails), bowls made of 12th-century Chinese porcelain and a dish that may also be Chinese were found there by the archeologist. . . . If local traditions are to be believed, this place was destroyed by a great tidal wave which diverted the river from its original bed. The town was later abandoned.

It is unclear whether this passage refers to the site of the stone well or to Nosy Lava. Verin, who surveyed these sites in 1969, believes it concerns the latter, but no other visitor to Nosy Lava ever reported a stone well. Would the people who lived here have carved *two* of these extraordinarily labor-intensive wells for places separated by less than half a mile? It may be that the two sites were once one and the same, for the owner of Nosy Lava told a French visitor in 1947 that the island had once been a peninsula; perhaps the "great tidal wave" retained in local memory was responsible for separating the two.* Regardless, the evidence further indicates the presence of an erstwhile settlement of some size.

Unfortunately, we cannot find anyone with a boat to take us across to Nosy Lava, but our guide says he'll take us to the third site. It's on a peninsula on the south side of the Mahanara, so we hike back to the road to have Marcelin drive us there. Dewar is especially curious about this third site, for it has the remains of what Verin believes may have been a stone mosque. Stone architecture dating from before the European arrival is extremely rare in Madagascar, and in all cases it is associated with the Islamic traders who populated the island beginning in the ninth or tenth centuries. The French trader Nicolas Mayeur, traveling along the northeast coast in 1775, discovered similar stone buildings "just before arriving in the south of Vohemar," which residents told him had been built by vazaha:

> The local people believed that there was at one time a foreland which stretched far out into the sea and formed [a] fine, spacious, and safe harbor where vessels were quite sheltered. This foreland was submerged and the harbor destroyed and filled in, however, as the result of a storm. A natural consequence of this disaster was that the settlement was abandoned and the white people left.

---

*Or was the wave said to have destroyed the town a folk memory of one spawned by Krakatoa, whose eruption sixteen years before Guillaume Grandidier's visit sent devastating tidal waves around the world?

Could Mayeur have stumbled upon the remains at Ambinanin'i Mahanara, as the peninsula site is known?

Marcelin drops us at a broken-down bridge over a stream; the road goes no farther. As we begin the hike north to Ambinanin'i Mahanara with a guide hired from a local village, a light rain begins to fall, though the sun is trying to burn through a white-gray cloud cover. There is no wind, and the air feels warm, so the walk is pleasant. Soon after starting we pass through a hamlet of two or three huts, where a man greets us effusively with hands full of dried, sticky vanilla beans. After that we enter uninhabited land. Even though the ocean cannot be far away, I can't hear it. The only sound is the occasional warbly, guttural cry of a cuckoo roller, an endemic gray and blue bird with a large head. The landscape changes from open country studded with traveler's trees to vanilla plantations beneath a light-filled canopy.

Then, quite suddenly, we enter a gloomy copse of thick, overarching trees. The path is soft underfoot, and it's quiet save for the now perceptible *whoosh* of the surf. The ocean lies just over a steep ridge to the east; we must be on the peninsula proper now. The trees in here are old and twisted, and so dense that the rain fails to penetrate the canopy. Most of the landscape hereabouts has been deforested, but no one has touched this stand.

Then I see why. Up on the ridge to my right rest dozens of tombs. None of us speaks as we glide past these vine-draped graves. They are of different sizes, from a few stones piled up to full-size wooden sarcophagi laid out on the ground. Some are unadorned, others bear names, dates of death, crosses, flowers. On a few, bottles and bowls rest—offerings for the ancestors. Huge orb-weaver spider webs dangle between branches, their multifaceted veils forming a kind of natural stained glass.

I glance at our elderly guide to gauge his reaction to our unwitting trespass onto his sacred burial ground, but his face remains inscrutable. He is a seventy-five-year-old man who introduced himself only as Monsieur Koto. In Madagascar, "Koto" is a common name for a little boy, but as Herbert Standing writes in his whimsical 1892 book *The Children of Madagascar*, "Often these names are retained all through life, the syllable *Ra* (for which the English Mr., Mrs., or Miss is the nearest equivalent) being prefixed. Thus you will meet quite old men bearing the name Rakoto . . . but use makes it familiar, and one does not think of the absurdity of the real meaning." Better "little boy" than some of the other names Malagasy parents can come up with, I think. "Sometimes the relatives seem quite at a loss for a

new name," Standing notes, "and so apparently call the child anything that happens to come into their heads . . . thus we meet occasionally with such names as Rubbish, Snail, Pumpkin, Sweet Potato,
Goldfish, Drygrass, Cannon."

"Cannon" would be more appropriate for M. Koto. His broken
nose, white stubble, and ramrod-straight posture lend him a tough-
ened look that belies his age. Experience emanates from him like a
pheromone. We're not in the presence of an old man but a venerable
elder, whose heavily lidded eyes seem to look right through you. His
people, the Anjoaty, are known for their occult power—early European travelers to the island often used the terms "Anjoaty" and "ombiasy" synonymously—which makes our presence on his ancestor's
consecrated turf all the more unsettling. Yet because of his people's
Arabic lineage, M. Koto is perhaps the closest thing we'll come to a
living incarnation of the Rasikajy. It also makes him, incidentally, a
living symbol of Madagascar's singular creolization.

Like the Vezo, the Anjoaty are not one of the eighteen official tribes,
even though people claiming to be Anjoaty are found right around the
island's coasts. Their name means "people of the river mouths," and
they pride themselves on a potent connection with places where
rivers meet the sea. This bond is both physical and metaphysical, as is
hinted at in the French word used for the Anjoaty culture, *embouchure*,
which refers to both the mouth of a river and a mouthpiece.

As the Ambinanin'i Mahanara graveyard shows, Anjoaty bury
their dead in the littoral, close to the mouth of a river. This enables
ancestors to mediate between the land of the living (the earth) and
that of the dead (the sea). Anjoaty like M. Koto exercise their supernatural power (*hasina*) through their ancestors buried in the
embouchures. In the northeast, this power is manifest in what Grandidier called *masim-bava* or "sainted-mouths," an ability to speak the
truth or even self-fulfilling prophecies. The ethnographer David
Hurvitz describes how an Anjoaty from Vohemar told him that
embouchures were the most powerful places in the landscape:

> When I asked him why, he answered, "Just because" . . . I think
> I can now add: because that is the place where water from the
> interior passes into the ocean, representing the passage from life
> to death (which is why ancestors are buried there). If that is not

convincing, let me say this: The power of an Anjoaty mouth is like the power of a river mouth, and in the same way, a second meaning of the Malagasy word for embouchure, *vinany*, is prediction or forecast.

The Anjoaty's legends of migration from Arabia and down the east coast of Madagascar embody these beliefs and also provide clues to the movements of the long-dead Rasikajy. Many Anjoaty believe their forebears originally came from Mecca, though they retain few vestiges of Islam, and each group of Anjoaty has a different legend of precisely how they came. The Anjoaty of Vohemar, for instance, maintain they arrived via a legendary sunken island called Mijomby. While the specifics of the legends differ, they all share the notion of traveling from land over water to land. In fact, as Hurvitz has written,

> . . . the entire corpus of mythology is animated by an opposition between land and water, with constantly recurring themes of creation from the sea; return to the sea at death . . . ; islands sunk into the sea or villages transformed into lakes, drowned ancestors transformed into crocodiles or survivors saved from drowning on the backs of fish; fish, amphibians, canoes (and even coffins) serving to ferry people across water; and in general . . . common reference to a world of people beneath the sea.

Anjoaty of the northeast also believe in the legend of Darafify, a kind of Malagasy version of the American legend of Paul Bunyan and his blue ox Babe. An Anjoaty, Darafify was a giant with cheeks the color of the *dara*, the red fruit of the Madagascar date palm. He came to Madagascar from Arabia with a red ox bearing eight teats, which begat Darafify's vast herds. He landed on the northern tip of the island and eventually worked his way down the east coast, performing various legendary feats along the way, including killing a giant snake that devoured men and cattle. "He cut it into little pieces and scattered the remains of its body far and wide, so that they could not find each other to join together again," wrote a French traveler to Madagascar who collected a version of this tale in the 1880s. When Darafify died, it is said, his two wives cried so much that their tears formed two lakes.

The Darafify legend differs in its particulars as one moves down the east coast of Madagascar, but its overall nature is the same. Grandidier proposed that the Darafify legend is nothing more than a mythological history of the Islamic colonization of the island. Darafify personifies the Arab colonists who settled in Madagascar beginning in the ninth or tenth century. His travels from north to south represent the Arab migrations along the same route, and his struggles with the giant snake and other opponents signify the Arabs' conflicts with local Malagasy along the way.

I've been wondering if we'll have a conflict with M. Koto, and sure enough, before we can reach the other side of the burial ground, our Anjoaty guide waves Darsot over with that insistent downward motion of the hand that Malagasy use to mean "Come here." He's standing by a kind of totem pole bristling with zebu horns. What's he up to? Will he demand we leave the cemetery immediately, even turn around and go back?

Darsot turns and calls me over.

Me?

"He says this marks the line between traditional and Christian burials," Darsot says, pointing to the totem.

Suddenly, inexplicably, M. Koto, master of the occult, living representative of the mysterious Rasikajy, bursts out laughing. Covering his mouth with the back of a hand, he goes on chuckling, his lidded eyes twinkling with amusement. I can't fathom what set him off, but his delight is my relief. The somber air of a minute ago clears in a flash. He couldn't care less that we're in his sacred burial ground!

With a newfound lightness in our step, or at least in mine, we continue on and soon reach the end of the peninsula. A raised finger of land, it juts north into the mouth of the Mahanara. To the west of it is a sheer drop; to the east, down off the lofty central ridge, lies a narrow plain of secondary forest backed up by the sea. Clearly the peninsula began life as a sand dune, but now it hosts a wood of thin-trunked trees. The sun has begun to break through the clouds to the northwest, backlighting the Mahanara, whose surface glitters as if strewn with a million diamonds.

Perhaps it is shock. Perhaps it is Dewar's unbreakable silence and the Malagasy's natural courtesy. But no one says a word when we realize what has happened here. The entire tip of the peninsula has been dug up. Recently. Deep, gaping holes gouge the ridgetop. One of them could swallow a house; another is narrower, perhaps eight feet on a side, but is at least twenty feet deep. The whole looks like a construction site, or Darafify's sandbox. Some of the holes have stacked tree trunks wedged into their sides to keep the sand from spilling in.

Someone has gone to one helluva a lot of work here. For what?

A handsome, muscle-bound Betsimisaraka youth in shorts and machete, who followed us from the one hamlet we passed, tells us through Darsot that a group of vazaha did this last year, and that they would be back. He didn't know what they were looking for. Dewar does, I'm sure, but as usual, he's not making me privy to his thoughts. He's squatting down next to one of the holes, holding his blue baseball cap in one hand, whispering quietly to Darsot. It's hard to read M. Koto, too, who stands off to one side, his face as solemn as it was before his curious outburst. Apparently he didn't know of this rape of his people's land until we arrived.

I head down to the shrub-filled plain to have a look at the stone building that Verin believes may have been a mosque. The looters haven't touched it. Vines shroud its two remaining walls, the south and the east. The walls are about twenty-five feet and thirty feet long, respectively, and about a foot and a half thick. Verin reports that in the 1920s, a French official looted the structure's stones for a road he was building from Sambava to Vohemar. Another Frenchman reported in 1947 that "the district administrator demolished in particular a door or an arched window, which was the only opening in the high stone wall that was well known to him." High stone wall no more: These remnants stand but three feet high. And there are no distinguishing architectural features—say a door or an arched window— that might give important clues to the builders.

I climb back up to the ridgetop, where Dewar is staring into a pit.

"Your nemesis," I say.

"Yeaaaaah," he replies, running his fingers through his hair.

"Will this stop you from excavating here?"

"This? This might be a problem. We're going to file reports locally and in Tana, try to stir up some trouble. The problem, of course, is if it pisses off enough local people, vazaha coming here and ripping holes all over the graveyard, it does make it difficult for others to come back." He looks around the torn-up mound. "It's like, holy shit."

"Were they looking for tombs?"

"Yeah. Islamic graves don't contain grave goods, and these people were supposed to be Islamic, but they did bury their dead with grave goods—Chinese porcelains. What the French dug up here when they found their quote 'treasures,' they're all tombs. If these are like the tombs found at Vohemar, they are probably fourteenth, fifteenth, six-teenth century."

He looks down the slope toward the "mosque."

"Those walls down there look like those of Mahilaka, which is a twelfth-century site. But nobody in Mahilaka was being buried with anything that would cause this much trouble. They must have found something."

The hike back is sadly reminiscent in mood of the hike back to Bemanevika village. At the delapidated bridge where Marcelin dropped us off, we find the Renault and a large pile of white bird feathers. This guy is going to single-handedly depopulate the avifauna of Madagascar, I think. While Darsot heads up the road calling for our driver, I spend the time looking for *Brookesia* chameleons, without any luck. It's late in the afternoon, and the light is getting bad. Half an hour later, Marcelin saunters back, rifle in hand and birdless, thank goodness. We cram into the tiny car. I'm delighted to realize that, for the first time, I'm sitting on the left side; my inflamed right shoulder will get a break.

But it's not for long. As we pause in the nearby village, Darsot turns around in the front seat and says, "Want to try a local drink?" We're to wait here while Marcelin drives M. Koto back to his village, and a local man has invited us to his home for a cocktail. As dusk falls, we meander through the village, a sizable hamlet of several dozen thatched huts on stilts, all the same size and shape, spread out willy-nilly on the packed earth.

Our host, an elderly man with a beaten face, steps into his dark-ened house, and we follow behind, Darsot first, then Dewar and Richard, and finally me. There's not much room, and we sit on the floor, knees to our chins, in a row against one wall. The hut has a bed with old sheets and a mosquito net, mats on the floor, a new fishing net hanging in one corner, and a side table with a cassette player and batteries scattered about. That's it. While Darsot chats with our host, translating choice bits into French for Dewar, I gaze out the door of the hut at the village settling down for the night. Chickens flap into a bushy mango tree or walk up the angled trunks of smaller trees to spend the night. A boy chases a loose zebu to the guffaws of villagers. Smoke wafts from rooftops, filling the twilight air with the nostalgic fragrance of burning wood.

While I'm musing, a young man suddenly clambers past me and plops down in the only available floor space. His eyes are crossed from too much palm wine. Which we are treated to shortly, out of a discolored kerosene jug stuffed with thatch. Our host pours some into a metal dish like a dog bowl and passes it around. Not wishing to be disrespectful, I make the motions of taking a sip, but keep my lips tightly sealed, imagining a vicious bout of hotely belly several days hence. Perhaps thinking the same thing, Richard watches me closely from a foot away. Does she notice? In case she or our host has doubts, I smack my lips approvingly and say "Good!" then pass the bowl to Dewar, who does drink.

We're saved from a second round by Marcelin, who races into town, beeping his horn, then driving between the houses of the village itself looking for us and beeping some more. Once we're back in the car, he rockets back to Sambava, eager no doubt to cook up that flock of birds he shot. No one's up for conversation, least of all Dewar, so I spend the time watching for the flicker of firelight between the slats of huts we fly past in the dark. And trying to keep pressure off my right shoulder, which is jammed once again against the right side of the car.

On the way, I think back on Raxworthy telling me, while we waited for that ancestor ceremony that never happened, how you go with the flow in Madagascar. Dewar has shown over the past week that he adheres to the same philosophy. Okay, so maybe you have to drink palm wine out of a filthy jerry so as not to offend your host. Maybe you have to cross greased poles for bridges as part of your fieldwork. Maybe you find that not one but two of the archeological sites you've come 10,000 miles to study have been destroyed. So be it. You go with the flow. *Moramora*, the Malagasy say. Take it easy.

I recall, too, what Burney told me about Dewar. "I've learned a great deal from Bob. He's the type of archeologist who has the necessary patience to do the work. The kind of people who typically get in trouble in places like Madagascar are those who are insensitive to local cultural concerns, like burial customs and attitudes about where foreigners should go and shouldn't go."

It's a good lesson in dealing with Dewar's impenetrable silence, I decide. Go with the flow—or in this case, flowlessness. Considering what he's had to put up with over the past few days, it's nothing. Besides, what else can I do?

13.

IF GOING WITH THE FLOW is the best way to get along with the Malagasy, the best way to get along *famously* is by speaking in proverbs, which lie at the heart of who the Malagasy are as a people. "I find that oftentimes I can express things in French or even in English using a Malagasy-type proverb," Paul Siegel told me. "I'll see an analogy to something in my mind, and when I verbalize that analogy, I find that it goes across absolutely clearly. Sometimes my colleagues will say, 'God, you're speaking like a Malagasy,' which I take as a personal compliment."

The Malagasy, it is safe to say, are addicted to proverbs. They come spilling out on all occasions: when sitting around the family hearth, bargaining in the marketplace, giving a funeral speech, or orating before the National Assembly. Indeed, nothing will energize a Malagasy audience faster than an aptly quoted proverb. "They will forgive much in the way of want of logic, and even paucity of thought, to say nothing of a bad choice of words and indistinctness of expression, if what they do get be only served up with a few terse and racy sayings that are already familiar to their ears," wrote John Alden Houlder, a chief admirer of Malagasy sayings. For Westerners who prefer getting to the point, this national *idée fixe* can be exasperating. As Houlder says, "many speakers are tempted to attach more importance to the ornaments of speech than to the purpose for which they are used, and, for the sake of making a point, pay more attention to the flowers of rhetoric than to its ultimate fruits."

What such speeches lack in unambiguous argument, they make up for in the wit of the proverb. For the corpus of Malagasy sayings, assembled over the centuries, is one of the world's most robust. "There are [proverbs] applicable to virtually every emotion, state, and condition known to mankind," writes Leonard Fox. "In that respect they constitute a universal philosophy of life that transcends its Malagasy context and merits admiration as one of man's noble attempts to construct a valid moral and philosophical framework for his existence." Many of the proverbs in common use today in Madagascar likely date back many centuries, encapsulating the age-old wisdom of the Malagasy people. As one Malagasy wrote in 1846, more or less summing up his people's worldview, "Much money and many cattle may be completely consumed quite easily, but wisdom is a possession that lasts forever; the truth is not destroyed, it is eternal."

Like those in other cultures, Malagasy proverbs are often untranslatable in their original form. They contain words or concern customs or concepts long vanished, or they deal with uniquely Malagasy cultural traits that would require preliminary explication before a foreigner would have a chance of understanding the proverb. Even those sayings that embody universal truths often tend in translation to lose their "piquancy and relish," as one commentator put it, which is why the Norwegian missionary Lars Dahle published his 1877 *Specimens of Malagasy Folk-lore*, still the classic in the field, in Malagasy. The reader of translated versions also misses many of the poetic elements inherent in Madagascar's proverbs, including alliteration, onomatopoeia, caesura, punning, and occasional rhyme.

Even so, many Malagasy sayings readily speak volumes about their creators. "A people is known by its proverbs," Houlder declares, "for what is nearest and dearest to the heart of a nation, the aspect with which they contemplate life, how honor and dishonour are distributed amongst them, what is of good and evil report in their eyes, will surely be apparent in their proverbs." These come, he says, from "the one old book of human nature," which is the same the world over. "The cover may be white, brown, or black, and the leaves more or less discolored, but there is no essential difference in the characters written by the finger of God upon them."

Some Malagasy proverbs have a familiar ring. Just as we don't count chickens before they're hatched, so the Malagasy "don't decide before God, like the hatchers of fowls." Just as we don't make waves, they "swim without disturbing the water." They know that money can't buy love ("friendship cannot be bought") and that there is safety in numbers ("cross in a crowd, the crocodile won't eat you"). For the Malagasy, a rolling stone "stops not till the bottom is reached."

Many Malagasy sayings, in fact, are miniature cautionary tales concerned with good and evil. "The thing done is like a loincloth," holds one saying; "good, it goes all round; bad, it goes all round." Lies may be "plump out of sight," but they are "thin face-to-face." And one should remember that "shame changes your face" and "obstinacy makes you a dwarf" (literally, "little as to your father"). In short, say the Malagasy, "man is his conduct." Resourceful people on Madagascar know that "long feet will find food," whereas lazy people are reminded that "if you do not want to go to the forest, do not expect the wild boar to come to you." One should be generous of heart; after all, "clapping won't skin the hands." A selfish person is "a butterfly in the sun: he won't give up his warm place." Particularly if he is well-off. "How is it," asks a Malagasy proverb, "that only the rich are nig-gardly?"

Poverty, of course, is a recurrent theme in Madagascar. "Delight is not for the poor," say the Malagasy, and "only in sleep do we resem-ble the rich." The poor and hungry on the island are "like an ox's stomach in the winter: only half full." But whether rich or poor, "men are like the rim of the pan: they are part of one circle." So it is worth remembering that while there may be "many cocks within the fence, and they all want to crow," the end result is the same: "One thousand words, one hundred speeches: it will end in stubborn silence." And few would dispute the contention that "it is better to die tomorrow than today."

Many Malagasy proverbs embody such simple truths. "If a roost-er does not crow, it is because it has been eaten," they say. Why cry at a funeral? "A dead man cannot see your tears." When asked shortly after independence whether he planned to cut ties with France, Madagascar's first president replied with a well-known proverb: "Do not kick away the canoe which helped you to cross the river." The Malagasy will advise you not to try to be someone you're not, because "a hare-lipped person who whistles makes life hard for himself." And for goodness sake, "if you are just a dung beetle, don't try to move mountains."

Humor is rife in these sayings, as it is in the hearts and minds of the Malagasy. "When men fight by the light of the moon, the bald ones are sure to be hit," they say. Better to have "a pig fight: a grunt or two settles it." Even better is to avoid confrontation altogether. Be like "rats about to pass each other on a narrow beam [who] are both courteous." Still better, take note that "when old women meet, they overwhelm each other with congratulations." Speaking of older women, one would be wise to keep in mind that "the wife is liked, but the mother-in-law is loved." The Malagasy love children most of all — most of the time. "Nothing is more precious than your baby," they declare, "except when he bites your breast." And, of course, "other people's children cause your nostrils to flare."

Proverbs are so intrinsic to Malagasy thought that they underlie many hainteny, the traditional poems of the Merina. Indeed, the links between these two types of "wisdom literature" are so inextricable that even today, Malagasy confuse the two, Fox says. A number of poetic forms exist in Madagascar, and many other tribes besides the Merina write their own poetry, but hainteny are the most highly refined. Usually between four and ten lines though sometimes much longer, they deal with love in all its forms as well as with wisdom, character, and moral issues. As with proverbs, we as foreigners and readers of translations can only hope to approach a surface understanding of these highly complex poems, which rely heavily on metaphor and allusion. Like proverbs, they are of indeterminate though clearly considerable age; as Fox notes, "poetic techniques of such great refinement do not appear with sudden spontaneity."

The depths of meaning in hainteny can be extraordinary. Take the following example:

> *Whish-whish, shaking of the winnowing-pan,*
> *thump-thump, pounding of the pestle.*
> *The lid is in tears up to the neck.*
> *The ladle stirs.*
> *But it is only the little spoon that enters the royal courtyard.*

This poem can be interpreted in at least three ways, writes Fox in *Hainteny*. On the most obvious level, it describes the island-wide practice of preparing and eating rice. On another level, one may view it as portraying a woman's suitors, men respectively ostentatious, macho, sensitive, crafty, and unassuming, the latter being the one who wins the woman's favors. Finally, the poem can be seen as a fairly unambiguous depiction of the sexual act.

The degree to which the Malagasy can infuse metaphor and meaning into a few well-crafted lines approaches that of Japanese haiku, as in this hainteny:

> *The fringe of my lamba is damp:*
> *into the water I was about to drink*
> *a frog jumped.*

In this poem, the speaker appears to be the older wife of a polygynous man. (Some Malagasy tribes still practice polygyny.) On one level, it appears that she has been crying, dampening her wrap, because the younger wife (the frog) preempted her just as she was about to enjoy the favors of her husband (the water). On another level, ashamed at having wept, she pretends that her lamba became wet after a frog splashed water on it.

Though the bulk of hainteny focus on love, others deal with the moral issues of greatest importance to the Malagasy. These include a respect for one's elders:

> *That road, there in the north,*
> *is shady, like an old road;*
> *it seems to meander here and meander there,*
> *but in the west it circles the home of my parents.*

Good and evil:

> *Broad fingers, large fingers:*
> *if a thief gives you his hand,*
> *reject it like a dead beast.*
> *Broad fingers, large fingers:*
> *If someone of good character offers you his hand,*
> *think of it as moonlight permeating your heart.*

And wisdom:

> *Three ears of corn with roots*
> *and thin bananas with many sprouts.*
> *Even if the rice is exhausted*
> *or the year is a hard one,*
> *one cannot show reluctance with beloved friends:*
> *for there are still the three ears of corn with roots*
> *and the thin bananas with many sprouts.*

Intriguingly, scholars have found inherent similarities between hainteny and a poetic form from the Malayo-Indonesian region known as *pantun*. Like hainteny, pantun, which dates to at least the fifteenth century, is distinguished by a signature use of allusion and metaphor. As in hainteny, too, there is also a close connection between pantun and proverbs. Indeed, the *Hikayat Hang Tuah*, a seventeenth-century compilation of older materials dating back to the fourteenth century, uses the word "pantun" to mean both a "figurative saying" and also the type of poem known by that name. Pantuns most closely resemble hainteny when dealing with love in all its manifestations, as in this example:

> *Red ants within the bamboo,*
> *the flask is filled with rose water.*
> *When desire seizes my body,*
> *you alone bear the cure.*

In this pantun, the bamboo represents the male sexual organ, inflamed as if by red ants, which can only be soothed by the female sexual organ and its secretions—the flask filled with rose water. As in many hainteny, the metaphorical beginning of the pantun sets the tone for its more literal end.

Again, as in all aspects of their culture, the Malagasy took the rootstock of Eastern poetry that they brought with them and fashioned it into something endemic to Madagascar. Hainteny, as the most advanced of the various forms of Malagasy poetry, can be so allusive as to be utterly obscure to the Western reader, but this does not mean that one cannot appreciate them. As Fox has written, "the artistic refinement of hainteny causes them to transcend the cultural specificity inherent in their metaphors and allusions, and permits their beauty to be perceived even on the simple level of sheer lyricism." Then again, some hainteny are charmingly clear:

> *"I laid my eggs," said the bird,*
> *"there, in a tall tree;*
> *the tree was blown down by the wind;*
> *the wind was stopped by a promontory;*
> *the promontory was bored through by a rat;*
> *the rat was eaten by a dog;*
> *the dog was flogged by a man;*
> *the man was overcome by a spear;*
> *the spear was overpowered by a rock;*
> *the rock was passed over by water;*
> *the water was walked in by a little bird."*

The syncretism of Malagasy poetry, blending the endogenous and the indigenous into something unique, finds its apotheosis in the work of Jean-Joseph Rabéarivelo, widely considered Madagascar's greatest poet. Sadly, this syncretism, which bore such rich fruits in his work, played out tragically in his short life. Rabéarivelo was born in 1901 to an aristocratic though poor and unmarried Merina mother. His formal education ended at the age of fourteen, but through voracious reading he mastered French by his early twenties, when he came under the sway of the French symbolist poets. His early rhymed poetry, written in French, owes as much to the likes of Baudelaire and Mallarmé as to his homeland, as Rabéarivelo reveals in "Influences" (in which Iarivo is Antananarivo):

My song is saturated with thy light,
its soul so long has felt the influence,
Iarivo sky, encircling, blue, immense,
of all thy shifting shades of sound and sight.

But I would have my song's course wedded quite
to thine, fair river, sky-betrothed, that thence
its suppleness attain the elegance
of thy dim margins which the sun makes bright,

and I may honor more that foreign speech
which with delight my instinctive soul can reach,
which I adopt, nor grieve upon that head,

when from high terraces I soothe my heart
and, gazing moved, name over every part
of thy perfected grace, land of my dead.

Scholars consider Rabéarivelo a precursor of Negritude, the midcentury poetic movement born among French-speaking African and Caribbean writers as a protest against French rule and the policy of assimilation. Yet, as "Influences" intimates, he was more in thrall of the French than in opposition to them. Indeed, he had no political agenda—his poetry is concerned with the contrast between the light of reality and the darkness of the imagination—and he even allied himself more closely with his French idols than with his own people. "[T]his is what I am: imperiously, violently, *naturally*, a Latin among the Melanians . . . with the latters' features," he wrote in his journal.

His devotion to those Latin "roots"—and to his favorite daughter, who died suddenly in 1933, plunging him into a depression from which he would never recover—ultimately led to his death. In 1937, his dream of finally visiting France seemed about to come true, with friends in the Malagasy government supporting his application to join the Malagasy contingent to the Universal Exposition in Paris that year. But when a group of basket-weavers was sent instead, Rabéarivelo broke down. On June 22, after dispatching his manuscripts to various friends, he killed himself by taking cyanide. He was thirty-six years old.

A burgeoning interest in his native land and heritage eventually eclipsed Rabéarivelo's devotion to things French. He had an intense love of the hainteny, which he said "responds to the need of the race, whose supreme elegance consists in an eloquence at once consuming and imperceptible, like a flame kindled at high noon." In his mature work, collected in two books of poems, *Presque-songes* (Near-Dreams) of 1934 and *Traduit de la nuit* (Translations from the Night) of 1935, Rabéarivelo owes much to the spirit of hainteny. As Fox notes, the poetry in these works, which Rabéarivelo wrote first in Malagasy and then translated into French, "does not attempt to reproduce the style of hainteny in French; rather, it expresses the poet's assimilation of that genre and its spirit, which he casts in a new and entirely personal form. . . ." This poem comes from *Translations:*

> *Slow*
> *as a limping cow*
> *or a mighty bull*
> *four times houghed,*
> *a great black spider comes out of the earth*
> *and climbs up the walls*
> *then painfully sets his back against the trees,*
>
> *throws out his threads for the wind to carry,*
> *weaves a web that reaches the sky*
> *and spreads his nets across the blue.*
>
> *Where are the many-colored birds?*
> *Where are the precentors of the sun?*
> *—Lights burst from their sleep-deadened eyes*
> *among their liana-swings,*
> *reviving their dreams and their reverberations*
> *in that shimmering of glowworms*
> *that becomes a cohort of stars,*
> *and turns the spider's ambush*
> *which the horns of a bounding calf will tear.*

In his final years, Rabéarivelo took to translating and adapting existing hainteny. His last collection, *Vielles chansons des pays d'Imerina* (Old Imerinan Songs), published posthumously in 1939, consists of hainteny that he presents as prose poems, a form well-loved by the French symbolists. As such, these poems, which combine indigenous oral literature of Madagascar with a form borrowed from overseas, perfectly symbolize the creolization of Malagasy culture. Some of them seem to sum up Rabéarivelo's, and perhaps Madagascar's, dilemma:

There in the north stand two stones and they are somewhat alike: one is black and the other is white. If I pick up the white one, the black one shames me. If I pick up the black one, the white one shames me. If I pick them both up, one is love, the other consolation.

14.

I FIND LITTLE OF EITHER love or consolation in the work I've been doing with the provost of Yale all morning, my last in Sambava. Crouching in the hot sun outside Dewar's bungalow at the Coconut Palm, Richard and I rinse the potsherds we gathered in a bucket of water, then lay them on Dewar's sifting screen. Once each is dry, we paint a line of nail polish remover on it, on which Dewar will later write an identifying number. "I'm sorry, it's *really* exciting," Dewar said as he pressed us into service. "But if you don't do it, they disappear into . . ." He let his words trail off. Neither of us is disappointed when Dewar and Darsot finally come out of their bungalows two hours later, where they've been writing up their notes, and join us for a final lunch in the restaurant.

"You'll have this song in your notes forever," Dewar says now as he finishes the last bite of a salad he wouldn't touch a week ago.

He's giving me the promised summary of his findings, and so I've pulled out the tape recorder, not willing to entrust my one chance to get the goods to my personal shorthand. Even though it's painfully loud, the music has actually been enjoyable this time, with catchy Malagasy pop giving way at one point to a welcome set of Stevie Ray Vaughan. But the current song is a rap number with obscene lyrics— "Fuck that!" repeated over and over again. If our Malagasy hosts knew what the rapper was yelling, I'm sure they'd be deeply offended. They're more used to lyrics like that on Richard's lamba, which, beneath a clock showing a time of 3:00 P.M., has a line in French enjoining, "Don't be late for our rendezvous, my love."

Dewar is like an ON/OFF switch, and right now he's ON. And so am I, absorbing his speech like a sponge. His words are not Rabéarivelo, alas, but they're arguably going down just as pleasantly after a week of waiting. He recaps each of our journeys in depth, giving me his opinion about the significance of what we found at Benavony, the rockshelter, Bemanevika, Mahanara, and a few other places Marcelin flew us to along the Vanilla Road. We found no signs of the earliest Malagasy, much less clues that might help penetrate that "most beautiful enigma." But all in all, Dewar got what he came for.

"I'm quite pleased, in fact I'm *very* pleased," he says, blurting out the "very," that the zone between Benavony and the Mahanara River valley is potentially a place where he can set up an archeological reconaissance project and collect useful data to compare with other parts of the island.

I realize with relief that the music has stopped. It's quiet in here, save for the muffled *thwump* of bullish waves down on the beach and a lone cock crowing even though it's midday. Perhaps he's protesting the clutch of dead chickens I just saw go past the window in the hands of one of the restaurant staff.

"The final piece of work we've been doing here is making contact with the political leaders in all the areas we visited," Dewar says. "We were trying to establish a sense in them that we're interested in their history and not in looking for Arab gold. And also in developing the kind of relationship that will permit us to come back at a future date and begin some more serious research."

"A collaboration with local people rather than suspicion," says Richard, who, like her husband, is smoking a thin cigar now that our meal is finished. "And you can see there's a basis for suspicion. We saw it with our own eyes. You have to put some distance between yourself and the bad guys."

Dewar continues as if he hasn't heard. His face takes on a liveliness I didn't know it could possess.

"So this morning we were off at the *fivandrony*, which is the equivalent of the prefecture level. The president, who had received us quite warmly when we first arrived, was gone, but his assistant was there. Turns out his assistant once upon a time was a history student at the University of Madagascar. So *not only* did he know about archeological sites like Irodo to the north of here, but after we described for him the kind of graverobbing we saw at Ambinanin'i Mahanara, he was the one who characterized it as an act of piracy and told us that he was going to establish an immediate inquiry into who was responsible and what kinds of steps can be taken to prevent this kind of destruction and looting of archeological sites in the future."

For all his animation, Dewar might be Lawrence declaiming before the Arabs.

"He was very explicit and completely without prompting on our part in saying, 'This is a matter of the destruction of the national patrimony. If this is allowed to continue without check, there will be no way for our children to ever be able to answer the question of what are the origins of the Malagasy? What is our history in this corner of the island? What *is* our past?'"

For the better part of an hour, Dewar's spigot gushes without ceasing, slaking my thirst for his thoughts on a week of protracted physical and mental exertion. He's on a roll, and I get a heady draft of what Jolly calls his "really original mind."

As I watch his expression morph like the surface of a lake in a gusty breeze, his balletic hands threatening to make early sherds of the lunch dishes, I realize that I should have given him the benefit of the doubt earlier on. Burney considers him "a very careful scientist, rather conservative in his interpretations," and it's clear to me now that Dewar was simply taking everything in back at all those sites. It's not his style to blurt out whatever comes into his head at the moment, but rather to moramora and sift his thoughts before making a carefully reasoned statement, at the proper time and place.

How Malagasy of him.

Island of Dreams

When you plant one rice seed, you harvest a hundred.
—*Malagasy proverb*

Without the forest, there will be no more water,

without water, there will be no more rice.
—*Malagasy proverb*

1.

"You have all the luck!" says Patricia Wright, standing in the dappled sunlight of the rain-forest floor, arms crossed, beaming. With her slightly Asian eyes and relaxed pose, Wright, a primatologist at the State University of New York at Stony Brook, seems as serene as a Buddha, though her outfit of green field pants, tan shirt, and binoculars runs more toward Smokey the Bear. She is leading me and an Earthwatch team on an orientation hike through her research site at Ranomafana National Park in southeastern Madagascar, and she's just acknowledged our good fortune at having a troop of rare golden bamboo lemurs, a species she codiscovered in 1986, suddenly appear in the trees over our heads.

Ranomafana is a remnant stand of rain forest covering more than 150 square miles of steep ridges and sunken valleys on the eastern edge of the escarpment that runs like a spine down the center of Madagascar. It is the home of two tribes, the highland Betsileo ("the many unconquered"), the country's preeminent rice-terrace agriculturalists, and the lowland Tanala ("those of the forest"), leading practitioners of *tavy,* or slash-and-burn farming. It is also home to no less than twelve species of lemur, a dazzling number that epitomizes Ranomafana's ranking as one of the world's most biodiverse regions. The newly arrived team is here to further Wright's behavioral study of the Milne-Edwards's sifaka, a beautiful brownish-black lemur with crimson eyes that, at nearly fourteen pounds, is the second-largest living lemur.

Wright's hope on this morning walk is to introduce the team to *Propithecus diadema edwardsi,* as her research subject is known scientifically. But now, only a half hour after leaving the research cabin on the edge of the Namorona River, we suddenly come upon the goldens in a stand of thin bamboo. They move unrushed yet rapidly through the trees. The size of house cats, with long, furry tails, they offer me only fleeting glimpses of their golden fur. Then one runs down a branch arcing over the trail a few feet from my head. I rush to get my camera in position, but after a quick, round-eyed stare into my eyes, the lemur is gone, across the trail and into the trees. Not for nothing does "lemur" come from the Latin word for ghost.

Compared to the Milne-Edwards's sifaka, the golden bamboo lemur is undistinguished-looking. But as I scramble up a ridge and admire it spotlighted in a patch of sunlight, I gaze in awe, as if catching sight of the Dalai Lama, because it is an animal of almost mythic status. Known only from this rain forest and another a bit farther south in Madagascar, it is one of the rarest primates on Earth. Before Pat Wright and German biologist Bernhard Meier, working independently, discovered it in these forests a little over a decade ago, the golden bamboo lemur was unknown to science. For primatologists, finding a new species of primate is like an explorer stumbling upon another Grand Canyon.

But what they found out about the lemur was even more surprising. For one thing, it eats bamboo. It's rare enough for animals to eat bamboo; these lemurs are the only primates known to do it. But the park already had two bamboo-eating species, the greater bamboo lemur and the gray bamboo lemur. How could three species live on the same food source in an area only about the size of five Manhattans? Careful scrutiny of eating habits revealed that the three species consume different parts of the plant: The greater eats the tough husk; the gray dines on the tips; and the golden prefers the tender new shoots and leaf bases. Most remarkably, the parts the golden prefers harbor cyanide. In fact, each day it consumes more than twelve times the amount of cyanide that should kill an animal of its size.

"One stalk would kill us, yet it eats it all day!" Wright says now, throwing her arms up and laughing in a deep-toned, almost masculine way. "Nobody knows *how* it does it." Her oval eyes dance over each of us in turn, as if seeking tacit agreement of yet another rain-forest mystery awaiting illumination.

Before Wright habituated a few groups of golden bamboo lemurs during fieldwork here in the late 1980s, it took months just to catch fleeting glimpses of this shy animal. (This trait is one reason scientists missed it for so long.) So we are indeed lucky to have such a close brush. Yet looking at Wright, her expressive face framed by frizzy brown hair cut in a long pageboy, proud as a mother showing off her babies, I feel that somehow this meeting is not serendipitous. She has achieved so much in this rain forest since she first arrived in the region in the 1980s that having her cherished golden bamboo lemurs come down to greet us on the team's first day only seems fitting.

Wright is almost singlehandedly responsible for the creation of this national park, Madagascar's fourth and most successful. Having a new species of primate to show off helped spark interest at the U.S. Agency for International Development, which funded the park's creation in 1991. Indeed, as a 1993 evaluation of the park's initial phase stated, "if the golden bamboo lemur had been discovered in another site, such as the eastern forest site of Andringitra [a reserve about 100 miles to the south, where goldens have subsequently turned up], USAID funds for development and conservation would now be channeled to Andringitra, not Ranomafana." But it was Wright's tireless efforts at all levels—international, national, regional, and especially local—that set the stage. "It is clear that [Wright] has been the predominant influence in the past, and will be in the future," the 1993 evaluation concluded. "This is because of the force of her personality and of her vision for the Ranomafana region."

If one were to ask *Can Madagascar be saved, and if so, how?* Ranomafana, more than any other protected area in Madagascar, would be a logical place to begin. For if you believe Pat Wright, minor miracles like our unexpected glimpse of the golden bamboo lemur occur here as regularly as the rain that sweeps in great sheets off the Indian Ocean. In many ways, Ranomafana, with its riches and its problems, is a microcosm of Madagascar. And if one wants to see a model of sustainable development—that is, a place where poor people can improve their lives without compromising their natural resources—one can hardly come to a better place than Ranomafana.

Or so Wright would have you believe. Forcibly.

I first met her at a talk she gave in Boston. She is the kind of speaker who, by letting her peaceful eyes linger long on as many people as possible, makes each person feel she is talking to him or her personally. She speaks with verve and humor, her expressive face transforming constantly, from feigned surprise to deep seriousness to wide, toothy grin. By the end of the lecture, I could see that most listeners had taken her argument that Ranomafana was that proverbial model hook, line, and sinker.

Not everyone has done so. Joe Peters, an adjunct assistant professor of natural resource management at Grand Valley State University in Allendale, Michigan, spent two years at Ranomafana in the early 1990s as a conservation technical advisor. In a recent scientific paper, one of several he has written that challenge the notion that Ranomafana is that model Wright speaks of, he maintains that the park was "flawed from the start because of exogenous impulses that projected it into national-park status in the first place." Foremost among those impulses, in Peters's view, was a goal to preserve the biodiversity at all costs, even at the expense of the rights and livelihoods of local people. Peters feels that the park has not adequately compensated the 27,000 Malagasy in the park's three-mile-wide "peripheral zone" for losing access to the forest their families have utilized for generations. "Is biodiversity preservation still good," he asks in another of those papers, "if it practically denies 27,000 of the poorest people in the world access to subsistence resources?"

I have come to Ranomafana to try to see for myself where the truth lies.

One thing is clear: The forest is still here. That cannot be said of areas farther east. Rain forest once blanketed the entire eastern portion of Madagascar, all 100 miles or so between the white beaches lining the Indian Ocean and the top of the escarpment that runs north-south down the island, just east of center. Save for a large patch in the northeast and progressively smaller patches as you head south, all of that has fallen to farmers, ranchers, and loggers. Scientists calculate that the original extent of Madagascar's eastern rain forest, estimated at over 27 million acres, dwindled to less than 10 million acres by 1985. Between 1950 and 1985, the most concentrated period of clearance, the annual rate of loss averaged almost 275,000 acres, which is roughly equivalent to a patch the size of Long Island lost every three years.

Ranomafana's rain forest survived because it straddles the escarpment's highest, steepest ridges, where the soil is unusually poor. But without protection, it was only a matter of time before the advancing front of farmers and loggers would wipe it out. Even now, with that protection, the forests lie to some extent at the mercy of the Betsileo and Tanala who inhabit this mountainous region 350 miles south of the capital.

We leave the goldens and continue hiking deeper into the forest on the trail known as Talatakely ("Little Tuesday"), named for a small village with a Tuesday market that once existed on this ridge. Skinks scuttle off into the dry leaf litter, and I can hear birds singing off through the trees. Wright points out some of the native plants along the path—tree ferns, their verdant fronds reaching out gracefully from a hairy trunk; the *ravenala* or traveler's tree, so named for the fresh water that thirsty travelers can find cached at the base of its fan-like array of leaves; and a wispy green plant known as *Rhipsalis*. Madagascar's only native cactus, *Rhipsalis* is what Wright terms a "biogeographic miracle," for its genus, which exists naturally all the way over in the Americas, somehow managed to get to the island. "It was one of the few friends I found when I got to this forest," she says, "because when I was doing fieldwork in Paraguay, I used to watch my little night monkeys come out and eat the fruit."

The trail rises through a forest harboring few of those towering buttressed trees so characteristic of, say, Amazonian rain forest, for the area was heavily logged in the late 1980s, even as Wright observed her lemurs. "We would write one thing in our notebooks and a tree would go down," she told me. "Write something else and another would go down." The lack of big trees and the time of year—it's October, the end of the dry season—allows a lot of light to penetrate to the forest floor. In fact, this rain forest is unlike any I've been in before. Those I've visited—in Peru and Brazil, in Thailand and Sumatra, even here in Madagascar—are hot, humid, richly odoriferous affairs. You languish in the stifling heat of midday and sprawl sweating in your tent at night. Everything you own is perpetually wet and smelly. It rains a lot, and heavily. But Ranomafana's forest is different. Last night, I could barely sleep—for the cold. Thinking *tropical rain forest* before I left the States, I deliberately left my sleeping bag at home, bringing only a sheet sewn to form a sleeping bag. The two thin airline blankets I pack for emergencies were of little help, and even when I put on my only warm clothes—a sweatshirt, blue jeans, and a pair of wool socks—the chill still got through.

Now, as I walk through the morning rain forest with Wright, I notice other disparate aspects. For one, I can see sky. It's just little patches through the trees, but it's there. Looking up, I can get an image of the whole sky—it's clear, with just a few wispy clouds—by assembling those patches in my mind, as if they were dabs of color on a pointillist painting. In typical rain forest, you're lucky to have more than a dim twilight at ground level, much less a glimpse out through the canopy. Another thing: It's dry. The air is dry, the trail is dry, the dead leaves littering the forest floor are dry. I know it's the end of the dry season, but since when is a rain forest dry? I could ask Wright, but she's too busy indulging her passion for teaching.

"This is a *Dombeya*," she says, stopping beneath a tall tree with broad, maple-leaf-shaped leaves. "It's the only tree I can hear before I see, because when it's flowering the bees buzz all around it."

She smiles and looks at each of us in turn.

"Why is it important to people? It's fast-growing, and people also strip off the bark and use it like rope."

She picks up a stiff, thick leaf.

"*Avahi* eat these leaves," she says, using the generic name of the woolly lemur. "Look at this leaf—would you eat this?"

Wright laughs that mannish laugh and goes on, savoring her rapt audience of fresh volunteers.

"Jorge Ganzhorn from Germany did an analysis of the niches of two nocturnal primates here that are both about the same size, *Avahi* and *Lepilemur*. He found that they eat totally different leaves. The leaves *Avahi* eats are very high in tannin, and *Lepilemur*'s are very high in alkaloids. Each of them has separately evolved a way to get rid of those different kinds of toxins. They can't do both at the same time, it's too hard."

An hour into our jaunt, just after we descend a sheer ridge and come out in a bushy clearing by a marsh, a group of Props suddenly materializes. "Props" is Wright's abbreviation of the Milne-Edwards's sifaka's generic name, *Propithecus*, pronounced with a long "o." The Props are less shy than the goldens. One settles down in a bush not ten feet into the forest and watches us as it languidly munches berries. Its thick fur is a dark chocolate brown, save for an off-white band around its midsection, and it has a long, thin tail. The lemur also has a red collar and a red tag for easy identification. "Red Red," as he is known, is an older male. But he is not dominant. That role is reserved for females, not just in this species but in all lemurs.

As the Earthwatch volunteers familiarize themselves with their research subject, which Wright will have them follow all day for five straight days, our leader tells us about a young male Prop that left its group to join a group of females.

"As the relationship was developing, he would get lost in the territory and get panicked," she says, sweeping her eyes conspiratorially over the group. "He would call and call and call. The females would sit there, and they knew he was really getting lost, and they could see how panicked he was getting. They didn't answer. He'd get more panicked, and they'd just sit there. Finally, when he was just becoming frantic, they'd give one answer and he'd come back, acting like 'Thank God I found you, I found you.' And they'd act like, 'Yeah, we're dominant, we're the leaders. Don't ever forget that.'"

Glancing at Wright speaking with easy confidence to the team, I can't help but think that that story might as well be autobiographical.

2.

PAT WRIGHT ALMOST DIDN'T come to work in Madagascar. In 1984, while attending a conference in Nairobi, a colleague suggested they pop over to have a look at the island. "It wasn't until I flew over Madagascar that I realized what deforestation was all about—all those barren, barren hills," Wright told me. "I burst into tears, and I decided I'd never return to Madagascar, it was too sad. Luckily, we have short memories."

Like Wright, when you fly over the island for the first time, heading to Tana in the central Highlands, you can't help but feel that the question *Can Madagascar be saved?* is moot. If you come by way of Nairobi, as I have on four occasions, you pass over the northwest coast, making a diagonal for the capital in the center of the island, though no matter which direction you arrive from, the view is largely the same. As soon as you pass over a band of coastal forest, whether rain forest in the east, dry forest in the west, or spiny desert in the south, you enter a barren land. Mile after mile after mile, you see below you bare, treeless hills or plains, often gouged by erosion gullies and blackened by brush fires. On a clear day, this view stretches to the horizon in every direction, and you feel as if you're looking at the surface of another planet—a dead planet. If you come at the end of the dry season, the smoke from those countless fires burning out of control below makes the view even more apocalyptic.

Madagascar can't be saved, you think. It's already been lost.

Fortunately Wright does not subscribe to that notion. She not only returned to Madagascar, she has returned annually ever since.

For fourteen years, she has done everything in her power to answer that question with a resounding *Yes!* Her answer is Ranomafana. To begin to examine that overriding question and Wright's response to it, it's instructive to look at the history of conservation in Madagascar, which set the stage for Ranomafana National Park.

Conservation has a long pedigree on the island. As early as the late eighteenth century, a prescient Andrianampoinimerina banned the cutting of live firewood. But Madagascar was still thick with forests then, and Malagasy slashed and burned with impunity, and not only for agriculture. Richard Baron once saw villagers who were dragging a tombstone through the forest chop down 25,000 trees just to make a path. In his 1881 "Code of 305 Articles," Prime Minister Rainilaiarivony tried to slow the felling by declaring that anyone caught cutting virgin forest for agriculture or charcoal would be chained in irons. But the felling continued, and a decade later, Baron for one was declaring the law a dead letter. "Probably more than one half of the original forest has been already cut or burned down," he wrote in 1892, "and in a few generations, at the present rate of destruction, this still magnificent mass of vegetation will have been swept out of existence."

Soon after the French took control in 1895, they began to clamp down. They started reforesting the island by establishing plantations of fast-growing, nonnative species such as eucalyptus and pine. They banned the killing of lemurs, and in 1927, they formed the first protected-areas system in the African region by setting up the first ten strict nature reserves, including Lokobe, which armed guards patrolled and protected. Building upon that foundation, the country went on to establish national parks, classified forests, reforestation zones, no-hunting reserves, and so-called special reserves, which protect specific plant or animal species. Today, Madagascar boasts more than fifty protected areas.

Unfortunately, the French also instigated major deforestation on the island. They planted the most fertile bottomlands on the eastern seaboard with coffee, bequeathing huge plantations to French and Chinese owners regardless of whether Malagasy already worked that land—which they had for centuries. Indeed, the geographer Lucy Jarosz links deforestation in Madagascar directly to colonial-era cash cropping of coffee as well as to logging and forest concessions, which she says resulted in the destruction of nearly three-quarters of the remaining primary rain forest in Madagascar in the first three decades of French rule.

Since Malagasy no longer had use of those bottomlands, they took to cultivating forested slopes for subsistence. French officials responded by banning the clearing of forests for tavy, ostensibly, Jarosz notes, to promote "rational forest resource management." If you toed the line, officials gave you free chemical fertilizers and coffee trees. If you didn't and they caught you practicing tavy, you had to either round up thirty men to work with you on a public works project or face a prison sentence.

The ban backfired, leading to even more deforestation. As Jarosz states, the ban "elevated the practice of tavy to a symbol of independence and liberty from colonial rule." First stripped of their ancestral bottomlands, then denied tavy, which they held as an integral aspect of *fombandrazana,* the "way of the ancestors," the Malagasy rose up in revolt. It's little surprise that the Rebellion of 1947 began in the Ranomafana region, where resident Betsileo and Tanala view tavy as their inalienable right. As an elder told the anthropologist Paul Hanson, who lived and studied in a village near Ranomafana, "We Malagasy are free. What makes us free is the search for fertile land."

Independence didn't help matters. Indeed, the immediate postcolonial era witnessed a radical downturn in conservation activities, mainly through lack of funding. Patrols decreased, and rangers could no longer bear arms. In 1961, the government drew up a threatened-species list and prohibited hunting of everything from the radiated tortoise to the blind cave fish. But enforcement of the new law proved woefully weak. The following year, President Philibert Tsiranana decreed that all Malagasy men had to plant 100 seedlings a year or suffer a tax, but the law was abolished a decade later. As a result, between 1960 and 1990 half the remaining forests in Madagascar vanished.

In 1970, an international conference on the environment, held in Tana and attended by numerous foreign scientists and conservationists, briefly inspired renewed field research and conservation efforts and made Madagascar front-page news. But the "May Revolution" of 1972, which the Malagasy viewed as a kind of second independence from continued French domination, triggered a series of political revolutions lasting until 1975. That year, Didier Ratsiraka established a socialist regime and booted most foreigners out of the country, leaving science and conservation at a virtual standstill. For the next decade, Madagascar truly seemed to be lost.

3.

One night on the porch of the research cabin she built high above the Namorona, Wright tells me about the chilly conditions she faced when she first began working in Madagascar in the mid-1980s. We're both feeling subdued after a big meal (fish with tomato sauce, rice, carrots, bananas, beer) and a day marching through the forest on the padded heels of Props.

"There was a lot of xenophobia, because the country had been cut off from the West for over ten years," she says, the unforgiving white light of a kerosene lantern casting her features in stark relief. "Everybody was afraid of strangers. I can't tell you how many times Malagasy told me, 'We don't need technological expertise. Why are you coming here?'"

Beyond her, flanking her like bodyguards, two traveler's trees she planted in 1989 rear into the night air, their fans of leaves also cast in the lantern's unnatural glow. The light fails to penetrate the tangle of forest behind them, though. Nor does the lantern's loud hissing drown out the thunder of the Namorona 100 yards down the hill. Our tents are down there, ranged along the forested edge of the river and its tributary, the Mariavaratra, whose quieter rush can also be heard. I have to lean forward to hear her.

"Now, for Americans there was none of that 'Yankee go home' stuff like there was in South America," she says. "It was the French whom they had all the hostile feelings for. But from the very beginning, we tried to be sensitive to the fact that Madagascar is not like anyplace else, and it's not our country. So we matched every technical adviser we hired with a Malagasy. This was a training period, where the technical advisers were to transfer information to the Malagasy so they'd be ready to take over in ten years."

She raises her eyebrows and smiles.

"It was a wonderful plan. The hitch was that the Malagasy chosen were somebody's cousin and not the most competent. So the first national director we got was one of the biggest crooks I ever met. He never came to work till the day he was fired. Then he sued us for firing him!"

A less dauntless person might have given up. But Wright has had plenty of experience in her own life to prepare her for the challenge of undertaking fieldwork in Madagascar, much less establishing a new national park.

Born Patricia Chapple in 1944, she was raised in Lyndonville,

New York, a small farming community of about 500. Though she graduated with a B.A. in biology from Hood College in 1966, little in the next decade of her life would indicate that she would become a leading primatologist. While working at Harvard Medical School in Boston, she became engaged to her high-school lover, the artist Jim Wright, and followed him to New York City when he entered the Pratt Institute of Art. They got married, and for the next ten years she was a housewife and social worker with the Department of Social Services, working in some of the city's poorest neighborhoods, including Brownsville and Bedford-Stuyvesant. "The only reason I wasn't killed was my naïveté," Wright says now. "Once I asked one of my clients, 'Is it safe here?' and she said, 'It is for you, because we know you're here to do good. But it ain't for others.'"

One day she walked into a pet store and, though she had no way of knowing it at the time, her life changed forever. Captivated by a furry little creature—a night monkey—she bought it for forty dollars and took it back to her apartment. The next morning she left for work as usual, leaving the monkey to fend for itself. When she returned home at the end of the day, there was no sign of the monkey. She searched in vain and finally decided it must have escaped through the open skylight. Distraught, she called the police to inform them that a nocturnal South American primate was on the loose in New York City. Then, around dinnertime, Wright suddenly heard a *click-click-click* sound. "Did you hear that, Jim?" she asked her husband and began looking around. Turns out the monkey had slept the day away in the springs under their bed. Herbie, as they called him, was a great pet, except that every time they left him alone in the apartment, he trashed the place. Wright decided Herbie needed a mate, and she and Jim made a special trip to Peru to bring back a female.

A year later, in 1973, they took a vacation to Costa Rica. During the trip, Wright fell horribly sick. "I didn't eat for a long, long time," she recalls. "I told Jim I had to leave or I was going to die." Back in the States, doctors quickly determined the cause of her malady: pregnancy. She had been suffering from morning sickness. Wright gave birth to a daughter, Amanda, later that year. Surprisingly, two weeks later the female night monkey, too, gave birth. "So we raised our infants together," she laughs. In hindsight, one is tempted, as with that brush with the goldens, to see more than serendipity at work. But it was not until two years later, when tragedy struck, that Wright finally decided to devote her life to primatology. When Amanda was two, Wright gave birth to another girl, but the baby died within days.

"As I lay there in the hospital, I made a decision," Wright says. "If I couldn't do what I wanted to do—have two kids—I would do what I second most wanted to do, study night monkeys."

Within two years, she found herself back in the Peruvian jungle with Jim and Amanda, ready to study *Aotus*, as the animal is known scientifically. One afternoon soon after their arrival, she went out alone to look for night monkeys. On the way back to the camp, where she had left her husband and daughter, she got lost and ended up spending the entire night in the forest. Early on, hearing a large animal crashing through the forest, she scurried up a tree, only to spend the rest of the night carefully removing from her hands and knees two-inch thorns she'd gotten on the climb. Toward dawn, as she sat leaning against a tree, she heard that now-familiar *click-click-click* sound above her and felt something land on her head. A male night monkey sat above her in the tree, urinating and defecating on her. "He was angry that I was in his territory," Wright laughs. "I was in heaven."

Six months later, she returned to the States and called the late Warren Kinzey. Kinzey was a biologist who, probably more out of the goodness of his heart than because he thought anything would come of it, had earlier encouraged the Brooklyn social worker who had contacted him to give her dream a whirl; he had even offered her a few pointers on field research. "He didn't believe I'd actually do it," she says. But when she described over the phone the data she had collected, he said, "I'll be right down." He drove to Brooklyn from his home in upstate New York, looked at her data, and said, "My God, you've got a paper." A year later, that paper was published in the prestigious primatology journal *Folia Primatologica*.

Wright was on her way. A workaholic by nature, she made up for lost time. Between 1976 and 1984, when she conducted her first fieldwork in Madagascar, Wright studied night monkeys, yellow-handed titi monkeys, dusky titi monkeys, and capuchin monkeys in various sites in South America. She trapped and censused six species of marsupials in Peru and analyzed the reproductive and foraging behavior of four-eyed and white-bellied possums in Paraguay. She censused Philippine tarsiers in the Philippines and investigated the behavior and ecology of western tarsiers in Malaysia. Amanda was with her on many of these forays, having begun to accompany her mother into the field at age three, though Jim by this time had left Wright (they later divorced). During that period, Wright also entered graduate school and in 1985 left the City University of New York with a doctorate in anthropology, almost twenty years after she graduated from college.

For her work in Madagascar, she would go on to earn a MacArthur "genius grant" in 1989 and, in 1995, the National Medal of Honor of Madagascar, the equivalent of a knighthood. She has remarried, spending a month of each summer in Finland with her husband Jukka Jernvall, a paleontologist and developmental biologist at the University of Helsinki, who also works in Madagascar.

Now, on the cabin porch at Ranomafana, Wright continues on, oblivious to the chill that descends every night.

"International agencies began to realize that if they were going to put a lot of money into Madagascar, which many considered a priority because of its biodiversity, then they had to do a lot of training and changing of people's minds."

She chuckles.

"We've come a *long* way since then."

Again, her words might as well be autobiographical.

4.

BY THE EARLY 1980S, conservation thinking among the world's environmental and development organizations had come a long way, too. A new ethic was beginning to take shape. Eventually that ethic would gain a name—integrated conservation and development—along with widespread official acceptance at the 1992 U.N. Conference on the Environment and Development in Rio de Janeiro. But initially it was nothing more than the idea that if you wanted to conserve biodiversity and ecosystems, you had to make it worth the while of the people who lived in the neighborhood. Simply declaring a new reserve and telling everyone around it to keep out did not work; people, as anyone would who felt cheated out of his or her birthright, found a way in. It may seem painfully obvious to us today, but even as recently as the 1970s, the notion of giving local people a stake in the conservation of their own natural resources was all but unknown to large development agencies such as the World Bank.

In Madagascar, things had begun to improve in the 1980s following the conservation black hole of the previous decade. Enthusiasm sparked by the 1970 conference resumed, and in 1985 the Malagasy government convened a second International Conference on Conservation for Development. Even President Ratsiraka appeared willing to listen, and Britain's Prince Philip, international president of the

World Wildlife Fund (WWF), wasted no time in commenting to him at the meeting, "Your nation is committing environmental suicide." Following the conference, environmental organizations stepped up activities in Madagascar. WWF alone increased its expenditures there tenfold between 1983 and 1993. Foreign researchers, too, enjoyed new leeway, and it was these scientists who brought the novel thinking to Madagascar.

The new approach had its first test at a reserve formed in the scorching spiny desert of the southwest in the late 1970s. Beza-Mahafaly—Beza means "many (or big) baobabs" and Mahafaly is the tribe that lives in the area—came about in a way that seemed to perfectly accommodate the new ethic. It all began in 1975, when Alison Richard began a search with two other scientists, anthropologist Robert Sussman of Washington University in St. Louis and biologist Guy Ramanantsoa of the School of Agronomy at the University of Madagascar, for a site in the southwest where they could establish a new reserve. The reserve would protect fast-dwindling spiny and gallery forest, and it would serve as a training ground for the next generation of Malagasy scientists. Moreover, it would differ from reserves created during the French colonial period in that, from the beginning, it would directly involve local people.

Along the Sakamena ("Red River from Far Away"), they found just what they were looking for: a 250-acre patch of intact gallery forest and, a few miles farther up a dirt road, a 1,200-acre spread of pure spiny desert. Little did they know how ideal this site was. For as soon as Richard and her colleagues began cautiously asking around, they found to their astonishment that local elders were more than receptive to the idea of forming a reserve. In fact, concerned about how rapidly their ancestral forest was disappearing, they had long sought for a way to protect it. It was a perfect match. The Mahafaly would donate the land for the reserve, and the researchers would raise funds to undertake development projects to improve lives in the hardscrabble villages surrounding the reserve. In 1985, Beza-Mahafaly became the first new reserve established in Madagascar since the 1960s.

Over the next eight years, with WWF funding, staff demarcated the reserve boundaries, fenced the gallery forest, hired park guards from local villages, built simple research facilities, and established a field school for students from the School of Agronomy. It wasn't until 1985, when USAID kicked in a big grant and WWF increased its funding, that long-promised development activities began. Reserve personnel helped build a new school, dug wells, repaired a road to the nearest market town, worked with farmers on agricultural and agro-forestry projects, and began restoring a dysfunctional irrigation canal. In 1990, Alison Jolly was able to write optimistically that Beza "has become a model for increasing support to many forests in Madagascar's 'necklace of pearls' " — those remaining forest stands ringing the coasts.

Unfortunately, Beza has not been an overnight success, for reasons ranging from inadequate funding to an incipient distrust between local Mahafaly and the Merina from the capital who administer the reserve. Despite its rough edges, however, the reserve has served as a valuable test case of integrated conservation and development. Today, except for the director, who is a Merina from the School of Agronomy in Tana, the entire reserve staff, including park guards and naturalists who monitor target plant and animal species, consists of Mahafaly recruited from local villages.

The idea launched at Beza took greater form in Madagascar in the mid-1980s. Spurred by increasing outside interest in protecting the island's biological treasures, the Malagasy government launched a concerted effort to establish an integrated conservation and development system in the country. This led to a national conservation strategy in 1984, a forest policy in 1985, and the world's first national environmental action plan (EAP) in 1989. The fifteen-year EAP, designed to run from 1991 to 2005 and infused with roughly $168 million* in pledges from outside donors for its initial five-year period alone, has inspired more than twenty other nations to fashion similar plans.

To put the plan in action, two new bodies were formed: the USAID Grants Management Unit, which would supply the funds, and the nongovernmental National Association for the Management of Protected Areas, known by its French acronym ANGAP, which would serve as the national park service and oversee management of the island's protected areas. Working together, ANGAP and USAID selected nine protected areas — including Beza and Ranomafana — as sites for so-called ICDPs, or integrated conservation and development projects. USAID supplied the cash, and ANGAP brought in "operators" such as the Missouri Botanical Garden and Duke University to undertake research and help get the ICDPs off the ground. All parties agreed on the overall goal of the ICDPs: to place high priority on conserving biodiversity and training conservation profession-

---

* Malagasy conservationists were thunderstruck by such amounts. In 1987, the national investment budget for the entire protected-areas system was less than $1,000.

als, while also involving local communities in managing the protected areas for mutual benefit and sustainable use.

In other words, Beza writ large.

And Ranomafana. In essence, Beza has passed the baton to Ranomafana, which, again if you believe Wright, has become the large-scale model of integrated conservation and development that Beza first promoted on a small scale.

The ICDPs ended with the conclusion of the EAP's first phase in 1996, and many conservationists felt they failed. Some maintained that they were too local and too reliant on the traditional concerns of conservationists and development experts, making them largely irrelevant to the serious issues facing Madagascar and its environment. "According to this critique," write Alison Richard and Sheila O'Connor, both long-time researcher-conservationists in Madagascar, "ICDPs are at best a palliative, but certainly not sufficient and should not be viewed as the central thrust of a conservation strategy." ICDPs also made painfully clear the disparities between local people and the outsiders who were ostensibly trying to help them. As Richard and O'Connor write, "all too often those who benefit most bear little of the cost, which is shouldered by those who can least afford to pay and for whom the benefits are least." Such asymmetry, they note, is "a strongly marked and pervasive feature of ICDPs: Differences in formal education, authority, opportunity, and access to resources distinguish local communities from those who bring the conservation message, whether foreigners or Malagasy nationals from outside the community."

Joe Peters, for his part, feels Ranomafana's ICDP simply missed its target. "The underlying assumption of the ICDP approach is that these development interventions will reach those whose activities threaten the protected areas, and that these interventions will motivate the target populations to modify their activities in keeping with conservation goals," he has written. However, most interventions during his tenure at Ranomafana were limited to a small number of pilot villages along the main road, he says. "[T]he sheer demographic and geographic scope of the project . . . predisposes against reaching even a portion of those persons whose activities threaten the park."

Wright concedes that, like Beza, Ranomafana—which continued as a park after the ICDP ended, of course—has had a long ramp-up time and its success must be measured in small steps. "[W]e learned," she has written, "that long-term success is a day-by-day, year-by-year process of negotiation, discussion, and compromise." From the start, she and her colleagues strived to work with local communities around Ranomafana, she says. Her approach was "bottom-up," not "top-down" as so many large-scale, World Bank–funded development projects had been in the past. The idea was to work up from the needs and wishes of the local people, rather than down from the preconceived theories and methods of the funders.

In the design and planning phase of Ranomafana National Park, for instance, Malagasy teams trained in education and health services began monthly visits to all villages in the peripheral zone, finding out how residents lived and getting them used to the idea of the park. To understand impacts on the forest and its resources, researchers investigated the ways in which villagers traditionally use the forest. They were not surprised to discover that local Betsileo and Tanala make intensive use of local resources. They burn the forest to grow hill rice, chop down trees for house construction, harvest reeds for mat-making and tree ferns for use as flower pots, graze cattle in the forest to hide them from rustlers, and gather firewood, honey, and medicinal plants. Some illegal timbering also took place, they found.

Those early years of research identified the problems that the Ranomafana National Park Project hoped to address, Wright says. But thinking up solutions, particularly establishing alternative sources of income for villagers suddenly prohibited from tapping the forest's bounty, was no easy task. To further the work of those initial years, Wright, as international director of the Ranomafana National Park Project, worked with Malagasy and Western colleagues to form six special teams. These focus on six areas that adherents of integrated conservation and development see as critical: park management, ecological monitoring, biodiversity research, health, economic development, and education and ecotourism.

5.

IN MY EFFORT TO SEE how each of those six areas is functioning, I start with park management, driving down one morning with Wright in the project van to pay a visit to the park director in Ranomafana's lower reaches. The road, part of a paved highway that stretches for 150 miles from Fianarantsoa, the Betsileo capital in the southern Highlands, to Mananjary, a port town on the east coast, winds down through rain-forest-cloaked hills. It's a clear day, but wisps of mist still hug the hillsides.

Four miles from the park entrance, we come to Ranomafana Town, a ramshackle village of several thousand souls spread out on either side of the road. Managing the park along this road must be the easy part, I think, for most of the park consists of steep, jungled ridges and hidden valleys far from any habitation, much less road. Yet as we drive up the hill above Ranomafana Town to the park headquarters, I see smoke from a fire high on a hillside, like a taunt to the park director. Can he not even manage the park within his sights?

The park headquarters is a long, one-story building with four or five rooms facing east toward the ocean far below. It's prefabricated log, just like Wright's research cabin. After making us wait on the wide porch just long enough to remind us who's boss but not long enough for us to get annoyed, the park director appears and greets us with a smile. Dressed in the standard business attire of the professional Malagasy—tailored slacks, button-down shirt, nice shoes— Jocelyn Rakotomalala seems brawny and fit, though he's not a big man. With his round, slightly beefy face, he looks like a handsome Mañuel Noriega. He graciously ushers us into his office.

The room is small and spare, with little on the walls, and his wooden desk is all but bare. Wright settles in the corner of a sofa across from him, while I grab a chair between them facing an open window, through which wafts a heady scent of blossoming bougainvillea. It occurs to me that I'm sitting between the two most important players at Ranomafana National Park, the president and prime minister if you will.

Thinking of that fire up the hill, I ask Jocelyn about his greatest challenges with managing such a vast spread. He pauses before answering.

"We're concerned about the carrying capacity of the park," he says, speaking in a soft, surprisingly high voice. His English is good, but I have to lean forward to hear him. "Nobody"—he glances at Wright and smiles—"can say now what that is either in tourists or researchers."

He looks down at his desk, still smiling, then goes on. Tourism to the park is increasing by 25 percent annually, he says, with 6,500 visitors in 1996, and dozens of foreign researchers and their graduate students descend on Ranomafana every year. Again he looks at Wright and smiles.

"Everybody has an opinion about research and tourism," he says. "We need zoning, and to close Talatakely at certain times of the year." Talatakely is Wright's "Little Tuesday" research trail.

In the polite, unassuming way of the finest Malagasy diplomat, Jocelyn is staking his ground, challenging Wright in a completely unchallenging way. Wright and her Ranomafana Park Project ceded full control over management of the park to ANGAP and its ambassador Jocelyn only a few months ago, in June 1997. He has his own goals—increase tourism to 20,000 a year in five years and make the park financially self-supporting, for instance—and those goals may not entirely jive with Wright's. Wright was responsible for bringing Jocelyn here in the first place, but their relationship by the very nature of it—figuratively speaking, he's now raising Wright's baby—is delicate.

If there's one area in which Jocelyn's park management and Wright's biodiversity research mesh seamlessly, it's with ecological monitoring. And this is where I head next, to visit with the TSEs, a French acronym for Technicians for Ecological Monitoring. The five TSEs and their assistants, all of whom are Malagasy from local villages, monitor the health of the park by checking periodically on five target subjects—birds, insects, chameleons, small mammals, and plants—to see how the park and its visitors affect them over time. They supply these data to Jocelyn to improve park management.

We met the TSEs the first day. Wright ushered the Earthwatch team into the laboratory, a prefab log building a short walk from the research cabin. It seemed crowded enough in the peaked-roof lab with the Earthwatchers, but then the TSEs and other Malagasy research assistants began filing in. Ten, twenty, close to thirty smiling, self-deprecatory Malagasy were soon standing shyly against one wall. "Without this team, Ranomafana would not be Ranomafana," Wright declared proudly, standing before her carefully assembled team, some of whom she has worked with for more than ten years. "All speak Malagasy, English, French . . . and Latin!" she said, looking around the room with eyes ablaze and laughing her deep laugh. She had two of the TSEs come forward for special recognition. Emile Rajeriarison and Loret Rasabo, she said, served as her guides when she first came to Ranomafana in 1986.

Later that day, over a Three Horses beer, Emile told me his story. With his mildly pointed face and unshaven cheeks angling down to a bushy goatee, he looks more Arabic than either Indonesian or African. He was born and raised in Majunga on the west coast, and perhaps some of his ancestors in Majunga were Arabian merchants. Emile came here when he was eighteen to work on a hydroelectric dam that the Japanese were building across the Namorona. He fell in love with Loret Rasabo's sister and decided to stay; they are now married with six children. Emile told me that in the early 1980s, when a French scientist became the first foreigner to try to work here, the villagers fled from him. Locals were still wary of the French, and by extension all foreigners, more than three decades after the 1947 Rebellion, which French authorities brutally suppressed, burning Ranomafana Town to the ground and killing hundreds of local people. Emile, not being a native of the area and having had plenty of contact with foreigners in Majunga, had no such fear and led the French scientist around. When Wright arrived in 1986, he led her around, too.

Now, as the sun reaches midway through the sky, I join Wright and the Earthwatch team to see firsthand what Emile and his fellow TSEs do. We crowd into the project van and drive up the serpentine road toward the Highlands. Thick rain forest pushes down onto the road on one side, while on the other, the frothing Namorona crashes through a gorge far below. During cyclone season, from January through March, this stretch can be treacherous. Two years ago when I first visited Ranomafana, a landslide triggered by the heavy rains had taken out an entire hillside, and the road with it. Because this is the only road to the coast in the region, porters ferried goods between trucks dead-ended on either side of the slide, and a veritable village of food shacks sprung up on both sides to cater to the truckers. To get to the park entrance, I had to scramble across the landslide. With every step I sank up to my knee in gluey mud. It was pouring, and as I crossed I kept glancing up at the avalanche of rain-riven mud above me, wondering if it would suddenly come down.*

Now, in the dry season, the steep slopes are stable, and we reach our destination without incident. It's a metal bridge over a Namorona tributary, where the TSE team, by prearrangement with Wright, waits to guide us to its current monitoring site. Loret Rasabo leads the way into the forest. He is tall for a Malagasy, with a quick smile and languid eyes. Wright told me that now that the ornithologist Olivier Langrand, author of *The Birds of Madagascar*, no longer lives on the island, his protégé Loret has become the best birder in Madagascar, eagerly sought after by ornithologists and birdwatchers alike. So I stick close to him, hoping for a good sighting. But there are few birds around, and we hike in silence.

---

* Not long after that 1995 visit it did, sending the bulldozer that kept the mud slope crossable tumbling down the hill, where it burst into flames and killed two Malagasy.

We're at a higher altitude here, and the forest canopy is lo\
only forty feet high or so. This is montane woodland as opposed to
the lowland forest around the research cabin. Many of the trees have
lost their leaves, and the tall bamboo grass that covers the ground is
all dead, killed by a rare hail and frost several weeks ago. It's a tinder-
box in here, and so unlike typical rain forest that it's eerie.

Before I have a chance to work up a sweat, we stop at a fencelike
black-mesh net strung between two trees. Here Ernestine Raholima-
vo, a short, wide-eyed Malagasy woman who heads up the team,
explains its mission. Twice a year in the cold and hot seasons, she
says, the TSEs descend on sites like this one to monitor the five target
subjects. There are seven sites within the park, each representing different habitats, elevations, and degrees of human disturbance, and
each is paired with a companion site in the adjacent peripheral zone.

"They also patrol the park and ask villagers about what they take
from the forest for medicine, food, mat-weaving, and so on," Ernestine says in perfect English. She is one of a handful of Malagasy
whom Wright has helped to get advanced degrees in the States.

Each of the five TSEs then gets up to say what he does. Most are
endearingly shy about speaking in English before a dozen vazaha, but
not Emile. Wearing a ZOOLOGICAL SOCIETY OF SAN DIEGO sweatshirt
and green Ranomafana cap, he looks the part of professional field
researcher, and he plays it with aplomb. He saunters over to the
black-mesh net and explains that it's a malaise trap for catching flying
insects at night. They fly into the net, which rises vertically from the
ground to about ten feet up, then instinctively fly upward, he says,
where an overhanging portion of the net stops them. This flap arcs up
to a light attached to a branch. When they fly toward the light, they
end up as specimens in a bottle affixed to the corner of the net. He
also uses pitfall traps—small plastic containers sunk in the ground—
to capture ground-dwelling insects. He takes all specimens back to
the lab, where he identifies them as to family.

Peering into the bottle, Emile sticks his finger in and begins vigorously scraping out ants that have crawled inside.

"I like ants, but I don't like them to eat my specimens," he says
quietly, and is surprised when he gets a laugh out of his audience.

He then yields the forest floor to the others. Loret explains how
he censuses Ranomafana's 114 bird species by walking a transect and
looking for birds or hearing their calls. Pierre Talata, a middle-aged
man with large, watery eyes, tells how he searches for Ranomafana's
diverse set of chameleon species by following a transect at night,

amp to locate them in the trees on either side and

r species name, height off the ground, and other infor-

ar stories come from the other TSEs.

at these modest men, I think if anyone will save Madagascar it will be these people. Not foreign scientists or development experts or conservationists, even ones as well meaning and hard working as Pat Wright. Not even university-trained Malagasy from the capital like Jocelyn Rakotomalala. But these regular people, these local people. "The assumption has always been that wildlife conservation is the prerogative of naturalists, trained ecologists, and foresters, and that modern biological science is the only discipline needed to carry it out," write a trio of Indian conservationists in a recent issue of the scientific journal *The Ecologist*. "Lacking formal training, local people are deemed to have nothing to offer conservation."

These men challenge that assumption. Trained on the ground over the past decade by Wright and an international cadre of specialists, they have added Latin names and modern theories to their self-taught knowledge of their island's natural history. Their enthusiasm for their work and their implicit understanding of its value shines through in these brief speeches. Who better to take the pulse of a place than the people for whom it is home?

When the speeches are over, Emile leads the way back to the road down a different path. The unobstructed sun we've had for several days now shares the sky with some threatening cumulus clouds. Will we finally see some rain in this rain forest?

We traverse a ridge covered in low, scraggly trees. Emile points out several with small holes gouged out of their trunks. They were made by that most unusual of lemurs, the aye aye. After using its bat ears to pick up the sound of a grub or other insect moving about inside, the aye aye chewed the holes with its beaverlike teeth, then jabbed its twiggy middle finger inside to spear its victim.

I chat with Emile about the Malagasy wildlife he knows so well. When I tell him that the leaf-tailed gecko is my favorite animal in Madagascar, he turns, flashes a knowing smile, and nods.

"And how about the leaf-nosed snake, with the females having those wonderfully pointed noses?" I ask.

Emile says nothing. I follow him down an incline and over a narrow stream, somewhat disappointed that our banter has ended. Did he hear me?

"Actually," he says quietly, almost as an aside, "I think the male has the pointed nose and the female has the leaflike nose."

I have to smile. He's right, of course. Emile may not have a university degree, or any formal education past the third grade for that matter. But he arguably knows as much about his country's natural history as any Western-trained naturalist. How many specialists with his depth of knowledge would be so self-effacing? When I share the incident with Wright later, she laughs and says, "He was deciding whether to be polite, but the scientist in him won out."

6.

ECOLOGICAL MONITORING IS but one piece of the pie when it comes to biodiversity research at Ranomafana. Scientists have flocked to the park, for even within Madagascar, one of the world's top five "megadiversity" countries according to Conservation International, it is extraordinarily species-rich. Botanists from the Missouri Botanical Garden, surveying in one-hectare plots in Ranomafana, counted 37 families and 105 species of trees, for instance. That is more diverse than lowland rain forest in Madagascar's vast neighbor, Africa, where botanists in Gabon counted 29 families and 99 species.

Fauna within Ranomafana show as great a diversity as flora. Besides the three bamboo-eating species and the Milne-Edwards's sifaka, day-active lemurs include the red-bellied, red-fronted, and black-and-white ruffed lemurs, while nocturnal lemurs feature the woolly, greater dwarf, and rufous mouse lemurs as well as the *Lepilemur* and aye aye. There are eight species of carnivores, all of which are viverrids, the line of small meat-eaters that predates canids and felids (of which there are none native in Madagascar). Among them are the fossa and the strikingly marked ring-tailed mongoose, which is notorious for devouring lemurs facefirst. The park has eleven species of tenrecs, including the rare aquatic tenrec, which had not been seen for a quarter century when it was discovered here. There are numerous other curious creatures, from red diurnal rats to suckerfooted bats.

Ranomafana is famous not only for its mammals. The park also harbors close to half of the island's complement of 265 bird species. Birders can find all five of Madagascar's endemic avian families here, and eight out of ten of the park's birds are endemic at the species level. These include such internationally recognized rarities as Pollen's vanga, the brown mesite, the short-legged ground roller, and the yellow-bellied sunbird asity. If you go for snails, you're in for a treat at Ranomafana. Within its borders you can find the world's largest tree snail and its most diverse assemblage of so-called bird-egg snails. You might come upon a curious species of "trapdoor" snail that lopes in a bipedal fashion up trees, or a dead leaf that turns out to be a slug with hidden tentacles. Not surprisingly, many of Ranomafana's creatures, like those in Madagascar as a whole, serve as tantalizing reminders that the island once abutted all the other southern continents. The park's six species of gray-green *Astacoides* crayfish, for instance, count their nearest relatives in Tasmania.

Sustained research at Ranomafana began with a single rain-soaked campsite that Wright and a few other researchers shared in 1986, the year she codiscovered the golden bamboo lemur. Today, the research domain at Ranomafana includes the cabin and lab, two satellite research camps at sites deeper within the park, and three extensive trail systems, with more planned. Hundreds of foreign researchers have conducted fieldwork here, and dozens of Malagasy scientists got their start working out of Wright's research cabin. In fact, more fieldwork has taken place in this small hill of rain forest than in any other protected area in Madagascar. From the start in 1987, Wright and colleagues pushed a team approach to science, with researchers who study lemurs, herps, birds, bugs, snails, fish, and ecological conditions sharing findings in an effort to understand the interrelationships of species and habitats. Fieldwork has ranged from taxonomic work on specific species to long-term ecological research designed to plumb the secrets of the ecosystem and how its denizens evolved.

Scientists have also helped to quantify the human impact on the forest. For Wright has always strived to link biodiversity research with the needs of local people, she says. One of the animals you'll come across in Ranomafana's rain forest is, curiously enough, the domestic cow. Wright told me that local Betsileo and Tanala have concealed their zebu in the forest ever since nonlocal Malagasy, brought in during the mid-1980s to help cut timber, stole virtually every zebu in every village. "People who had fifty cows before the timber exploitation had none at the end of it," she said. During initial planning of the park, villagers pleaded with Wright to allow them to keep their livestock in the woods. Wright agreed on a five-year moratorium to study bovine impacts on the forest. "Beth Middleton, a Southern Illinois University professor, checked all these cow patties— yeah, pleasant job, right?—in order to identify seeds and leaves," she said. Middleton determined that the zebu hang around marshes and eat very few plants in the forest proper. So Wright and the villagers agreed to allow all zebu currently in the park to stay, but not to add any more. "So everybody's happy," she said. "Research said it was okay."

And Wright said it was okay. As matriarch of Ranomafana, particularly the research domain, she has come under fire for what some deem an authoritarian control over who gets to do fieldwork there, and of what kind. Janice Harper, an independent anthropologist who came to Ranomafana to undertake a health and socioeconomic study in peripheral-zone villages, maintains that "From the very beginning, I was informed that any attempt to interview any local residents prior to approval by [Wright] would result in the termination of my research. 'If you get on [Wright's] wrong side, then you're out of here,' I was warned by the local expatriate director. She reminded me that every social scientist to come before me, with one exception, had their research terminated when their views conflicted with those of project administration." Harper chose to leave the project rather than have to work under these conditions.

The problem, in Peters's view, is that Wright's primary concern as a biologist is protecting lemurs and other wildlife. Even the 1993 independent evaluation of the Ranomafana ICDP's first phase implies this, he says, when it states that "preservation of biological diversity is the goal, while the other aspects are the proposed means toward that goal." For Wright, Peters believes, people come second, including local Betsileo and Tanala and the social scientists who study them. "Only with experience have biologists begun to realize that technical and social problems are inseparable," he writes. "Unfortunately, the biologist in charge of the Ranomafana ICDP has yet to learn this lesson." Citing Harper's experience at Ranomafana but without naming Wright, Peters says those in charge of conservation projects like the Ranomafana ICDP must "improve the flow of information throughout the policy-making system, which will aid in early detection of sources of perceived inequity."

7.

THE ONLY INEQUITY I perceive between me and the lemur I'm watching at the moment is that he can sleep while I have to stay awake. To get a taste of Wright's research at Ranomafana, I'm helping volunteer Ursula Brandon observe the behavior of Yellow Silver, a young Milne-Edwards's sifaka who has been doing precisely nothing for the past hour.

He's perched on a vine far off Little Tuesday, stretching horizontally from one thin tree to another only two feet away, so he can lean against one while putting his feet up on the other. One hand is draped lazily over a twig. His two-foot-long black tail hangs down beneath him, and occasionally he curls it up chameleon-style, then lets it unfurl. Though he's twelve feet off the sloping ground, he allows me to approach within six feet; I'm close enough to see his fingernails and his eyelids, which are currently shut, as they've been for most of the past hour.

Every five minutes Brandon takes note of what Yellow Silver is doing. Sleeping . . . sleeping . . . sleeping. Her beeper goes off and she scribbles in her notebook, laughing that it's the same: sleeping. Lejean Rakotoniaina, the young Malagasy assigned to ensure Brandon doesn't lose Yellow Silver when the lemur decides to move, laughs, too. Brandon can't note "nearest neighbor" because the other Props in the troop are more than 5 meters (or 16.5 feet) away. In fact, they're out of earshot on the other side of the hill.

Male Milne-Edwards's sifakas breed at six years, and Yellow Silver, at four years old, is beginning to look outside his group for a potential mate, Wright says. Groups have only three to nine members and defend a territory against other groups. Yellow Silver's group is Group I; three other groups live in the area. "He'll be the one that's on the edges all the time, seeing and smelling who's out there," Wright told us when we first came upon Group I this morning. "I predict that this group will hit all of the borders now that Yellow Silver is in the mood for moving around. Which group should he move into? Where will he get the best breeding position?"

Wright's study of the demography, behavior, and ecology of the Milne-Edwards's sifaka began in 1986, and it's the first continuous long-term study of a Malagasy rain-forest primate ever done. Like all lemurs, the Milne-Edwards's sifaka has a severely restricted range — in this case, the moist eastern Highland forests of Madagascar. The species has gone extinct outside a small range centered on Ranomafana, but here in the park itself the species is well protected, and researchers can easily study it. Wright and her helpers have observed the behavior of thirty-three individuals from three groups. (Group IV was only recently identified.) They also captured twenty-one of those thirty-three by shooting them with tranquilizer darts, then, before releasing each one, they fitted it with a colored collar and accompanying tag for quick identification. (Yellow Silver has a yellow collar and silver tag.)

As Jane Goodall did with chimpanzees and Dian Fossey with apes, Wright is gathering valuable information on population density, territory size, and habitat needs for this little-known species. People

like Jocelyn Rakotomalala seek such information to estimate needs
for a viable population. Wright has learned much about infant development, reproduction, and predation. A particular interest is how
female dominance affects group dynamics and reproductive success.
She's now writing a book on the topic entitled *When Females Lead,* an
area she knows something about, I dare say.

The Earthwatch volunteers are furthering Wright's long-term
effort. Theoretically, each team member is supposed to follow a
selected lemur all day for five consecutive days, taking note every five
minutes of what his or her charge is up to: grooming, calling, eating,
playing, scent-marking, and so forth. But so far, only Brandon and
one or two others have shown the tenacity of research guides like
Lejean to stay out all day, rain or shine. And the team is supposed to
do this twice, once this week and once next.

Of course, our job is pretty easy just now. Or I should say Brandon's job; I'm just observing. For all Wright's talk of Yellow Silver's
desire to move on, he's presently showing not a hint of it. He's sleeping . . . sleeping . . . sleeping. It's about seventy degrees, and a light
breeze keeps the understory pleasantly cool—perfect, really. Through
the thin tree trunks and lightly leaved canopy, I can see a distant hillside of rain forest as well as discrete patches of blue sky that, again,
collectively add up to a lot of sky seen and a lot of light coming in. My
notebook is brightly lit by direct, warm sun.

Its warmth, and the monotony of watching that lazy lemur, are
making me sleepy. The soil—dark, moist, richly aromatic—is bed-
soft, too. The only thing that keeps my eyelids open is a pretty little
paradise flycatcher, a small orange-breasted bird with blue circles
around its eyes that is flitting about in a nearby garden of ferns, vines,
fallen branches, and dead leaves. Short, dark tail: it's a female; the
males have those impossibly long, sweeping tail feathers.

Well, it's no use. Breathing in the oxygen-rich, slightly humid air,
I allow myself the luxury—one Brandon doesn't have, poor girl—of
dozing off.

Six hours after waking from that nap, I find myself in a research situation that could not be farther from my "work" with Yellow Silver.
It's now long after dark, it's pouring a cold rain, and I'm alone and
lost in the rain forest.

What kind of idiot would want to go out in the middle of the night, in a downpour, and scramble off-trail through brambly rain forest, searching for a metal cage with a chicken in it? My kind, it seems. Wanting to get a taste for what other kinds of research are taking place here besides Wright's, I have volunteered over the past several days to help a young biologist named Luke Dollar catch fossa using chickens as bait. One of Wright's graduate students, Dollar is doing a study of the island's largest mammalian meat-eater, and he needs to put radio collars on them in order to track their movements through the forest.

Damn, damn, *damn*! I pick myself up yet again after losing my footing and sliding downslope on my butt for a few feet. The slopes are forty-five degrees here, and it's all but impossible to keep my footing. Slipping a few feet is one thing, but I've already had two major falls in the hour I've been out here by myself. Both ended in slams into trees that left me seeing stars. I keep grabbing branches for support only to find them break off in my hand, the rotten bastards. Wire-thin vines I can't see even with my headlamp keep slicing my face and arms, and others along the ground trip me up. Every time I look down at my ankles there's a new leech there; if my socks and pants weren't already caked with smeared mud, they'd be stained crimson with blood.

The trap I'm looking for is one of eight that Dollar has set up every 100 meters (330 feet) in a rough square deep in the forest. At the closed end of each golf-bag-sized, steel-mesh cage, Dollar has pinned a live chicken behind a row of thin wooden stakes he has jammed down through the mesh and into the ground. This keeps the chicken in but won't stop a fossa—though the trap door that the animal triggers on the way in will. *Cryptoprocta ferox* is Madagascar's premier lemur-eater, and it's not for nothing that its specific name means "ferocious." For his doctoral research, Dollar is studying the animal's behavior and ecology at Ranomafana. So far he has caught and collared four fossa, which several Malagasy helpers are now radio-tracking around the clock in eight-hour shifts, to learn how far they travel in a day, how big their home ranges are, when they're active, and so on. Already Dollar has found evidence that fossa, previously thought to be strictly night- or twilight-active animals, are cathemeral—that is, active day and night, right around the clock.

Which is why we've been checking the traps not only in the early morning but late at night as I am now. The only difference this time: Dollar is not here to find them. He left this morning for Fianarantsoa ("the Place where Good is Learned"), the Betsileo capital two hours up the road, and did not return in time to check the traps after dinner. So, assuming the role of that proverbial idiot, I offered to inspect them without him.

"You sure you can find them in the dark?" Wright asked me back at camp, her eyes smiling in the light of the kerosene lamp. I should have taken that as a warning. I guess I thought it would be an adventure, and lucky for me, Will Wilson, a twenty-five-year-old Australian volunteer, must have felt the same, for he agreed to join me. It clouded up late this afternoon and, by the time we set off after dinner, it was raining quite hard, which only added to our sense of adventure.

Having helped Dollar in previous days, we located the first three traps quite easily. No fossa, alas. Even though the rain made it quite chilly, the chickens seemed fine. We sprinkled a little uncooked rice down through the wire mesh and watched them go at it vigorously. Then, feeling confident, I offered to track down numbers four through six, which were 330 feet apart off "D" trail. A few days ago, I got lost searching for just the first of those and had to call cravenly for Dollar to come off the trail and find it for me. And that was in full daylight. But I didn't have the compass then; this time I did. I remembered the bearing Dollar took and headed off, leaving Wilson waiting on the trail in the rain.

"I'll be back in a few minutes," I said.

Yeah, right.

I found the first trap, Number Four, quickly. But then it was only seventy-five feet off the trail, and I was able to follow a faint path made by our previous passings. Again, no fossa. I gave the chicken some rice, took a bearing, and moved on. The hills rise and fall and twist and turn here, undulating like waves in a bay. It's easy to be fooled into thinking your route follows the lay of a hillside or that you can eyeball straight. So I took my bearings every ten feet or so. But in close-packed rain forest, on undulating hills, getting off-bearing by a degree or two even for a few steps can lead you astray real quick. And my cheap plastic compass, which is not calibrated for the Southern Hemisphere, seems to jump between *two* settings a few degrees apart. I'm not sure which one is correct. Several times I had to make my way back to Number Four and start over, having not located the neon orange ribbons Dollar has tied every seventy-five feet along the transect. Time slipped by. After half an hour, I finally came upon the second one, Number Five.

The chicken was dead. A lump of wet feathers heaped against the wooden stakes. I felt awful. I was mad at Dollar for not being there and for letting that poor chicken freeze to death in the pouring rain. Or did it die of fright? Perhaps a ringtailed mongoose came sniffing around, even took tiny bites of the bird through the mesh. . . .

Things went downhill from there. Now here I am, wondering

where the hell Number Six is. I just crossed over "W" trail, where earlier I stumbled upon a Betsileo *vatolahy,* or sacred stand of memorial stones, overgrown with vines. I wouldn't want to come upon the stones now—too spooky. I just saw an orange tag, but it looked old and tattered. Was it one of Dollar's? I left Wilson an hour ago; I'm sure he's getting worried. A while back I thought I heard him calling through the rain, but I'm not sure. My sweat-soaked back feels cold whenever I shift my pack, the damn compass is fogging up inside, and bugs attracted by my headlight keep flying into my eyes.

"Fuck it."

I'm surprised to hear the sound of my own voice after stumbling along so long in silence. I take a bearing, 160 degrees, and turn to head back. Buggered as it is, the compass, I know, is the only thing that will keep me from getting completely lost. So every second or third step, thinking only of getting out of this bloody rain forest and into my dry, warm sleeping bag—which I borrowed from Dollar, of all people—I stop and check the bearing. Then I just smash through the forest, little caring if I am on the exact path I came out on. When I think I am in range of Wilson, I begin yelling.

"Will! Will!"

I hear a faint response. Crash on, yell again. The response gets louder. I keep the bearing at 160 degrees, plunge ahead. Finally I see a twinkle of light between the tree trunks. I fight my way through the final stand of thicket and am on the trail.

"Where the hell you been, mate?" Wilson asks kindly, clearly relieved to see me.

"Couldn't find the damn third trap," I say, winded, wiping sweat and dirt and bits of leaves from my brow.

We spend only a few minutes searching for the final two traps. In vain, as it turns out. It's late—10:30—and we fear that Dollar may set out looking for us. So we head back along the trail. Forty-five minutes later we reach the camp. Everyone has long since retired, except for Dollar and Wright, who are chatting quietly on the porch.

Standing there soaked to the bone, deeply chilled, and beyond exhausted, I tell Dollar about my endless search for the third trap off "D" trail.

"Third trap?" he says. "Peter, there are only two traps there."

I look at him in disbelief. I want to strangle him, sitting there all smug and dry and warm. But it's my mistake. We missed a trap between what we thought were Numbers Two and Three. I muster a feeble chuckle, mumble goodnight, and stagger down to my tent.

8.

THE NEXT MORNING, the focus of my effort to come to terms with Ranomafana begins to shift from wildlife to people. In the few days I have left here, I want to try to assess the project's other three components: health, rural development, and education and ecotourism. Peters's opinion notwithstanding, Wright says that from the start, the Ranomafana project has focused on identifying the needs and improving the welfare of the local population. Not an easy thing to accomplish in a place as poor as Madagascar, which a 1996 World Bank list of 133 nations put at the twelfth poorest.

Indeed, the first thing that strikes you as you head out into the countryside is the extreme poverty of most rural areas. In 1994, the average Malagasy earned $210. That is hardly enough to feed one's children, much less properly house, clothe, or medicate them. As a result, 51 percent of Malagasy children under five are chronically malnourished, according to the USAID, and about 16 out of every 100 Malagasy children die before they reach five. Even so, the population is exploding, with a full 45 percent of Malagasy under the age of fifteen. "Madagascar," Alison Jolly stated bluntly in 1989, "is rapidly heading to a confrontation between the human population and the supply of fuel and food, with fuel running out first."

Despite the seeming richness of its setting, the Ranomafana region is no exception. One development expert who had spent his career working in some of the poorest developing countries, including Nepal, Ghana, and Sri Lanka, told me the poverty at Ranomafana was some of the worst he'd ever seen. "Nothing prepared me for this," he said. "The level of poverty here is a stark and grinding reality for people." And the population is soaring. Over the past three decades, the population in the Ranomafana region has grown by 111 percent.

It was to assess the living conditions, and particularly health, of local villagers and their children that Lon Kightlinger, a graduate student at the University of North Carolina, arrived at Ranomafana in 1990. When he first talked to Wright about doing a health and socioeconomic study at Ranomafana for his doctoral thesis, Wright asked him if he spoke French. No, he said. "Well, I hate to tell you this, but if you're going to work in Madagascar, you've got to speak French," she replied. "I speak Malagasy," he said. Turns out Kightlinger had spent a dozen years as a missionary on the island. In early 1990, he, his wife, Mynna, a Malagasy medical doctor, and nine Malagasy nurses set off into the peripheral zone with packs on their backs; a year later, they had completed a seminal study of village health.

Kightlinger's group witnessed that "stark and grinding" poverty in the 516 households in eighteen villages they visited in the peripheral zone. They learned that nine out of ten people were farmers and household size averaged six people, with some families having up to fourteen children. None of the villagers practiced contraception; only one out of ten had even heard of contraceptive devices. Over those twelve months, Kightlinger's team also studied a fixed group of 1,292 children. They discovered that 62 percent of the children were underweight, with 17 percent severely malnourished when measured against World Health Organization standards. Those children also suffered inordinately from disease. More than half of them had malaria, and eight of ten had intestinal parasites, especially roundworm and whipworm.

In 1991, Wright says, the Ranomafana project's newly formed health team began delivering health-related services requested by the villagers that Kightlinger's group had visited in the peripheral zone. These included medical care, basic first aid, pointers on nutrition and sanitation, and coordination with government immunization programs. Three years later, the health team added a family-planning program. Project workers built two new clinics in the peripheral zone and renovated the hospital at Ranomafana. The health team also began monitoring the impact the Ranomafana project has on the lives of villagers. "This team has continued to bring news, health care, and assistance to the residents every other month for over five years," Wright says. "This is a tangible benefit from the park, and it has helped local residents to accept the park."*

Not surprisingly, the project has its own ideas about improving economic conditions, which in a region as poverty-stricken as

---

* Not everyone agrees with Wright's take. According to Christian Kull of the Yale School of Forestry and Environmental Affairs, who conducted a study of the evolution of conservation efforts in Madagascar, "The project has been criticized strongly for its adverse effects on child nutrition (as forest access is restricted). . . ." Janice Harper contends that project administrators blocked her efforts to have the health team assess a crisis taking place in one of the project's pilot villages. Within thirty-six hours of Harper's arrival in the village, a resident died, and over the following year she worked in the village, villagers had continued to die at the rate of at least *one per week.* According to Harper: "After nearly 10 percent of the village had died, I sought assistance from the project. . . . Rather than respond to this request in a positive fashion, I was first met with profound indifference, and then regarded as a potential danger should I actually publish the fact that children were starving to death and adults were dying of unexplained fevers and respiratory disorders in a pilot village of the project. . . . Any attempt to discuss the realities of death and sickness, and possible problems with project strategies affecting health or economic status, was condemned as criticism."

Ranomafana is as vital as improving health. Since most peo⌐ farmers, that means agricultural development, and agricultural o opment has to start with offering alternatives to tavy.

Tavy farmers clear-cut and burn a hillside in order to plant h rice. They can farm that hillside for about three years before the soil's nutrients are exhausted; they must then leave that hill fallow for at least fifteen years before it becomes profitable to repeat the process of cutting, burning, and planting. Residents have practiced tavy here ever since their forebears first settled the region in the late eighteenth century. While wet-rice farming in paddies was an individual family affair, dry-rice farming on hillsides was communal, with what Joe Peters terms "an implicit code of social reciprocity." Traditionally, the local king or *mpanjaka* (literally "the one who rules"), with the consent of village elders, allocated tavy fields to newly established households, which would have enough land to properly rotate planting.

When the population was low and the forest stretched as far as the eye could see, tavy was actually an efficient use of resources, which comes as a surprise to many Westerners, who have long been taught to think of slash-and-burn farming as the murderer of the world's rain forests. According to the French scholar Jean-Louis Guillaumet,

> Outsiders have been shocked and have accused the peasants who practice tavy of practically every evil. However, this type of agriculture is well adapted to the environment for farmers with limited technical means. Fertilization by ash, fallow with forest regrowth, and the forest in turn rebuilding the soil is a simple and judicious use of natural processes. This system of traditional agriculture gives maximal yield for minimal human effort. Only when the human effort is subsidized by dead forests of the coal age, in modern oil-based fertilizers and herbicides, and oil-run machinery, can we achieve more yield per man-hour than the peasant who uses live forest as his subsidy.

Since Ranomafana's farmers cannot afford modern agricultural equipment or products, tavy is the only way they can tease life out of their nutrient-poor soils. Bruce Johnson, who conducted a soil survey at the park, claims the soils here are some of the poorest in the world. "This is the primary reason why farmers must practice slash-and-burn agriculture," he writes. "The standing biomass of the mature forest releases a quantity of nutrients that is unavailable from the soil."

the recent spike in population around Ranoma-
formerly sustainable tavy system into disequilib-
increase deforestation in the vicinity of the park.
who have lived there for generations typically
lowed fields, new migrants must slash-and-burn standing
to create their own tavy fields. Most of the remaining woodland
in the region lies on precipitous slopes, a poor place to practice tavy.
As Richard and O'Connor declare in a recent paper, "In Madagascar,
we can confidently predict that forests cleared for agriculture on
steep hillsides will not become transformed landscapes in the service
of the local people"—as are, for instance, the great rice fields in the
flatlands around Tana, which are inundated by annual flooding of the
Ikopa River—"but rather . . . will be the first step in a slide toward
soil loss and sterility."*

As an alternative to tavy, Johnson has suggested providing villagers with fertilizer in exchange for leaving the forest intact. "Given
the population pressures within the region," he has written, "agriculture without man-made fertilizers is absolutely unsustainable without
the loss of forest and/or human lives." But Peters holds that Johnson's recommendation insufficiently takes into account villagers'
powerful cultural ties to tavy. As Paul Ferraro, another Ranomafana
researcher, put it:

> The residents of the Ranomafana region view the forest around
> them through the notion of *tanin-ðrazana,* which literally means
> "land of the ancestors." As they see it, their ancestors gave them
> the land, and therefore they are the rightful owners. Despite
> more than a century of attempts by external powers to exert
> their will over the management of these resources, many people
> in the region continue to assert that it is their right to use the
> resources as they please.

To learn about some of the alternatives to tavy and other types of
forest exploitation that the Ranomafana project offers local farmers, I
stop one afternoon by the office of the economic development team,
which sits just down the hill from the park director's office in
Ranomafana Town. As luck would have it, Norman Uphoff, who has

---

* The World Bank estimates that degradation in general costs the Malagasy Republic,
which has been called "the world champion of erosion," $100 to $290 million annually, or
about 5 to 15 percent of the country's gross domestic product.

run the program since 1992 but now oversees it from afar, is visiting for a few days from Cornell University, where he directs the Cornell International Institute for Food, Agriculture, and Development (CIIFAD). A tall, fit-looking man bursting with energy, he reminds me of a college football coach. I don't know if it's the fact that he is about to catch a plane back to Tana, but in the few minutes we have to chat in an airy, undecorated room, Uphoff speaks so fast my head spins.

"We have nine programs with CIIFAD in different countries, and at all those we're talking about *farmer-centered* research and extension," he says. "How do we get away from this linear model where the scientists develop the technology, give it to the extension agents to give to farmers and get them to accept it? It's a one-way street. We try to say it's two-way, but it's 90 percent one-way."

He chops his hand down on the table before him.

"We're saying: Let's start with farmers' own sense of need and bring to bear whatever technology we can, but experimentally. Not here's what you should do, but let's try these two things, or you try one and your neighbor tries another. Let's not look over each other's shoulders, let's satisfy ourselves that it's good. We're trying to change the relationship between holders of scientific knowledge and holders of local knowledge, which is important for finding an efficient solution."

One way Uphoff's group has changed that relationship is by employing Agents for Conservation and Development, or ACDs. These are local people recruited from villages in the peripheral zone, who train with CIIFAD experts before heading off into the hills to work closely with farmers. Since the ACDs are native sons, residents are more willing to work with them than with foreigners or nonlocal Malagasy. The ACDs assist the farmers in testing various alternatives to tavy, which keep farmers from succumbing to the urge to establish new tavy fields and help them get more out of existing ones.

In his rapid-fire fashion, Uphoff describes some of those alternatives, which are just getting under way. Rather than just sowing hill rice on those steep slopes, the ACDs help farmers plant mixed crops, including fruit trees and vegetables other than rice. These plants help retard erosion and provide crops that farmers can sell in the market in the lean months before the rice harvest. Uphoff's team is looking at off-season crops, such as beans and potatoes, that will allow farmers to get another sixty-day crop in and also augment their nutrition and income. The ACDs help the farmers build ponds to raise tilapia, carp, and the meaty local crayfish, which are as big as half-pound lobsters.

They provide them with seedlings of fruit and coffee trees, and with the knowledge of how to grow them.

"Turns the notion of a nursery on its head," Uphoff says, his eyes burning into mine. "Don't just give people plants, give them plants and knowledge so they know how to propagate them themselves and will teach others to propagate them."

Teaching has been that two-way street Uphoff mentioned. During a prolonged drought a number of years ago, local cultivators discovered rice intensification, a technique that allows farmers to double or even triple their rice production. A brief rain fell early in the planting season that year, and the farmers, fearing no more rain would come, tried transplanting the young plants far earlier, at roughly eight days rather than the customary two or three months. To their surprise, the plants did not perish but, stressed by the unfavorable conditions, grew even faster. A Catholic priest, recognizing the value of his constituents' discovery, helped systematize it, and Uphoff's group has improved upon it while working with local farmers. For instance, instead of leaving standing water in paddy fields, which reflects the heat of the sunlight that helps plants grow, farmers keep the fields well drained most of the time.

"We developed these things experimentally with the farmers," Uphoff says, beginning to pack things feverishly into his briefcase. "So clearly we're growing with them."

Suddenly he pauses and speaks slowly for the first time.

"I remember when I first went to the CIIFAD board with this opportunity to work in Madagascar. One board member said to me, 'Just remember, Norman, Madagascar is the graveyard of dreams.'"

He looks at me and smiles without mirth.

"That's the reputation it has, and I can see why. We've had our disappointments. But we're making a lot more headway than we ever expected."

Just as suddenly, he returns to his staccato delivery, his hands once again rapidly slicing the air.

"There's this farmer we work with. He's got fish ponds, fruit trees, mixed cropping. I wish I had my photo album to show you where he lives. It's a squalid little house, but this guy's got ideas, he's got plans. . . ."

9.

I RISE AT 5:15 TO raindrops on my tent roof. I've made arrangements to meet that farmer Uphoff mentioned and, perhaps because the farmer can't afford to lose the work time, he agrees to see me at the delightfully unbusinesslike hour of seven in the morning. I eat a quick breakfast with Wright (does she ever sleep?), then hike the half mile down the hill, across the new bridge over the Namorona, and up to the park entrance, where Wright's Malagasy driver is waiting for me.

We drive a mile down the road to the Domaine Nature hotel, the only park hotel outside of Ranomafana Town. Here, two of Uphoff's staff—an American technical adviser and a Malagasy translator—greet me and introduce me to a mustachioed Betsileo named Paul Rakotonirina, the ACD who works with the farmer. With the volunteer Will Wilson and a Wright staff member also joining us, we set off in single file, straight uphill along a muddy path, a cold, silent party of six.

For the first time since I arrived, this rain forest actually looks like a rain forest. Mist filling the river valley behind us seems to absorb the light of the dawn, making for a melancholy scene. Drops of water hang suspended from leaf tips. Above us, pendulous clouds press down on the higher ridges of rain forest, obscuring them. With every step on the rain-slicked path, our boots seek out the security of a rock or root; brushstroke smears on the ground mark our many slips. It's chilly enough for a sweater, though I'm warming up from the steady uphill march. Each of us is lost in his or her own thoughts.

As we pass through a banana plantation laid out willy-nilly on the slope, I notice wild ginger growing along the path, its light-green leaves twitching with every drop of rain. Suddenly I hear Uphoff's brisk voice in my head: "I was walking along the road one day and saw all this weed, wild ginger, growing along the roadside. It's beautiful, with a nice fragrance and flowers. I thought: Maybe it knows something other plants don't. I took it back to the States and analyzed it chemically. Turns out it has a lot of phosphorus, which means the plant has a special knack for extracting phosphorus from the soil. Phosphorus is the limiting factor in getting bigger yields here. So we tell farmers to cut down wild ginger, chop it up, put it on their compost. I'm trying to get some hard numbers on this, but Samuel, an ACD we work with, says compost can double your maize and bean yield, and wild ginger can *triple* it."

Again, as with rice intensification, Uphoff displays a willingness as a foreign expert to learn on-site, to dispense with the paternalistic attitude of "we know it all already, let us tell you how it's going to be."

After half an hour, we leave the banana plantation behind and come upon a series of terraced fields. They step end-to-end up the hill to the edge of rain forest, which protrudes from the base of heavy cloud. Ranomafana National Park starts right behind this man's farm. Above the fields but before the trees sits a one-room house on a rise, the first I've seen since we left the road. It's of mud and wood, with a thatch roof, and it's compact enough to fit in the bed of a dump truck.

The farmer comes out of the house. He's a small, scraggly man. Despite the chilling rain, all he wears is a woven straw hat, threadbare shorts, and a T-shirt showing a pair of bearded black musicians below the logo SLY AND ROBBIE WORLD TOUR. Dried mud cakes his bare legs and feet. Full of bad teeth, his mouth lies twisted open in a leer, an expression that gives him the look of a simpleton. Can this be the farmer Uphoff raved about — "this guy's got ideas, he's got plans"?

Through the translator, Paul the ACD introduces him as Rakoto Pierre, using the surname first as Malagasy do. When I tell the farmer my name is Pierre, too, he smirks nervously and looks at the ground. Paul explains to him that we've come to see his farm and learn about the techniques the project has taught him. Pierre stands with his arms crossed against the cold, and I can see goose bumps on his biceps. I wonder what it would feel like to trade places with him right now. How deep a chill would I feel? What would I be thinking? Would I be screaming hungry?

Suddenly Pierre speaks. He welcomes us to his farm and apologizes that his house is too small to accommodate us, so that we have to stand out in the rain. And he is sorry that his wife and children cannot come out to greet us, but several of his children are sick. He gestures back toward the house, where his wife peers from a tiny window, a baby perched on the sill. I say *Salama,* and she returns my hello with a smile. Turning back, Pierre says he would be happy to show us around and answer any of our questions. He looks at me for the first time, letting his eyes linger as if to say "Fire away."

Instantly I regret my ungenerous thinking about him. Humbled, I begin asking questions of the translator, who either lets Paul ask Pierre or asks him herself. Standing in his rain-slicked mud yard, Pierre answers in brief, almost whispered responses. The rain smears the ink on my notebook until Wilson hands me one of the Rite-in-the-Rain pens Wright distributed for the *Propithecus* study. I feel guilty

standing here in my expensive rain gear and waterproof boots while
our host shivers in half a layer.

Born in Ranomafana Town, Pierre has lived in this house for five
years, he tells us. He's thirty-nine years old, and he and his wife have
six children. When I ask if he cleared the forest himself, he says yes.
Then, perhaps fearing we are connected to persons of authority—for
whom there is a general distrust among locals, possibly as a result of
the fallout from the 1947 Rebellion—he quickly adds that he had a
permit to do so. As a general rule, the individual who first clears and
cultivates a piece of land in the Ranomafana region gains exclusive
agricultural rights to it. But maybe Pierre is just playing it safe. Looking out over the terraced fields, he says his farm is about five acres.
Would we like to see it now?

Pierre leads us around his neatly tended fields. He strides barefoot along their raised edges as comfortably as I would on a sidewalk,
but I step gingerly, not wishing to end up on my back in the mud like
that ill-fated missionary. I glance down the hill. It's still shrouded in
fog, with only nearby trees standing out against the white like
specters. If it was clear, it would be a view to vanquish all views:
steep, rain-forest-garbed mountains falling off toward the east, where,
on a clear day, one can see the Indian Ocean shimmering ninety miles
away.

All the things Uphoff talked about are here. We walk along the
edges of rice paddies, where Paul notes that Pierre practices rice
intensification, and along fields of beans, corn, and cassava. Breadfruit, lychee, banana, and other fruit trees hug the hillside between
fields and in their own discrete plantations. Pierre shows us his fish
pond, which is about the size of a family pool in the West, and above
it, a smaller pond for crayfish. I see neither fish nor crayfish, but he
assures me they are there. Both ponds bristle with healthy stands of
the spiky, dark-green reed that villagers use to make mats. As we
walk, Pierre points out other plants that, trying to take it all in, I miss
entirely—lemon grass for making tea, and a bushy plant with pink
flowers that retards erosion. Though I don't see them, he says he
grows medicinal plants, a species of tree that honeybees prefer, and
tree ferns—all so he doesn't have to go to the forest to get them. And
he chops up wild ginger for his compost.

I can see why Uphoff holds this man up as the model farmer—
he's willing to try all Uphoff's techniques, and he says all the right
things to a visitor. An uncharitable thought crosses my mind: Might
they be paying him off to use his conveniently located farm as a show-

piece? But I quickly abandon such thinking. I am truly astonished at how much Pierre appears to accomplish here, totally on his own. The farm is diverse, well maintained, even beautiful. Paul, who visits the farm two to three times a week, tells me through the translator that it is not hard to motivate Pierre.

Wondering about Uphoff's two-way street, I ask Paul whether the project has learned anything, in turn, from Pierre. The ACD talks with his client for a moment, then says yes. Pierre taught the project how to grow avocado and breadfruit in the nursery, and how to improve one's yield of cassava: Farmers normally don't till the soil when growing cassava, but Pierre does, and it turns out to be more productive.

While Pierre's actions are taking pressure off the forest, one has to wonder how much all this is really doing to improve his life. Clearly he remains poverty-stricken, barely able to provide for his family. Like most people around the park, Pierre has no electricity, running water, or material wealth, and he or his wife must collect fuelwood at least every other day. Because of the region's poor soils, rice yields for both upland and paddy rice here are far lower than the national average, and Pierre admits that he can't grow enough to feed his family. He says he must supplement the income he makes from selling his rice in Ranomafana Town and his vegetables at the hotel down the hill by working on another farm. Daily wages in the area are pitifully low, running between 2,000 and 4,000 Malagasy francs, or forty to eighty cents. The flowerpots Pierre can make from the tree ferns he grows sell for 1,500 to 5,000 Malagasy francs, but tree ferns grow very slowly, and it takes on average five days of work to produce eight to twenty pots. Who knows how long it will take for his fish and crayfish ponds to return the investment of time and effort he put into them?

Before we leave, I ask Pierre if I can take his picture. He looks at me shyly and prepares to pose before his house. Struck by a thought, he shuffles past me to a banana tree, where fifteen or twenty seedlings of coffee plants sit in pots fashioned out of black plastic. He picks one up and returns to stand in front of the house. Looking through the camera's viewfinder, I realize with a start that, having composed himself for the photo by closing his mouth and removing the cold-induced grimace from his face, Pierre is strikingly handsome. He stands confident and proud and even manages an impish-looking smile, the first I've seen on him.

At once I feel overcome with emotion for this man. The struggle he has to make every day of his life is heartbreaking. Again it strikes me, as it has on other occasions in other lands, how close and yet how immeasurably far apart this man and I are. I can touch him, he can touch me, we both eat, sleep, talk, raise families. Yet our lives could not be more divergent if we came from different planets. Seeing him shiver in this bone-chilling rain, I think of giving him my rain pants as a gift, but wonder if it would be inappropriate. I settle on a few pieces of candy for his kids, and with velomas and *misoatras* (thank-yous) all around, we set off back down the hill.

10.

AS PIERRE'S SITUATION suggests, aid in health and agricultural development may improve lives, but to offer opportunity for a better life, or even a different one, many development experts feel that improving education and encouraging income-generating alternatives such as ecotourism are crucial. Again, Wright says the Ranomafana project has keyed in on these two areas since the beginning.

Education in Madagascar is about what you'd expect in a nation so poor. According to recent figures from the United Nations Development Program, public expenditure on education as a percentage of gross national product was just 1.9 percent in 1993–94, which is significantly lower than the 2.8 percent average for the world's least-developed countries. The adult literacy rate in 1994 was about 46 percent. In 1992–94, the number of students enrolled in primary or secondary education was 42 percent, and the number of book titles published for every 100,000 people in the country was just one.

Education in the Ranomafana area was not much better when Wright arrived in the 1980s. Kightlinger's socioeconomic study revealed that while 75 percent of adult men claimed to be literate, fully 70 percent of six- to nine-year-old children had no formal education whatsoever. Wright told me that a third-grade education, which is what Emile and Loret had when she first began training them, is the minimum necessary for such training to take hold; she has had no success with those who had no education at all.

Wright and her colleagues formed an education team to try to reverse this situation. In the early years of the project, they constructed four new schools and renovated seven others. They donated blackboards, chalk, and other school supplies; the national government supplied the teachers. The team built and stocked small libraries in six villages and taught children to plant trees and gardens, sing conservation songs, and listen to stories about Malagasy animals and plants. Today, Wright says, the education team fans out periodically into the peripheral zone bearing posters and other Malagasy-language conservation materials created by the WWF. In villages several days' walk from the road, they set up a portable projector and show slides about Ranomafana's wildlife and habitats on a white sheet. In many cases, this is the first that most children and many parents learn about the full range of riches their forest holds.

The project encourages villagers to leave home to learn more. One afternoon, I saw a vanload of schoolchildren empty out at the park entrance for a tour, and another afternoon I spent an hour in the Ranomafana National Park Museum, an exposition center established by the project in 1991. For a tiny museum this far off the beaten path, it is impressive. There are exhibits on the history of the park and the region, local culture, and flora and fauna, including live boas and native fish. Locals and foreigners alike can enjoy the exhibits, whose clear, simple label copy appears in Malagasy, French, and English. The English is not the broken, English-as-a-second-language style found in so many rural museums in developing countries, but clearly was composed by a native English speaker. On weekends, scientists give lectures, village musicians play, and the museum shows videos.

Local Betsileo and Tanala villagers seem eager to learn. Wright reports that more than 300 villagers visit the museum each month. And a few years back, after village elders attended a scientific conference at Ranomafana, they surprised everyone by requesting that the results of scientific enterprise in the park be made available to local villagers. Today, every other month, each peripheral-zone village receives an illustrated newsletter in Malagasy that describes recent research results.

Educating villagers to take an active role in conserving their forest—and make a living while doing so—has also taken place on a wide scale. Wright's team of paid biodiversity research assistants now numbers about thirty, and there are also the ACDs and other project teams involved in socioeconomic development in the villages. To learn about and participate in biodiversity surveys elsewhere, Emile, Loret, and several others have traveled to other parks in Madagascar and, on a special trip that Wright arranged, to Kenya and Tanzania. Now those men are returning the favor. The other day, Loret led the TSEs in teaching their techniques to a group of Malagasy officials visiting from Isalo National Park.

Wright has coupled on-the-ground training of locals with the university training of both local and nonlocal Malagasy in research, conservation, and management. To date, Malagasy have completed fifty-seven master's theses based on research at Ranomafana, with many more in the works. Wright and her colleagues have helped ten Malagasy, including Ernestine, the head of the TSEs, receive advanced training in biodiversity and the environment in the States, and four are currently earning Ph.D.'s there.

"All these Malagasy graduate students both here and in the States are my way of getting the future to go my way," Wright told me. "That's my hidden agenda. Keep training all these really competent people, who come back to Madagascar and take all the top research and conservation jobs. They're totally capable, and they have an understanding of two worlds."

Understanding two worlds—Malagasy and Western—is essential for the final component of the project's development concerns: generating alternative sources of income for villagers. Catering to foreigners may ultimately prove to be the most lucrative means of earning a living in the area. Again, the project has tried to help locals reap the benefits of ecotourism by, among other initiatives, providing the initial infrastructure to lure tourists here. They built the museum, a welcome center at the park entrance, a campground along the Namorona, and a marked trail system complete with a brochure describing curious creatures and plants to see during a self-guided tour.

Today, village committees help decide on new projects, using their portion of park entrance fees. As in all of Madagascar's protected areas, the entrance fee, which at Ranomafana is the equivalent of about ten dollars, is split between the national park system and peripheral-zone villages. Currently, the ninety-three villages share about $10,000 a year this way. A committee of eighteen village representatives elected for three-year terms decides on new projects. In 1995, they planned to use the money to buy seeds and build campgrounds, an artisanal training center, and small dams to expand existing rice paddies. Since these funds are available in perpetuity, Wright points out, the park itself has become integral to village economics.

Peters concurs. "The sharing of entrance fees is the most immediately conceivable and practical manner to distribute ecotourism revenues to remote populations and/or disadvantaged ethnic groups that otherwise would not benefit from tourism to the park," he says. In his estimation, before the entrance-fee sharing began, only about 65 people at Ranomafana, or 0.002 percent of the 27,000 people in the peripheral zone, benefited from tourism to the Ranomafana region.

Peters says a similar lack of equity showed up in project hiring. While the number of Malagasy employees grew from 20 to 114 during his tenure, the majority Tanala were largely left out of the equation, representing only 7 percent of hirees compared to 58 percent Betsileo and 24 percent Merina. (The remaining 11 percent belonged to other tribes.) Even more striking, fully 18 percent of employees came from a single village, the Betsileo community of Ambatolahy, which exceeds the number hired from all other peripheral-zone villages combined. Moreover, nearly eight out of ten employees were men. Such biases, Peters notes, can only cause conflicts among villagers as well as between villagers and project personnel.

Alison Jolly says such inequities are hard to avoid. "Although it's true that most hiring started from Ambatolahy, there is a real problem for foreigners in a foreign land, or indeed for anybody launching a new business or project," she told me. "It is much safer to hire friends of your present employees, or at least someone who has a reference for being reliable. Add to that geography: Many villages, especially Tanala ones, lie very far off the road. Add to *that* the differences in education: Villages near the road are much more likely to have some literate people. Ideally, of course, the park should hire one person per village, balanced by sex, but practically, Joe Peters must know that's impossible."

Wright, for her part, has not mentioned these inequities to me, nor whether they exist in the other park-related activities that she claims benefit Ranomafana communities. Project members, for instance, worked with a local nongovernmental organization to establish training workshops to teach villagers how to make and market handicrafts. They built a craft shop near the museum, which also has its own craft shop. These stores sell local wares, with 85 percent of profits going back to the artisans, Wright says. Finally, they have trained locals to be park rangers, campground managers, and tourist guides.

Guide services, park-entrance fees, handicrafts, a bimonthly newsletter—all are designed to strengthen that link between conserving nature and improving local people's lives. As Wright says, "The challenge for the next few decades is to reconcile the richness of the biodiversity with this overwhelming poverty of the human populations."

11.

On my last full day at Ranomafana I join a guided hike to a Tanala village. I am eager to visit one of the peripheral-zone villages and to see ecotourism in action here.

For we are certainly going on a tour. To meet our guide for the day, we gather on the side of the road just east of Ranomafana Town. We have to wait while the two project vans make several round-trips to the park entrance to shuttle our entire party down. Besides the twelve Earthwatchers, there are twelve college students newly arrived from the States. They are part of the University Development Linkage Project, another of Wright's initiatives. For two months, they will learn biodiversity research techniques here at Ranomafana, then conduct research on a particular theme for one year; Malagasy counterparts do the same in the States. One of the American students, a naïve, in-your-face kid with a goatee, would not shut up when his group arrived in camp last night. I see that he has already latched onto the guide. Maurice Rakotovao is an angular, thirtysomething Malagasy who speaks perfect English. He keeps his white button-down shirt open to his navel, and he's too quick to smile. I smell an operator.

I'm not partial to tours or large groups, much less to smooth-talking guides or know-it-all kids. As we head off on the purported two-hour hike, I make a mental note to keep enough distance between myself and everyone else to make it at least feel that I'm doing this on my own. I can't say I'm looking forward to watching this herd of camera-wielding vazaha descend on a tiny village. Of course, I am one of those camera-wielding vazaha.

Just how large a group we are becomes apparent five minutes into the hike. We come to a river with no bridge. The river is only a stone's throw across for someone with a good arm, but it's waist-deep and fast-flowing. Hearing we were on the way, a villager has built an arrow-shaped raft out of newly felled reeds. For 1,000 Malagasy francs (twenty-five cents), he poles people across, two at a time. A writer of the last century, after making use of a *zahitra,* as this kind of craft is known, wrote that "of all the rude, primitive, and ramshackle contrivances ever invented for water-carriage, commend me to a zahi-tra." But the raft has already upset once, sending two volunteers into the drink, and at this rate I can see we'll be here for an hour. As the group masses on the near bank, each person waiting his turn, I notice locals wading across upstream, where some ripples indicate the river is a bit shallower. Making note of their precise route, I make it across in two minutes.

Not wishing to wait for the herd, I head up the trail behind several of Wright's assistants, who have also joined the group. (She herself forgoes the journey, having another meeting scheduled with the park director.) It would be hard to get lost, as dozens of Malagasy are heading in the same direction. Are they coming to see the vazaha? Make a party of our visit? Many of them are dressed in colorful lambas and are singing and otherwise making merry. In spite of my misgivings, I begin to look forward to the event.

It dawned foggy and drizzly this morning, but by ten or so it begins to clear off, blue sky showing promisingly through the cloudbank. The route becomes clear shortly after I leave the river behind. Here in the peripheral zone, residents have cleared the jungle, and I can see the red-tinged trail snaking its way up through low secondary vegetation to a ridge a mile away. Our destination lies somewhere on the other side of that ridge. Malagasy in their brightly hued clothes dot the green hills like dabs of paint in an Impressionist painting.

All about me rise rounded hills, not so steep here halfway down the escarpment. According to one of Wright's colleagues, this area bore thick rain forest as recently as 1992. Today, it is a tapestry of tavy and paddy-rice fields, banana plantations, newly burned hillside, secondary scrub forest, and discrete hamlets of four or five thatch houses apiece. Wild ginger grows in profusion, its light-green color standing out from the darker green of hillside shrubbery. Looking around, I have to admit that such a landscape has more to interest the eye from afar than dense rain forest, discrete patches of which I can see still bristling the highest ridgetops like mohawk haircuts.

The sun has dried the trail, and there is good footing. The path ascends quite precipitously in places. I try to imagine climbing it in the pouring rain; one would need cleats. Now out in full, with white puffy clouds blocking it only briefly, the sun soon dries my river-dunked clothes. I get into that satisfying hiking rhythm—step up, deep breath, step up, exhale, step up, deep breath, step up, exhale. How many places around the world have I done this? I pass the time thinking back on great treks I've taken—to the base of Everest on the Tibet side, into the Brooks Range in Alaska on cross-country skis, through a tiger sanctuary on the flanks of Sumatra's highest mountain. I stop many times to let my mind wander over the spectacular view, and most of the group passes me.

An hour and a half after leaving the river, I gain the long-awaited

ridge. Down below, nestled in a bowl valley of rice paddies and ⸌
ested slopes, sits a village of perhaps fifteen houses. A crowd of M.
gasy fills its central arena, and I can hear drums, flutes, and wi
hoots. The party has already started, it seems. I make my way downhill into the village. Feeling self-conscious but not wanting to miss the
opportunity, I begin taking photos of people in the crowd: two men in
straw hats bundled despite the heat in homespun gray wool lambas; a
knot of children whose Sunday best cannot hide their filthy under-
clothes; an old woman in a yellow print wrap dancing barefoot inside
a circle of onlookers to the raucous music of a village band.

An Earthwatch volunteer taps my arm.

"This isn't the village," he says with an apologetic smile, and
points to other team members heading out of the hamlet.

It turns out that our group has stumbled upon a circumcision ceremony, to which all those Malagasy I'd seen on the way up here were
apparently headed. Our group arrived just after the circumcision had
taken place and had watched as the bewildered children exited the
place of cutting.

Suddenly I feel rotten. Even as I regret missing the event while I
contemplated my navel back on the path, I feel tormented by guilt for
invading these poor peoples' lives, during one of their most intimate
moments at that. It's clear that we vazaha are not the center of attention. Like all Malagasy, these people are too polite to even hint that
our presence is not wanted. As one visitor once put it about the Malagasy, borrowing another's comment about the Arabs, "They are most
unwilling to appear unwilling." But I notice that few adults smile.
Feeling like just one more insensitive tourist, I hustle away and fall in
line behind the group.

Things don't improve at Ambatovory, our destination. The name
means "group of stones," but there is not a stone in sight. Nor do
there seem to be any people. A scatter of eight or nine houses on a
muddy rise, the village rests in the same bowl valley as the circumcision village, several rice paddies away. Maurice assembles our throng
of about thirty Westerners and Malagasy in what passes for Ambatovory's main square, a flat of packed earth between houses. He introduces the mpanjaka, the king of Ambatovory.

As Maurice translates, the king welcomes us to his village in kindly yells. A slight man with a graying goatee like a ball of steel wool
clinging to his chin, the king appears to be about seventy. He wears a
General Custer hat and a blue trench coat several sizes too big. I can't
believe how loud he's speaking; we're right in front of him, and it's

aybe he's gone deaf. He gestures dramatically, and his
nce with the joy of it. He apologizes that few of his
are here; they're all at the circumcision ceremony.
y he yells: filling a gap. There he stands, a thin wisp
man, surrounded by thirty listeners, some of whom are twice his
size. The cowherd and the cows.

When the king finishes his welcome speech, Maurice leads the
group on a tour through the village.

"This is coffee," he says, pointing to a six-by-six-foot square of
red-brown beans spread out to dry. "They harvest it when it turns red
and sell it in the market for 6,000 Malagasy francs a kilo."

Everybody crowds around. Maurice moves a few steps away.

"This is rice," he says, indicating another square laid out on the
ground. "They will dry it and put it in the granary there."

He points to a peaked-roof structure, like a toy house, on stilts to
keep the rats out. Everyone swings his or her head that way. A few
steps more brings us to a house.

"This house is made from ravenala, the traveler's tree," Maurice
says, "the frame from the trunk, the walls from leaves, the roof from
thatch."

He steps between the house and another not three feet away and
enters it. The crowd follows, pushing in. Maurice's voice is muffled
for all but the eight or ten who make it into the house with him. I
don't even try, but lean against the outside wall in the sun. There is a
crisp white sheet of paper tacked there—letterhead from the local
tourist agency, with a typed message welcoming visitors to the village.

Next we hike en masse up a slope to the king's house. The king, it
seems, is not only mpanjaka but also ombiasy, and he will now show
us the instruments of his trade. Perched by itself on top of a hill, his
house is about half the size of Pierre Rakoto's. Built of rough thatch
dried gray in the sun, it resembles an enormous shaggy sheepdog.
Inside, the king accommodates us in groups of four or five, which is
all his hut can handle.

Pleased to have the king virtually to myself for a moment, I step
inside. Like all the houses we've seen, it is raised a foot or so off the
ground. The doorway is not made for a man of my size. I clamber in
and crouch by the door, letting my eyes adjust to the darkness
(there are no windows). I'm glad to see there are only two vazaha,
along with Maurice, one of Wright's Malagasy assistants, and the
king.

I look around and begin to take notes. The house is even tinier

than it appears from the outside. It is seven feet long by three
half wide, with a low-slung peaked ceiling of smoke-stained pa
fronds. At the end near the door, a stove of three stones cradles a po
on the boil. Smoke wafts up through shafts of sunlight piercing the
none-too-tight walls and sneaks out through the frond roof. On the
floor lie several plastic bowls and plates, a kerosene lantern, and some
woven mats, while on the wall above the king, I can see another piece
of that spanking new letterhead.

The king sits on a mat in one corner, surrounded by the tools of
his trade—smoke-blackened baskets, two large wooden spoons, a
pair of black zebu horns. I see a necklace with a Christian cross and,
oddly, a toy wooden house proportionally of a size one might see in
the States but not here.

The king smiles and waits while Maurice explains that our host
will now "call" the ancestors to tell them that he will describe the
materials he uses and direct a ceremony. The king then picks up an
object like a whisk broom. "If a zebu is killed, you have to keep its tail
in the house," Maurice translates. "When treating you, he blesses you
with the zebu tail dipped in water. There is toaka gasy in the zebu
horns."

"Wait, does he bless you with the toke gas or the water? Does he
drink the toke gas?"

I realize with a jolt that one of the Americans, whom I hadn't recognized in the dim light, is the punk kid with the goatee. Instantly the
ambience is shattered.

After answering the student, Maurice goes on.

"Now he's holding the *hazamanga*, a special blue wood he uses for
the ceremony."

"Ah, hold on. Haz mung? How do you spell that?" Goatee says.
He leans forward out of a dark corner and looks at me. "Can you
write that down? I wanna make sure I got this right."

I can't deal with this. At the risk of offending the king, I slip outside and sit on the blackened trunk of a huge tree that just missed the
king's house when it fell. The sun is blazing hot, and I find it hard to
think. This tree is all that remains of a stand of rain forest that recently stood here. It has fallen amid an entire hillside blackened by fire.
Below me spreads a graveyard of stumps, none more than four feet
high. It is a tavy field in the making, and I can see rice plants growing
in the ashes. The tree I'm sitting on is about five feet around at the
base—modest compared to some rain-forest giants but massive in this
context of forestlessness. Villagers have chopped off all of its smaller

elwood, and someone has even hacked at the trunk to

heat-induced torpor, I try to imagine what all this
like when this tree was a seedling. The young tree
have stood in an impenetrable rain forest, struggling to find the
stray shaft of sunlight that manages to penetrate the canopy and
reach the forest floor. All around would have been "trees, trees, nothing but trees!" as one traveler declared a century ago on a visit to the
eastern rain forest. "In these woods," wrote another, "the growth of
the trees is so dense that it is difficult to get a palanquin through in
many places, and there is a deep gloom below even at mid-day." Still
another described trees "of an extraordinary bulk and height, [which]
seem almost coeval with the creation." There would have been
lemurs, and flocks of forest birds, and chameleons leading their quiet
lives.

Now it's all gone.

I look around at the surrounding landscape. In the valley floor,
paddies. Moving upslope, fields of hill rice and discrete stands of
planted fruit trees. Above that, for a thousand feet or so, empty hillside covered with dull, whitish-green bunchgrass, which is so
unpalatable that even zebu won't eat it. On the ridgetops, those remnant stands of rain forest, the straight white trunks of outlying trees
standing out against the general dark green. No other woodland
exists anywhere in sight, nor any recently burned forest that I can
see. Was the tree I'm sitting on part of a sacred stand that the king
slashed and burned out of need? Did that "stark and grinding" poverty force him to do what his better judgment argued against?

Perhaps that's why I feel so uncomfortable with this tour.
Here's this venerable old man, infinitely more knowledgeable than
any of us of the ways of the world, not to mention bare survival,
opening his house and sharing his most sacred customs with a
crowd of total strangers. Was this, too, forced on him by poverty?
Even if he chose this willingly, how long will it be before he
becomes jaded, before he can't take the lack of respect shown by
tourists like that punk kid? And what happens to his most cherished traditions when he shares them with people like that? How
soon will they erode in the face of the juggernaut of modernization
that we all represent?

Even though we've been invited, I can't help but feel we're imposing. And I can't stem the thought, because its truth is incontrovertible, that we're hastening the end of an indigenous culture.

After the king has welcomed into his house everyone who has an interest, we hike to the other side of the village, where the Tanala have set up six peaked-roof, open-air pavilions for tourists to sit under. The exchange students crowd together under one, while the Earthwatchers and Malagasy research assistants spread out under four others. Under the sixth pavilion, a stooped old man with watery eyes appears with a valiha, the stringed bamboo instrument of Madagascar. In a high, soft voice, he says through the guide that he was born in 1914 and that he wants to apologize, but his wife, with whom he played, died two months ago, and he hasn't practiced since. He looks around the assembled crowd and smiles shyly, his toothless mouth agape.

The man begins playing. His fingers are amazingly nimble for a man of eighty-three, and his high voice carries well. And what energy. He plays and plays and plays, while we sit and listen and eat a lunch of beans and rice catered by the village and watch the king's teenage daughter dance a slow-moving, alluring dance. Four other locals — two men and two women — get up and dance, too, on a mat out in the sun. At one point, the valiha player joins the king and another man at a drum made of a length of bamboo. Singing as if to wake the ancestors, they build up quite a foot-tapping rhythm.

Then we, too, are dancing, vazaha and Tanala all strung out in a line, each of us with his or her hands on the shoulders of the next person. I'm reminded of the Malagasy game *dian-trandraka*, or "hedgehog steps," in which each player holds on to the person in front of him or her and sings and bends in imitation of the supposed movements of the tenrec. In any case, the mood is festive, especially with the toaka gasy going around, and everyone is having a good time.

Even I start to relax and enjoy myself. Who knows what the future holds? At the present, in these early days of ecotourism, things seem to be in some kind of order. These Malagasy have asked for this, and they are living it up for what it's worth *right now*. The underlying philosophy is, again, almost Buddhist. There is a snatch of Malagasy oratory that perfectly encapsulates the attitude of these Tanala, indeed of all Malagasy, an attitude that promotes a rigorous joie de vivre over dwelling on the profound and unrelenting suffering inherent in everyday life. It is titled "Take your fill of pleasure while you live":

Oh ye prosperous people, O ye well-to-do folks, take your fill of pleasure while you live; for when dead and come to the "stone with the little mouth" [the family tomb], it is not to return the same day, but to stop there to sleep; it is not to visit only, but to remain. The covering stone is what presses down over one, the red earth is above the breast, a temporary roof and tent walls surround one; no turning round, no rising up.

12.

AFTER HAVING HAD a taste of ecotourism and all the other pieces that make up the Ranomafana puzzle, I have to ask, Will it work over the long term? Wright told me she considers herself largely a figurehead at Ranomafana, now that ANGAP has assumed management, and all these programs, from biodiversity research to ecotourism, are in place and under the control of Malagasy, most of them locals. I don't believe her, and frankly I don't believe she believes herself. That's not to say Ranomafana would collapse without Pat Wright. But she has been the driving force behind most of the park's successes.

Even with Wright here—and she continues to approve every research project that takes place in the park—I sense the fragility of what she has accomplished at Ranomafana. One can't help but wonder, for instance: Could a change in the political climate, which is currently supportive of scientific research and environmental conservation, ruin everything here? Even Wright in all her optimism acknowledges the possibility.

"With Jocelyn as park director, I'm as happy as I've ever been," she says. "But what if we have to go back to Zeze?" The previous director was not only feckless but was eventually caught cutting timber inside the park and selling it. He resigned, but later ANGAP rehired him at a higher level. "If he assumed total control over Ranomafana, he could say, 'We don't like research, get out,' " Wright says. "Then this whole thing would be gone."

Even if her programs continue on track, are economic incentives such as establishing fish ponds and serving as park guides and hosting tourists in villages enough for such abjectly poor people? Some of these incentives have yet to pay off. Early on, for example, the project helped villagers plant plantations of *Harungana madagascariensis*, a native tree that bees use for making honey and people use for firewood and medicine and to make everything from utensils to houses. The trees are growing well, but since they take ten years to mature, no harvests have occurred yet. About fifty farms around Ranomafana now have active fish ponds, but as Pierre Rakoto's far-from-thriving ponds manifest, raising the fish is a tricky business, and even Uphoff admits the ponds' long-term value has yet to be demonstrated.

In a recent paper, Wright herself made a list of other current problems at Ranomafana. Communication—between project personnel and villagers, and between development experts and scientists—needs constant massaging. It is also hard to find qualified local people to fill administrative posts, and a lack of governmental priority has meant social programs such as family planning are progressing slowly. Despite its protected status, the park still suffers incursions from local people. Deep in the forest I came upon fresh stumps of trees cut solely to obtain honey, and farmers on the park's fringes continue to practice tavy. Honey collecting and tavy remain significant problems, partly because the success at Ranomafana has lured immigrants into the peripheral zone, and partly for lack of resources to adequately police the park.

"You know how many forest rangers there are for this district?" Uphoff asked me that day at the economic development office.

"I hate to ask."

"One. Whose motorcycle sometimes works." Uphoff shook his head. "He's a nice guy, but what can he do? What we're trying to do is promote 'social fencing,' where communities take some responsibility for managing the forest. It's very hard to argue that they don't have a stake in helping you protect it."

But would they have a stake in protecting something to which outsiders have now denied them access, even though the poorest households are, according to the anthropologist Sabrina Hardenbergh, another Ranomafana researcher, "extremely dependent on forest resources for household products and cash"? These people pay dearly for the park. As Michael Wright, a WWF senior vice president, notes in the book *Wildlands and Human Needs*, "In truth, rural communities end up paying most of the cost of these protected areas—by losing their access to resources and wildlife—while national governments, scientists, and the international community reap all the benefits in prestige and funds." Why would they have a stake?

Joe Peters opens one of his papers with mention of a satirical cartoon that appeared in the *Madagascar Tribune* in 1992 and that bears on this question. In the cartoon, a skinny Malagasy peasant speaks with a well-fed ring-tailed lemur sitting in a tree, from which two signs dangle. One says LEMURIAN DOLLAROS and the other displays the initials WWF. The man asks the lemur, "Why is it that there are stacks of money for you, while my children and I go hungry?" In the next frame, the lemur points a finger at the man and responds, "Wait until you're on the path to extinction!" Peters goes on to quote the 1993 evaluation report of the Ranomafana National Park Project (RNPP)—"preservation of biological diversity is the goal, while the other aspects are the proposed means toward that goal"—then adds: "It appears as if the peasant cultivator in the cartoon has a legitimate concern, as do the 27,000 peasant cultivators in [Ranomafana's] peripheral zone. . . . Even if we acknowledge the intrinsic right to exist of the twelve species of lemurs that inhabit the Ranomafana National Park, at what cost are we willing to enforce that right?"

By his own estimation, Peters guesses that during the first three years the project was under way, tourism benefited fewer than 100 people, barely 60 Malagasy from area villages received jobs, and improvements to infrastructure took place in under a dozen villages. All told, he says, less than 2 percent of the $3.2 million that USAID invested in the RNPP during that period went to "village projects."* Peters sums up: "As long as lemurs are given precedence over villagers, trust and confidence between local residents and park managers will not evolve, and conflicts will escalate."

Conflicts are bound to escalate because of rising tourist traffic as well. In five years, Jocelyn told me, he expects 15,000 to 20,000 visitors a year to his park. Yet the infrastructure is woefully inadequate to support even those that come now. "Many visitors choose to stay in

---

* A full 55 percent, meanwhile, went to university overhead and expatriate personnel. ("I know because I was one of them," Peters says.) To the uninitiated, this may sound absurd. But as Jolly noted to me, "This is an important case of a far more widespread dilemma — how to bring First World (U.S.) resources and expertise to help conservation. The fact that most of the budget does indeed go to expatriate salaries is shocking, but expatriates won't come without salaries comparable to what they have at home. Then you say, do you need expats? There are many brilliant Malagasy in Madagascar, and still more in France, who would be lured home if they got salaries comparable to those in France. But there are not *that* many trained Malagasy, and their salaries are so pitiful that any expat who tried to live on them would shortly turn to 'corruption'—at least petty corruption, to make ends meet.

"So Ranomafana's problems of allocation of resources, first to expats and then to the villages and villagers that the expats know and trust, are not individual misallocation of choices by the park managers, but the way that conservation effort is inevitably set up in a world increasingly divided between rich and poor. To fight this, and actually work from the bottom up, is more durable, but arguably would take so long there would be no forest left. Come to think of it, that is perhaps the heart of the dilemma: Defense of the forest is a hugely urgent need, even though, long-term, defense is useless, and one needs cooperative local support for the forest instead. A familiar problem, but with the park trying to do both things, with the rich/poor, foreign/local divide on top, there is just no way for the rich to be both the defenders of the forest and popular, too."

Fianarantsoa, because there are no comfortable hotels," he admitted. Even if they existed, no public transportation exists between the town and the park entrance, which lies four miles up a steep hill. Tourist guides are not officially a part of the RNPP and often do not adhere to fixed fees, which can cause resentment locally.

Wright feels the biggest problem might be securing long-term funding. Madagascar is not the place to find it, of course. "In the United States, there are no parks that are financially self-sufficient, and Malagasy parks will always need input from the international community," she has written. "When the Biodiversity Convention was signed in Rio de Janeiro in 1992, there was a commitment on the part of most of the developed world to assist the species-rich countries in the preservation of their biodiversity wealth. We hope the international community meets this monetary commitment."

When I ask Wright how the Malagasy government plans to spend funds currently coming in from the World Bank–funded Global Environment Facility, which provides funds for projects that protect the global environment and promote sustainable economic growth, she shakes her head.

"Will they use it to make five more Ranomafanas? That's not what we hear. They're going to spread it out all over."

"Too thin?"

"Too vague. When you make big money like that too vague, the pirates come out, and they take it. And nothing gets done."

13.

AFTER WE RETURN from the tourist village, Wright and I take a night walk up the trail from camp, scoping the foliage for nocturnal creatures. Suddenly she stops and trains her headlamp on something high in a tree.

"What's this? Hi, guys!"

"What are we looking at?"

"Avahi, two. They're in a *Dombeya* tree. They love *Dombeya* leaves."

I see them now: one clump of reddish-brown fur in the cleft of a tree, with two pairs of startled amber eyes aimed thirty feet down at us. *Avahi laniger*, the eastern woolly lemur. This species has very short ears almost lost in its head fur, and a very long red tail. But all I can see now is one rounded blob of fur. Woolly lemurs are nocturnal; these guys are sleeping in.

I know the feeling. It's been a long day. I got up at five to check the fossa traps, then did the trek to Ambatovory. After hiking back down to the road, we paid our second visit this week to the thermal springs that give Ranomafana its name, which means "hot water." Then, after dinner at a restaurant in Ranomafana Town, we drove back to the park entrance and hiked back through the dark to the camp. On the way, I found a leaf-tailed gecko glaring down at me from a twig hanging over the trail. That discovery, and the fact that it was my last night at Ranomafana, inspired me to ask back at camp if anyone wanted to take a night walk, even though it was about 11:00 P.M. Most people looked at me as though my stairs didn't quite reach the attic and shuffled off toward their tents without a word. But Wright didn't hesitate for a moment. Even after working in this forest for ten years, she is still fired up by the prospect of a night walk.

So here we are, on the path known as "X," scanning the trailside. The night is cool and clear, and I can see stars twinkling through the canopy overhead. Wright, the lemur lover, keeps her headlamp trained high in the trees. I aim mine lower down, in the bushes alongside the trail, searching for herps. She spotlights a mouse lemur scampering through the trees; I pick out two frogs.

I've been trying to get Wright to myself for a week, and it occurs to me that she has generously come along to give me that opportunity, since I leave tomorrow. So I ask her about her future plans. She's had such success here at Ranomafana; will she try to replicate it elsewhere in Madagascar?

"I have to decide if that's where I want to put my time over the next five years," she says. "The local people are the key."

"How likely is it that local people elsewhere in Madagascar would be as motivated as they are here?"

"I don't know," she says quietly and falls silent.

We slow-step up the trail, our bright lights casting nearby leaves in an unnatural whiteness. A rich aroma of leaves and dirt fills the air. After a minute or two, Wright speaks again, mentioning the officials from Isalo National Park and other areas who have come here to learn.

"I think the Malagasy are interested in replicating the success at Ranomafana. So even if funders aren't, it might happen anyway."

She doesn't sound convinced—or convincing. It's the first time I've heard her express anything but dogged optimism. Perhaps lack of sleep is finally catching up with her.

We continue on in silence, not finding much of anything. I guess the air's deepening chill has persuaded most creatures to keep a low profile, like those woolly lemurs. I'm beginning to feel chilled myself. The trail keeps twisting and turning, rising and dropping through a marshy, claustrophobic forest, like something out of a children's fairy tale. It's hard to scan the trees while also watching one's footing, and at one point when I step back to warn Wright about a hole in the trail, I plunge off the edge into the darkness. Luckily, I drop only about five feet before landing, to my surprise, on my feet.

But now I'm muddy and cold. Quite suddenly I feel utterly spent; the nineteen-hour day is finally taking its toll. I suggest we turn back, and Wright agrees. Irritated by exhaustion and cold, by seeing no animals, and by the doleful turn our conversation has taken, I ask Wright in a cynical tone about the "humanist/ecology" platform on which Madagascar's current leader, Didier Ratsiraka, won the presidential election in 1996. During Ratsiraka's first administration, a brutal socialist regime that lasted from 1975 to 1992, human rights and the environment lay at the bottom of the totem pole in Madagascar.

"What a joke, eh?" I spit out.

But Wright surprises me.

"You know what he told me?" she says in a high-pitched, almost incredulous voice, referring to a recent meeting she had with the president. "He said 'I've seen the light, and I *know* the environment is important. What I really want to do is make all of Madagascar a World Heritage Site.' " Wright says the last three words slowly, almost in a whisper. "I just listened, and I couldn't believe it. Because, actually, that's a really cool idea." And she laughs, shaking her head.

"Think he's sincere?"

"Oh, yeah, I think he's very sincere. It's a crazy idea, but, you know . . ."

She pauses and looks at me expectantly, her eyes glistening in the indirect light from my headlamp.

"Crazy ideas often change the world," I offer.

"Yeaaaaah," she says, nodding, and moves off along the trail.

14.

PERHAPS WRIGHT IS RIGHT about Ratsiraka. Earlier this month (October 1997), he signed into law Madagascar's fifth national park. Masoala National Park, which lies on a peninsula of the same name about 500 miles north of Ranomafana, will protect 840 square miles of virgin rain forest, part of the largest remaining stand of rain forest on the island, along with rare creatures such as the red-ruffed lemur and the serpent eagle. The Masoala peninsula was once a protected area, but the government degazetted it in 1964 for logging, and recently a Malaysian logging conglomerate sought concessions to exploit the entire peninsula. Ratsiraka, in one of the first acts of his new administration, put a stop to that. Even his notion of making the island a World Heritage Site has gotten some play.

"There is no country on Earth more deserving of world heritage status for its biodiversity," Russell Mittermeier, president of Conservation International and a primatologist who has done extensive fieldwork in Madagascar, said in a press release issued by his organization to mark Ratsiraka's announcement of the new park. "The declaration of Masoala as a national park is a wonderful, concrete demonstration of President Ratsiraka's commitment to the environment."

Ratsiraka's act is a promising sign for conservation in Madagascar. But the future remains hazy. Foreign logging interests continue to pressure the government to grant concessions to log forests throughout the island, including the peripheral zone around the new Masoala park. With the national population growing at nearly 3 percent a year, the Malagasy people themselves are an even greater threat. Deforestation, slash-and-burn agriculture, cattle ranching, and charcoal production continue to eat away at the island's tropical forest habitats.

Wright feels the next five to ten years will be critical for conservation in Madagascar. Not surprisingly, she has some thoughts on what should be done to further conservation efforts over the next decade, even as she acknowledges the challenge of securing the funding to do so. Many of her recommendations have to do with building the nation's capacity to look after its own, just as she has helped shore up the capacity of local people around Ranomafana to safeguard their forest.

In terms of increasing science capacity, Wright suggests building a national biodiversity institute, to offer centralized training in biology and technology to the future Emile Rajeriarisons and Loret Rasabos of Madagascar. Five long-term biodiversity research stations should be established, using as models existing ones such as La Selva Biological Station in Costa Rica or the Smithsonian Tropical Research Institute in Panama. Situated in protected areas, these stations should cover the major habitat types, including wet and dry forests, coral reefs, freshwater wetlands, and mangrove forests. At each station, four Malagasy biologists would oversee biodiversity research, ecological monitoring, protected-area management—all the things Wright has done at Ranomafana. More Malagasy graduate students and professionals should be given the opportunity to train overseas in everything from systematics to database management. Finally, Malagasy researchers should have the means to manage their country's biological riches with additional tools on a par with the national biodiversity database in Tana, wherein research results from all fields are collected and compared. Other such tools include geographic information systems, a video database of endangered species, and specimen reference collections.

Wright also has some ideas, naturally, about protected-areas management. She proposes reorganizing ANGAP into a bona fide national parks system, with standardized infrastructure, uniforms, rules, fees, and brochures. Newsletters, posters, field guides, television, and radio could increase efforts to raise public awareness of the need for conservation. Park managers should strive to improve both protection of their reserves and the tourist infrastructure around them, and they should train more local people to become research assistants, tour guides, and agents for conservation and development. Finally, while realizing that many scientists and conservationists feel that the initial ICDPs didn't work, she recommends learning from past mistakes and pressing on, maintaining the present, newly funded ICDPs and adding others.

Peters, for his part, would like to do away with the ICDP model altogether. In its place, he offers two suggestions, based again on his experience at Ranomafana. The first is to offer cash or in-kind distributions to peripheral-zone villagers, a strategy he considers "arguably the most ethical and effective way to preserve the forest. It recognizes that the poor pay the cost of biodiversity preservation through loss of access to subsistence resources, and compensates them for this loss." What if, he asks, the approximately $5.9 million that USAID had invested in RNPP by 1996 had gone instead to an offshore bank account or mutual fund? Assuming a 10 percent yield on the investment, $590,000 could be divided between the roughly 4,500 families in the peripheral zone. Each year, every family would get $131, which equals nearly two-thirds of the per cap income of $210. With this money, poor families could buy 440 to 880 pounds of rice, obviating any need to slash and burn forest to grow it. Or they could purchase fertilizer to increase rice production in existing rice fields. With just thirty-two dollars, for example, a Tanala household working the average of just over half an acre of paddy fields could, assuming minimum increases in yield, buy enough fertilizer to grow an additional 658 pounds of rice, Peters calculates.

As he acknowledges, however, past experience shows that giving cash away can create a dependency relationship and spur a decline in self-reliance, so as an option he proposes supplying villagers with in-kind distributions of rice seed or fertilizers. "Perhaps resident peoples would choose to abandon 'the ancestor's way' for a steady supply of rice," he writes. "This should be an issue which they determine themselves, if given the option. But are the development agencies and donors willing to make available such an option?"

If not, Peters has another alternative. The endowment above could go toward funding a public works program along the lines of Franklin D. Roosevelt's Civilian Conservation Corps of the 1930s. Program workers could renovate existing or build new infrastructure, which Peters notes appears to be mandatory for market-oriented agricultural development. With $590,000 a year, all 4,500 heads of households in the peripheral zone could work half-time year-round to build bridges, roads, trails, and so on. Peters:

> How does that translate into conservation? I have noticed crayfish hunters or fern pot gatherers who do not hunt crayfish or gather fern pots when they do temporary trail construction work for a dollar a day [the going rate]. They hunt crayfish and collect fern pots to sell because they need cash to buy food, and there are few income-generating alternatives. A public works program could generate income to purchase rice and simultaneously provide the infrastructure necessary for development.

I don't know if Peters's ideas make sense. Nor do I know if the ICDPs succeeded or failed. As for the Ranomafana National Park Project itself, I don't know if it has left most of the 27,000 villagers out of the ICDP equation, as Peters argues, or whether it's that model of sustainable development that Wright declared back in Boston. After all, I've spent only a short time here, and I've concentrated my efforts in and around Wright's research camp and Ranomafana Town, with only limited forays into the huge peripheral zone.

I only know what I've seen with my own eyes. Ranomafana's magnificent forests and wildlife are still here. In the areas I visited, research and rural development activities appear to be in full swing, Jocelyn Rakotomalala is providing strong leadership, and local people seem to be involved to an impressive degree. Is the project a model of integrated conservation and development? Perhaps not. But it represents a valiant attempt. In the concluding paper of *Natural Change and Human Impact in Madagascar,* a recent book that summarizes the latest scientific and conservationist thinking about Madagascar, Alison Richard and Sheila O'Connor deem integrated development and conservation projects "above all experiments, complex experiments trying to achieve difficult goals. . . . We must be prepared to admit when we have made mistakes, and to back up, start over, and try something new."

In my mind, it's the trying that's important. I remember what Richard told me on that long walk back to Bemanevika, referring to efforts like that taking place at Beza-Mahafaly Reserve. "We don't know enough to know that it's hopeless," she said. "In the absence of any certainty that it's hopeless, I think there is a kind of moral imperative to keep on trying. Without kidding yourself that you necessarily know what you're doing."

Wright may not be the perfect administrator, but she's putting her all into Ranomafana, which would not exist if it weren't for her tireless efforts over many years. By the same token, Chris Raxworthy is not only writing the herpetological history of Madagascar, but providing conservationists with the data necessary to determine which areas of the island are most in need of their attention and resources. Through his large, multinational expeditions, Dave Burney is both elucidating the island's paleoecological past and offering Western and Malagasy researchers the means to add their own findings to the scientific literature. Bob Dewar is willing to negotiate log bridges and sample dubious local hooch and otherwise adhere to the Malagasy philosophy of moramora in his search for answers to that most beautiful enigma, the Malagasy's origins, answers that give the people of Madagascar pride in their unique cultural heritage.

"I do have this notion," Richard told me at Bemanevika, "that if everybody contributes a drop, there will eventually be an ocean."

Will Madagascar get its ocean? Can this wondrous island be saved, and if so, how? Nobody knows the answers to those questions, but everybody familiar with the state of affairs in Madagascar would agree that finding answers constitutes a frighteningly monumental challenge, for which success is far from assured. Myriad factors militate against success, among them an exploding population, rapidly dwindling resources, and abject poverty exacerbated by a crushing external debt of $4.5 billion. The island also has many things going for it, among the most promising being a recent rise in Malagasy scientists eager to take over managing their country's scientific and environmental future. People like Achille Raselimanana and Toussaint Rakotondrazafy and Darsot Rasolofomampianina and Ernestine Raholimavo. "If these promising young scientists continue to be supported and encouraged," note Richard and O'Connor, "the next few years will see the emergence of strong national leadership in the environmental sciences."

There's no time to lose. John Behler of the Wildlife Conservation Society puts it bluntly: "The country will be either lost or saved during the working life of the current generation." Behler's words should be a call to action, not a cry of defeat. Richard and O'Connor perhaps say it best: "This is not a time for despair or for the fainthearted, but with the continuing commitment of our collective efforts, we believe there can be a bright future awaiting this island, its remarkable people, and its wealth of natural resources." All those in thrall of Madagascar, most especially the Malagasy themselves, can only hope they're right.

15.

IN MY FINAL HOUR at Ranomafana, I have an experience that is as much a symbol of why I wish to tell Madagascar's story as the time I picked up that snake in Isalo National Park. On the edge of the Namorona River, just downstream from Wright's camp, I join a future Malagasy king on a search for one of the most bizarre creatures in Madagascar, the giraffe-necked weevil.

I first met Pierre Talata, the TSE in charge of chameleons, during that visit to the ecological monitoring team a few days ago. He is next in line to become mpanjaka of Ambatolahy, the village closest to the park entrance and the unofficial First Village of Ranomafana. Emile Rajeriarison and Loret Rasabo live there, as does Pierre himself, of course.

When it was his turn to describe his work during our sojourn among the TSEs, Pierre got up before the Earthwatch group and crossed his hands in front of him. Perhaps because I stood right before him, or maybe because I was the only one taking notes, he latched onto me with his eyes and would not let go throughout his speech. I would have felt embarrassed if he hadn't had such an earnest expression on his face. As he spoke, he stared at me with his huge, wet eyes, wide and unblinking, looking as eager to have me hear his story as a dog begging for a scratch. He wasn't really seeing me; I was a focal point where he could rest his eye while he struggled to speak in English before a crowd of vazaha.

It was moving to see a future Malagasy king, dressed in a red sport jacket, down vest, and baseball cap, describing his scientific method. How he walked his 500-meter (1,650-foot) transect every night, scanning for chameleons within 5 meters (16.5 feet) of his headlamp and noting their species, sex, and precise location. "Fifty-six percent of the world's chameleons are here in Madagascar," he said then, pronouncing the word "chameleons" the French way and gazing at me with those unblinking eyes. "And Ranomafana has eleven species!" I had a sudden urge to hug him.

I watch now as Pierre, standing on a trail above the river, scans the vegetation with an intensity that has him leaning forward with the effort. Though I'm sure he doesn't even notice it, the view here over the Namorona is made for a postcard. Before me, towering green forest on either side of the river frames the gushing whitewater, while overhead a cloudless sky presses down, almost oppressively saturated with blue. Gusts of warm spray fly into my face. Casting my eyes alternately over plants and my companion, I try to imagine him as mpanjaka, the king of Ambatolahy. Would he yell his welcome to visiting tourists, as the king of Ambatovory did for us yesterday? I can't see it. He's too soft-spoken, too dear.

Or is he? Wright told me a story of how Pierre, for whatever reason, once raised his voice and banged on someone's door. Raising your voice is against the law in Madagascar, and he was thrown in jail. When Wright arrived that year to do research and learned of his fate, she drove to the jail and pleaded with the jailors: "He's a valuable researcher. Can I borrow him?" Only Pat Wright would try something as unlikely to succeed as this—and succeed. The jailors agreed to loan him out. "He worked for me for two weeks," Wright said, "and on the way out of Ranomafana I dropped him back at the prison."

Pierre *is* a valuable researcher. His specialty is chameleons, but like the other TSEs, he is an expert naturalist who knows more about the forest and its wildlife than many of the Western specialists who come here. Which is why he agreed to help me find the giraffe-necked weevil. He knows exactly which species of tree to find it on — *Dichaetanthera cordifolia,* a small tree in the melastoma family whose leaves are especially attractive to *Trachelophorus giraffa.* As soon as we cross the Namorona, he makes a beeline for two or three individuals of this tree growing on the bank above the river.

Of all the more than 650,000 beetle species so far identified in the world, the giraffe-necked weevil has to be among the most bizarre. The male looks like the insect world's stab at a cherry picker — imagine a lady bug with a boom. Above six black legs sits a round, crimson-colored body, from which a stout black "neck" as long as the body extends straight up. This would be peculiar enough if it ended there. But it doesn't. Beyond an elbowlike joint, the "neck" continues at an angle, and it is *twice* as long as the first part. It finally comes to a stop at a tiny black head with short black antennae. The male giraffe-necked weevil uses its outlandish appendage to roll a *D. cordifolia* leaf into a makeshift egg case, in which the female (which has a shorter "neck") deposits a single egg. I've seen these rolled-up leaves on the side of the trail for days, but as yet no weevils.

Giraffe-necked weevils prefer open spaces like the trail we're on, so Pierre steps delicately, carefully peering at the forest floor. Perhaps he's afraid of trampling a *Brookesia,* the smallest of the chameleon species he studies. The Betsileo maintain that one's hands and feet will swell up if one stands on a *Brookesia.* I doubt that Pierre, with his scientific training, believes in this superstition, or the general one Betsileo tell their children that harming or killing any chameleon will bring bad luck or even death. No, I'll bet this gentle man just doesn't want to inadvertently take a life.

Just when I begin to despair of seeing the beetle before I leave for the capital, Pierre suddenly exclaims and ushers me over with that distinctive Malagasy downward wave of the hand. When I'm at his side, he gently tips back a *D. cordifolia* leaf high over his head, aims those big, wet orbs at me, and grins. The leaf is large, and at first I don't see anything. Then, near the stem: a tiny insect. Pierre breaks off the leaf and lays it on the ground so I can take a photograph of my first giraffe-necked weevil.

I'm disappointed to find it's a female, and so small. I was duped into thinking the insect was bigger by what I now realize was a substantially over-life-size photograph of a male I once saw in a book.

But when I look up at Pierre, smiling triumphantly over me, I can only laugh at the absurdity of such a thought. Here I am, deep in a Madagascar rain forest on a picture-perfect day, sharing admiration of a creature straight out of fantasy with a future king.

Glossary

Aepyornis: the generic name of Madagascar's largest species of extinct elephant birds

agy: a Malagasy vine with stinging nettles

aloalo: tall, wooden carvings like small totem poles that are placed on *Mahafaly* tombs

Andriamanitra: the Malagasy God, creator of heaven and earth and all things in Malagasy cosmology

angady: the thin-headed spade of Madagascar

Anjoaty: "those of the river mouths," a subtribe that lives along the coasts, particularly in the northeast, and claims an Arab descent

Antaimoro: "those from the coast," a late-arriving Malagasy tribe that claims Arab descent, uses *Sorabe,* and practices divination and other magic arts

Antandroy: "those of the thorns," pastoralists who live in the spiny desert at the southern end of the island and build large stone tombs

Archeolemur: genera of extinct giant lemur thought to have lived primarily on the ground

Avahi: either of two species of nocturnal woolly lemur, with large eyes and great leaping ability

aye-aye: a bizarre nocturnal lemur with beaver teeth, bat ears, a foxlike tail, and a skeletal middle finger used to spear grubs from beneath tree bark

Babakotia: genus of extinct giant lemur with forelimbs far longer than its hindlimbs

Bara: reputedly the most warlike and one of the most African of Malagasy tribes, inhabit the south-central region in and around the Isalo massif

Betsileo: "the many unconquered," a tribe living around the southern town of Fianarantsoa that is known for its wood-carving and rice-terracing skills

Betsimisaraka: "the many inseparables," a group of tribes along the northeast coast that formed a confederacy in the early eighteenth century

bilo: an elaborate *Sakalava* ceremony to cure illness, during which the patient is placed on a high, wooden platform and made to perform proscribed rituals

celadon: a ceramic with a green glaze originating in China that was an early trade item in Madagascar and is commonly found in archeological sites

dady: the *Sakalava* custom of memorializing ancestors, particularly kings, by preserving their hair, fingernails, and other remains

Darafify: a legendary giant who migrated to Madagascar from Arabia with an ox bearing eight teats, as believed by *Malagasy* claiming an Arab descent

fady: taboos that are passed down through the generations via the *razana,* govern everyday behavior, and may differ from family to family

famadihana: "turning of the bones," the ancestor-honoring ceremonies during which the *Merina* and *Betsileo* disinter their dead relatives and rewrap the remains in fresh shrouds

fanorona: the national board game of Madagascar

fitenin-drazana: "sayings of the ancestors," which are comprised of riddles, oratory, proverbs, and poetry and are characterized by two-sided interaction

fokonolona: "the people of the village," local community councils

fokontany: part of local community government, a kind of people's executive committee

fossa: pronounced "foo-sa," the largest mammalian carnivore on Madagascar, which resembles a cross between a mongoose and a short-legged puma

Geochelone: genera of *Malagasy* tortoise that includes the extinct species of giant tortoise and living species such as the plowshare and radiated tortoises

hainteny: traditional poetry of Madagascar

Highlands: also known by the misnomer Hauts Plateaux, the central, mountainous region in the heart of Madagascar

hotely: small restaurants ubiquitous in Madagascar

hotely belly: the Montezuma's revenge of Madagascar

indri: a black-and-white lemur that sings an eerily beautiful song and is the largest of the living lemurs

kabary: stylized speech used in special ceremonies (compare *resaka*); in former times, an official gathering of the populace, during which the reigning monarch made proclamations

kalanoro: "wild men of the woods" in *Malagasy* legend

kidoky: the name given by west coast villagers near Belo-sur-Mer to an animal like a ground-dwelling giant lemur said to live in the region's dry forest (compare *kilopilopitsofy*)

kilopilopitsofy: "floppy ears," the name given by west coast villagers near Belo-sur-Mer to a hippolike animal reported seen in the area as recently as the mid-1970s (compare *kidoky*)

lamba: traditional togalike garment of Madagascar

lamba mena: silk shrouds used for wrapping corpses

lavaka: amphitheater-like erosion gullies that gouge hillsides throughout the Highlands

Lepilemur: also known as sportive lemur, any of several species of small, vertical-clinging and -leaping nocturnal lemurs

Mahafaly: "those who make taboos," a fiercely independent tribe of the southwest whose members build the country's most elaborate stone tombs

Malagasy: the name for the people and language of Madagascar

malgachisant: one who studies the natural and/or cultural history of Madagascar

megafauna: "big animals," which in Madagascar include all those extinct creatures weighing more than twenty-five pounds, such as the giant lemurs, elephant birds, giant tortoises, and pygmy hippo

menarana: hog-nosed snake that in the *Malagasy* Bible is the serpent that tempts Eve

Merina: "those of the Highlands," a tribe living in and around the capital city of Antananarivo; the most Indonesian of *Malagasy* tribes and the country's political and economic leaders

moramora: "take it easy" in *Malagasy*

mpakafo: "heart-stealer," refers to the widespread belief in rural Madagascar that white people kill *Malagasy* to eat their vital organs

mpanjaka: "king," refers to local chiefs of a village or region

ombiasy: medicine man who performs numerous ritualistic and healing functions in rural *Malagasy* society

omby: see *zebu*

Paleopropithecus: genera of extinct giant lemur; the most slothlike of the *sloth lemurs*, with arms twice as long as its legs and hooklike hands and feet

pantun: poetic form from Malaysian Peninsula distinguished, like *hainteny*, by use of allusion and metaphor

pirogue: hand-carved canoe, often with outrigger, found particularly along the west coast and brought originally from Indonesian region

Plesiorycteropus: extinct, aardvark-like animal of Madagascar, recently assigned its own mammalian order

prosimian: an early form of primate that includes the lemurs

raffia: an endemic palm used for building, basket-weaving, and other uses; the term is one of the few *Malagasy* words to enter the English language

raketa: prickly pear cactus, which was introduced to Madagascar and now infests the south

Rasikajy: Arab-derived civilization that thrived along the northeast coast between the ninth and sixteenth centuries

ratites: flightless birds that include the ostrich, emu, kiwi, cassowary, and rhea, as well as the extinct elephant birds of Madagascar and moas of New Zealand

ravenala: "leaf of the forest," an endemic, fan-leaved tree known colloquially as the "travelers' tree," for the freshwater found in the base of its leaf axils

razana: ancestors, particularly the most recent generations of one's forebears, whom *Malagasy* ritualistically consult on all areas of importance in their lives

reniala: "forest's mother," *Sakalava* name for the baobab tree

resaka: informal speech such as greetings, gossip, and other everyday talk (compare *kabary*)

Roc: giant bird in Arabian myth, so big it could lift an elephant in its talons; thought to have been modeled on Madagascar's elephant bird

Sakalava: "those of the long valleys," Madagascar's largest tribe, comprised of a loose confederation of subtribes found along the west coast from Nosy Be to Tuléar

sgraffito: type of early Near Eastern pottery with clay carved away to reveal a different-colored ground beneath; early trade item in Madagascar, commonly found in archeological sites

sifaka: any of several relatively large-bodied, long-legged, handsome diurnal lemurs

sikidy: divination techniques used to determine *vintana*

sloth lemur: name given to several genera of extinct giant lemurs thought, from bone structure, to have hung from trunks like sloths

Sorabe: sacred writings of the *Antaimoro* tribe, which constituted Madagascar's first writing

speleothem: general term used to describe all cave formations, including stalactites and stalagmites

spiny desert: unique ecosystem of southern Madagascar, characterized by octopus trees and other thorny plants

subfossil: a fossil in the making; that is, plant or animal remains that have not yet had time to become fully mineralized

Tanala: "those of the forest," a rice-farming tribe that lives in the region around Ranomafana National Park in the southeast

tavy: slash-and-burn agriculture

taxi-brousse: brightly colored small buses of Madagascar

toaka gasy: local rum

tody: Malagasy belief in the moral law of cause and effect, equivalent to the Hindu-Buddhist concept of *karma*

tromba: spirit possession, practiced most notably by the *Sakalava*

tsingala: water-borne insect that can kill cattle that swallow it

tsingy: from the *Malagasy* word for tiptoeing, limestone pinnacles found in the western dry forest that lie so close together that the *Malagasy* say one cannot put a foot down between them

tsy-aomby-aomby: "not-cow-cow," legendary beast with the body of a cow, uncloven hooves, and a propensity to eat people; may be lingering memory of the extinct pygmy hippo

valiha: zitherlike bamboo instrument of Madagascar

vazaha: foreigners, particularly Europeans

Vazimba: aboriginal people of Madagascar, according to *Malagasy* belief; the spirits of such; and a small *Sakalava* subtribe of western Madagascar

Vezo: a subtribe of the *Sakalava* who live along the southwest coast, use outrigger canoes, and are Madagascar's premier fishermen

vintana: destiny, which every *Malagasy* inherits at birth, may be good or bad, and is based on time, date, place of birth, and other, more complex factors

zebu: lyre-horned cattle of Madagascar, known as *omby* in *Malagasy*

Source Notes

AA refers to *The Antananarivo Annual and Madagascar Magazine*, a yearly scientific and literary journal published between 1875 and 1900, and is followed by the year indicating the issue, e.g., AA92 indicates the 1892 issue. For the reference in the Bibliography, see Sibree et al. (1875–1900). An *n* by a page number refers to a footnote on that page.

xxiv "a legacy from ancestors . . ." Jolly (1980), p. 223.
1 "May I announce to you . . ." Quoted in Lanting (1990), p. 28.
1 "a place where antique outmoded forms . . ." Attenborough (1981), p. 175.
1 "time had once broken its banks . . ." Jolly (1980), p. 10.
2 "This strikes me . . ." Diamond (1997), p. 381.
7 "the chiefest paradice . . ." Boothby (1647), p. 6.
7 "I could not but endeavour . . ." Quoted in Brown (1978), p. 46.
7 "The vast riches . . ." Ibid., p. 73.
7 "but scurvy School-masters . . ." AA92, p. 498.
8 "The independent Sakalavas . . ." Verin (1986), p. 399.
8 "Rice and I are one" Quoted in Lanting (1990), p. 79.
8 "Imerina has been gathered . . ." Quoted in Brown (1978), p. 130.
9 "waded to the throne . . ." Cousins (1895), p. 67.
9 "are not the customs . . ." Ibid., p. 96.
9 "the time when the land . . ." Ibid., p. 91.
9 "a heroine of Malagasy nationalism" Brown (1978), p. 167.
10*n* The commentator was quoted in McNeil (1996a).
12 "Tomorrow you will once more . . ." Quoted in Brown (1978), p. 271.
12 "all-out investment" Lanting (1990), p. 119.
13 "a tragedy without villains" Jolly et al. (1984), p. 211.

15   "Behave like the chameleon . . ." Ibid., p. 208.

17   "necklace of pearls" Quoted in Lanting (1990), p. 28.

17   "indefatigable drive . . ." From a letter to the editor published in *The Sciences*, July-August 1996, p. 7.

22   "one of the greatest unsolved mysteries . . ." Krause et al. (1997), p. 3.

22   "a vast natural laboratory . . ." Jolly et al. (1984), p. 105.

22   "a sanctuary for species . . ." Ibid.

23   "the particular conditions . . ." Ibid., p. 154.

26   "I do believe, by God's blessing . . ." Quoted in Brown (1978), p. 47.

26   "the contagion of the place" Ibid., p. 48.

27   "almost dead upon our poop . . ." Churchill (1732), p. 259.

27   "whereof one was myself . . ." Ibid.

27   "I told the merchant . . ." Ibid., p. 261.

27   "[T]he captain, having victuals brought him . . ." Ibid., p. 264.

28   "play'd the rogue" Ibid.

28   "three months together" Ibid., p. 268.

28   "naked as ever I was born . . ." Ibid.

28   "I was taken light-headed . . ." Ibid., p. 267.

28   "where I met with my father . . ." Ibid., p. 282.

29   "hammered in" Ruud (1960), p. 19.

29   "the taboos in all their many fields . . ." Ibid., p. 290.

32   "the anomalies of the mammal fauna . . ." Sibree (1969), p. 66.

33   "that sunken land . . ." Quoted in Wallace (1881), p. 438.

33   "Madagascar was no doubt . . ." Wallace (1876), pp. 286–287.

34   "I have gone into this question . . ." Wallace (1881), p. 426.

34   "not much inferior to Madagascar itself" Ibid., p. 424.

34   "comparatively easy stages" Ibid., p. 414.

35   "a land connection with *some* continent . . ." Ibid., p. 426.

35   "to our embarrassment . . ." Ibid., p. 417.

35   "We might as reasonably suppose . . ." Ibid., p. 420.

35   "all extensive groups have a wide range . . ." Ibid.

37   "Brazilian-Ethiopian continent" Hallam (1967), p. 226.

37   "Such was the need felt . . ." Ibid.

37   "sufficiently radical" Ibid.

37   "isthmian links" Ibid.

37   "Available evidence indicates . . ." Krause et al. (1997), p. 6.

39   "there is in traveling in the forest . . ." AA90, p. 197.

40   "one's hopes are only buoyed up . . ." Durrell (1992), p. 59.

42   "is not a buzz-z exactly . . ." AA90, p. 202.

42   "Don't be like the cicada . . ." Ibid.

42   "one cannot see one's way" Du Bois (1897), p. 68.

42   "When preaching at Ananalamahitsy . . ." AA84, p. 116.

42   "[T]hey eat all the herbage . . ." Du Bois (1897), p. 68.

43   "[L]ocusts are esteemed a great delicacy . . ." AA81, p. 72.

43   "If you cook them in lard . . ." Quoted in Standing (1887), p. 51.

48   "It would be better to trample . . ." Jolly et al. (1984), p. 208.

48   "the somewhat exaggerated fear . . ." Ibid.

48   "a kind of collective psychosis" Ibid.

48   "He who seeks to condemn . . ." Ibid., p. 207.

48   "cannibal monster" Ibid., p. 208.

49   "if a native happens to approach the tree . . ." Quoted in Copland (1822), p. 299.

49   "On perceiving a fly sitting . . ." Quoted in AA89, p. 15.

57   "a more or less continuous land bridge" Rage (1988), p. 266.

58   "no conceivable sea-level fall . . ." Krause et al. (1997), p. 15.

58   "if both Mount Antandroy and Mount Betsileo . . ." Ibid., p. 16.

59   "rafts with trees growing on them . . ." Quoted in Quammen (1996), p. 145.

59   "a number of great water courses . . ." Quoted in Krause et al. (1997), p. 6.

59   "twisted round the trunk . . ." Quoted in Quammen (1996), p. 145.

60   "if crossings of the Mozambique Channel . . ." Mittermeier et al. (1994), p. 20.

60   "only a small minority of the available fauna . . ." Hallam (1967), p. 206.

61   "Any event that is not absolutely impossible . . ." Quoted in Krause et al. (1997), p. 8.

65   "The ancestors come into our lives . . ." Lanting (1990), p. 74.

66   "every green plant . . ." Marcuse (1914), p. 149.

66   "A few very small tumble-down huts . . ." Barnard (1848), p. 7.

68   "The country about [St. Augustine's Bay] . . ." Boothby (1647), p. 5.

69   "Wee are increased in our number . . ." Quoted in Brown (1978), p. 43.

69   "many and well-built houses" Ibid., p. 46.

69   "old, ignorant, weake fellowes" Ibid., p. 45.

69   "she-cattle" Ibid.

70   "no other use but to destroy victuals" Ibid.

70   "of soe base and falce a condition . . ." Ibid.

70   "this most accursed place" Ibid., p. 46.

70   "After staring about us . . ." Barnard (1848), pp. 7–8.

70   "Tie up down there" Swaney and Willcox (1994), p. 218.

71   "the wives of the gun" AA92, p. 387.

72   "the father . . . takes a loaded musket . . ." AA83, p. 53.

72   "sometimes attended with very sad results . . ." Ibid.

72   "On one occasion a Sakalava boy . . ." AA81, p. 11.

72   "for sheer pluck . . .will compare favorably . . ." Marcuse (1914), p. 203.

72   "Usually, in such a case . . ." Ibid.

73   "They are not wanting in courage . . ." AA92, p. 388.

73   "many herbal quacks . . ." AA91, p. 323.

74   "All the things in the sacred ox-horn . . ." Ruud (1960), p. 201.

74   "They are not punished for *the stealing* . . ." AA84, p. 58.

74   "it is immediately known on both sides . . ." AA81, p. 13.

76   "I do not doubt that over the world . . ." Quoted in Gibbons (1996), p. 1497.

77   For more on the cichlid fish of Cameroon and the stickleback fish of Canada, see Ibid.

78   "the ring of wet and dry forest . . ." Jolly et al. (1984), p. 184.

81   "miasmatic vapours" AA84, p. 17.

81   "The first—the cold stage—is introduced . . ." Ibid., pp. 17–18.

81   "the intermittent fever of the ephemeral . . ." Ibid., p. 17.

81   "men with whom you dine . . ." Powell (1925), p. 366.

81   "He stood stretching out both his arms . . ." AA84, pp. 22–23.

83   "It can easily be understood . . ." AA90, pp. 131-132.

83   "sent to Fort Dauphin for some singing girls . . ." Barnard (1848), p. 96.

83   "a disease so strange . . ." AA89, p. 21.

83   "After complaining, it may be one . . ." Ibid., pp. 22–23.

84   "The strongest current of a four-cell battery . . ." AA92, p. 405.

85   "I found the woman on her knees . . ." Ibid., p. 442.

85   "Now, Ramatoa . . . I am going to cure you . . ." Ibid.

85   "more as a corrective . . ." Ibid.

85   "demoniacal spirit" AA82, p. 17.

85   "The remedy for her illness . . ." Ibid.

85   "food increases life" AA84, p. 65.

85   "This is the only rule of diet . . ." Ibid., p. 64.

86   "stout and strong" Ibid.

86   " . . . I told his mother . . ." Ibid., p. 65.

86   "The patient, whose face has been rendered hideous . . ." Marcuse (1914), pp. 183–184.

87   "They load their muskets . . ." AA82, p. 20.

87   "perfectly well and quite normal" Quoted in Chapman (1943), p. 126.

87   "This happened in one case . . ." AA92, p. 393.

89   "more than any other soul . . ." From a letter to the editor published in *The Sciences*, July-August 1996, p. 7.

89   "Head with lateral, orbital, and posterior crests . . ." Raxworthy and Nussbaum (1995), p. 535.

90   "By reason of the increasing power . . ." Marcuse (1914), p. 152.

91   "They have . . . supercilious airs . . ." AA81, p. 8.

92   "poorly through the worm" Sibree (1969), p. 285.

93   "a Caliban of a tree . . ." Quoted in Bradt (1992), p. 44.

94   "among the highest priorities . . ." Safford (1995), p. 118.

95   "The heat was grilling . . ." Marcuse (1914), p. 158.

95   "the air seemed to scorch the nostrils . . ." Ellis (1859), p. 114.

95   "sweltering in a heat beyond description . . ." Powell (1925), p. 381.

98   "I would take more care of her . . ." Brown (1978), p. 63.

98   "we lay down and were as happy . . ." Ibid.

98*n*   "just what one would expect . . ." Ibid., pp. 56–57.

99   "downright in love" Ibid.

99   "He was an old man . . ." Ibid., pp. 67–68.

100   "things did not answer my expectation" Ibid., p. 71.

100   "In his last years . . ." Ibid.

101*n* Regarding the fifty-four collected tortoises, see Webster (1997), p.61.

105   "Clearly, speciation in *Brookesia* . . ." Raxworthy and Nussbaum (1995), pp. 550–551.

107*n* "[o]ne African who is now free told us . . ." AA92, p. 392.

109   "Even to discuss questions . . ." Mack (1986), p. 64.

109   "the idea of the ancestors . . ." Ibid.

110   "to being condemned to eternal oblivion" Ibid., p. 65.

110   "The soul of the deceased . . ." Ruud (1960), p. 164.

110   "may justly be reckoned . . ." Ellis (1859), p. 376.

110   "the living are impoverished . . ." AA92, p. 411.

110   "The building of the tomb . . ." Ruud (1960), p. 141.

111   "One is willing to spend . . ." Lanting (1990), p. 114.

111   "dig no graves . . ." Richardson (1877), p. vi.

112   "After having washed the corpse . . ." Quoted in Mack (1986), p. 69.

112   "The first ceremony of drying the corpse . . ." Sibree (1969), p. 241.

113   "The whole family would feel . . ." Ruud (1960), p. 162.

113*n* "It is a source of vileness . . ." AA92, p. 416.

114   "Those who use such prohibited words . . ." AA83, p. 48.

114   "the ruler who is tolerant of many" Sharp (1993), p. 121.

115   "Give me my gun . . ." Richardson (1877), p. viii.

115   "For sitting or reclining . . ." Ibid., p. 6.

116   "Our gun and our spear . . ." AA76, p. 228.

118   "problematic" Raxworthy and Nussbaum (1996a), p. 375.

120   "everyone is a sorcerer . . ." AA87, p. 332.

122   "This spirit stood and began to evangelize . . ." Sharp (1993), p. 120.

123   "the quintessence of Sakalava religious experience" Ibid., p. 115.

123   "Discussions of the causes of spirit possession . . ." Ibid., p. 116.

125   "Unless regional biodiversity is documented . . ." From a letter to the editor published in *The Sciences*, July-August 1996, p. 7.

127   "Bullock in a crocodile's jaws . . ." AA82, p. 52.

137   "The Tretretretre, or Tratratratra . . ." Quoted in Battistini and Richard-Vindard (1972), p. 361.

138   "[T]his is a large bird haunting the Ampastres . . ." Ibid.

138   "According to the report . . ." Quoted in Brown (1978), p. 4.

139*n* "are not at all unlike an enormous quill . . ." Sibree (1969), p. 55.

143   For Martin's map of extinction patterns and discussion of the blitzkrieg hypothesis, see Martin (1984).

144   Budyko's equations are mentioned in Burney (1993a).

144   For Martin and Mosimann's model, see Mosimann, J. and P. S. Martin (1975). "Simulating Overkill by Paleoindians." *American Scientist*, vol. 63, pp. 304–313.

144   For a review of putative causes of Malagasy extinctions, see Burney (1993a).

145   "Got into the clutches . . ." AA81, p. 73.

146   "[T]he natives also swear . . ." Du Bois (1897), p. 60.

146   "live chiefly on stones . . ." AA75, p. 16.

146   "crawling saw-backed monsters" Sibree (1915), p. 297.

146   "I never saw such a dreadful snapping apparatus . . ." AA75, p. 25.

146   "one could almost cross . . ." Powell (1925), p. 418.

147   "the man who went fishing . . ." AA81, p. 73.

147   "I was somewhat inclined . . ." Sibree (1969), p. 58.

148   "The two males, rotund as Tweedledum . . ." Durrell (1992), p. 43.

149   "available soon" Quoted in McNeil (1996b), p. C4.

150   For Humbert's theory, see, e.g., Humbert, H. (1927). "Destruction d'une Flore Insulaire par le Feu: Principaux Aspects de la Végétation à Madagascar." *Mémoires de l'Académie Malgache*, vol. 5, pp. 1–80. For Perrier de la Bâthie's theory, see, e.g., Perrier de la Bâthie, H., (1921). "Le Végétation Malgache." *Annals du Muséum Colonial*, Marseille, vol. 9, pp.1–266.

151   For the 1974 challenge to the continuous-forest theory, see Koechlin, J., J. L. Guillaumet, and P. Morat (1974). *Flore et Végétation de Madagascar*. Vaduz: Cramer.

151   "it was under [man's] eyes . . ." Quoted in AA94, p. 150.

153   "whatever brought the first Malagasy . . ." Dewar (1984), p. 588.

153   "missing link" Ibid., p. 589.

153   "truly eerie" Ibid.

154   "a primeval savanna/forest mosaic . . ." Ibid., p. 590.

160   "the large and the distinctive" MacPhee et al. (1985), p. 469.

160   "Pollen zonation was discerned . . ." Burney (1987b), p. 132.

161   "a mosaic of woodlands, bushlands, and savanna . . ." MacPhee et al. (1985), p. 463.

164   "Walking under some trees . . ." Quoted in AA85, p. 118.

164   "ten times more virulent" Ibid.

165   "The Mkodos are a very primitive race . . ." AA81, p. 91.

165   "impenetrable forest" Ibid.

165   "May I never see such a sight . . ." Ibid., p. 93.

165   "a clear, treacly liquid . . ." Ibid., p. 92.

165   "six white almost transparent palpi . . ." Ibid., pp. 93–94.

170   "[e]very microfossil analysis . . ." Burney (1987a), p. 379.

175   "what can we expect in Africa . . ." Burney (1993a), p. 540.

179   "I had to admit finally . . ." Quoted in Godfrey (1986), p. 50.

179   "Strip the cumulative fantasy . . ." Ibid., p. 51.

181   "Oh, no, that's an elephant" Quoted in Burney and Ramilisonina (1998), p. 961.

182   "In 1946, when I was a girl of ten . . ." Ibid., p. 962.

184   "terrible animal" Ibid., p. 961.

185   "No, that's a sifaka" Ibid., p. 962.

187   "People constitute a great, broad mat" Fox (1990), p. 24.

191   "In the whole of the interior . . ." Quoted in Verin (1986), p. 33.

192   "May it not be that we have . . ." Sibree (1969), p. 116.

192   "the Malagasy agree closely with the Siamese" AA82, p. 3.

192   "He is inclined to suppose . . ." Sibree (1969), p. 105.

193   "We are also bound to ask . . ." Verin (1986), p. 35.

193   For the reference regarding "Malagasy culture is like a detective novel," see Haring (1992), p. 2.

195   "How is it that the well-used road . . ." AA81, p. 62.

197   "THERE IS NO GOD BUT GOD . . ." Verin (1969), p. 271.

202   "There are many plants . . ." AA81, p. 65.

202   "Rice is Andriamanitra" Quoted in AA84, p. 29.

205   "Ibn Lakis has informed me . . ." Quoted in Verin (1969), pp. 41–42.

206   "The (islands of the) Waqwaq . . ." Ibid., p. 42.

206   " . . . the Komr people, who have given their name . . ." Ibid., pp. 43–44.

207   "The people of Al Komr invaded Aden . . ." Ibid., p. 44.

208   "If we take such a work . . ." Sibree (1969), p. 105.

208   "However anomalous may be thought . . ." AA81, p. 102.

210   "I am confident that this deposit . . ." Dewar (1996), p. 481.

211   "Kalanoro are by far the most mysterious . . ." Sharp (1993), p. 138.

212   "We were informed by a trader . . ." AA89, p. 116.

212   "How is it your eyes are so red?" AA91, p. 366.

215   The reference for the scholar who argued Sanskrit loan-words came from Old Malay or Old Java is: Adelaar, K. A. 1989. "Malay Influence on Malagasy: Linguistic and Culture-Historical Implications. *Oceanic Linguistics*, 28.

215   "one of the greatest ironies . . ." Burney (1989), p. 8.

217   "represents a voyage of about 6,000 kilometers . . ." Quoted in Verin (1986), p. 39.

218   "to consider both the initial settlement . . ." Dewar (1996), p. 482.

218   "a further confirmation of the need . . ." Ibid.

219   "A most curious thing happened . . ." AA83, p. 117.

219   "clearly . . . there is a strong set of wind" Cousins (1895), p. 37.

219   "Was that chance?" AA87, p. 378.

220   "But they were not forsaken . . ." Ibid., p. 379.

220   "semi-savages though they were" Ibid.

223   "it should be stated emphatically . . ." Verin (1986), p. 34.

223n The reference for the Polynesian gene is Soodyall, Himla, Trefor Jenkins, and Mark Stoneking. 1995. " 'Polynesian' mtDNA in the Malagasy." *Nature Genetics*, Vol. 10.

224   "Savage and free on the far distant wilds . . ." AA81, pp. 67–68.

224   "The ox's horns go to the spoon-maker . . ." Sibree (1915), p. 184.

225   "I have never known a man . . ." Quoted in Jolly (1980), p. 175.

225   "Cattle are thus converted . . ." Mack (1986), p. 66.

225   "the zebu will lick bare stone . . ." Jolly (1980), p. 70.

225   "must go back to the very origins . . ." Verin (1986), p. 46.

226   "We know that the Vazimbas . . ." Ibid., p. 48.

226   "The end of their voyage . . ." Ibid., p. 42.

227   "Is it therefore possible . . . ?" Ibid.

227   "The Zanj have no ships . . ." Ibid., p. 43.

227   "the process by which Malagasy culture . . ." Mack (1986), p. 17.

228   "These ombiasses are very feared . . ." Ibid., pp. 35–36.

229   "insistent" Southall (1986), p. 415.

229   "These powers certainly provided . . ." Ibid., p. 416.

229   "mysterious ritual potency" Ibid., p. 413.

230   "the oldest one of the days" Ruud (1960), p. 31.

230   "To white one's neighbor" Ibid.

230   "to spit white" Ibid.

231   "says white words" Ibid., p. 32.

231   "If the birthday is at Asorotany's beginning . . ." Ibid., p. 59.

232   "The top of the heap of earth . . ." Chapman (1943), p. 137.

232   "the most horrid and execrable feature . . ." Copland (1822), p. 74.

232   "in societies where literacy is unknown . . ." Kus and Raharijaona (1990), p. 23.

232   "There is almost as much symbolism . . ." AA99, p. 326.

232   "The Malagasy conception of destiny . . ." Mack (1986), p. 40.

233   "The diviners watch carefully . . ." Ruud (1960), p. 61.

235   "corner of prayers" Fox (1990), p. 26.

236   "The place is holy . . ." Verin (1986), p. 266.

239   "One of these bridges . . ." AA92, p. 435.

239   "deepish streams by bridges . . ." AA99, p. 313.

239   "it's wrong to say there's a bridge . . ." AA81, p. 61.

239   "These rickety structures beggar description . . ." Ibid.

240   "has combined diverse imported elements . . ." Haring (1992), p. 4.

240   "The ethnic map of Madagascar . . ." Ibid., p. 23.

240   "The relative meaninglessness . . ." Southall (1986), p. 425.

240   "Malagasy culture may be . . . a hybrid culture . . ." Quoted in Dewar (1985), p. 306.

241   "It's the extremists I'm afraid of " Haring (1992), p. 26.

241   "Un, deux, trois: c'est gai!" Ibid.

241   "a very soft and musically sounding language . . ." Sibree (1969), p. 148.

242   "a fine, rackity-clackity, ringing language . . ." Durrell (1992), p. 32.

242   "swallow as many syllables as you can . . ." Swaney and Willcox (1994), p. 84.

242   "voiceless and voiced blade . . ." Fox (1990), p. 15.

243   "Bis(marck) him!" AA81, p. 44.

243   "I most certainly wash my hands . . ." Ibid.

243   "*vely* . . . struck . . ." Standing (1887), p. 80.

244   "to spend the night away from home . . ." Bradt (1992), p. 27.

244   "not to be eaten by an heir . . ." AA89, p. 103.

244   "the black soot" Sibree (1969), p. 159.

244   "the becoming good by spreading a mat" Ibid., p. 160.

244n   "everything would have gone well . . ." Mullens (1875), p. 285.

245   "inheritance-not-removing" Ibid., p. 248.

245   "fight at writing" Standing (1887), p. 82.

245   "a fought-out sale" Sibree (1969), p. 162.

245   "the finishing of counting" Ibid., p. 149.

245   "lost" Ibid., p. 162.

245   "the eye of day" Ibid., p. 157.

245   "flowers of the grass" Sibree (1915), p. 207.

245   "the whales are turning over" Ibid., p. 275.

245   "the *not* terrible bird" Sibree (1969), p. 157.

245   "offspring of the silk" Ibid., p. 151.

245   "darken the mouth of the cooking pot" Bradt (1992), p. 27.

245   "when the wild cat washes itself " Ibid.

245   "Halving of night . . ." Sibree (1915), pp. 93–94.

246   "It is most desirable that any Traditions . . ." Quoted in Haring (1992), p. 10.

247   "life is sweet . . ." Fox (1990), p. 24.

247   "A society and a morality . . ." Ibid.

247   "Sir, I eat (your) illness" Ruud (1960), p. 171.

247   "It is as if one simulates . . ." Ibid.

247   "one must say that he is seriously ill . . ." Ibid., p. 142.

247   "He/she who never will . . ." Ibid., p. 172.

247   "But finding himself, after a little time . . ." AA87, p. 375.

248   "The villagers said . . . that the high tide . . ." Quoted in Jolly (1980), p. 76.

248   "I . . . found a crowd of tens of thousands . . ." From a letter to prospective volunteers in the 1993 *EarthCorps Briefing*, published by the Earthwatch Institute, Maynard, Mass.

248   "any foreigner, be he of whatever nationality . . ." AA84, p. 2.

248   "wrote their own culture . . ." Haring (1992), p. 7.

248   "sayings of the ancestors" Ibid., p. 14.

248   "fighting like a drunk" Ibid., p. 29.

249   "to talk and dispute for the mere purpose . . ." Ibid.

249   "What then is this?" AA90, p. 174.

249   "If you try to lift it, you can't . . ." Standing (1887), p. 67.

249   "God's little lake . . ." Ibid.

249   "At night they come . . ." AA90, p. 174.

249   "The dead cry out to sell the living" AA76, p. 248.

249   "Which do you prefer . . ." Haring (1992), p. 56.

249   "Often have I sat by the village or camp fire . . ." AA99, p. 322.

249   *"Filokobokobo"* AA77, p. 375.

250   I paraphrased the tale of the wild hog and the chameleon from a version in Standing (1887), pp. 137–138.

251   "When driving their cattle . . ." Haring (1992), p. 28.

251   "speaking in tongues" Ibid., p. 25.

251   "Every thing has its place . . ." AA89, p. 38.

251   "Desolate one forsaken by friends . . ." Ibid., p. 33.

251   "[T]he Malagasy are ready and fluent speakers . . ." Sibree (1969), p. 193.

253   "concrete packed together" Quoted in Verin (1986), p. 264.

253   "made by ropes" Ibid., p. 261.

253   "M. Jully has revealed . . ." Ibid.

253   "a wall made of stones . . ." Ibid.

253   "exposed a surrounding wall . . ." Ibid.

254   "just before arriving in the south . . ." Ibid., p. 265.

254   "The local people believed . . ." Ibid.

255   "Often these names are retained . . ." Standing (1887), p. 38.

255   "Sometimes the relatives seem quite at a loss . . ." Ibid., p. 40.

256   "sainted-mouths" Quoted in Hurvitz (1986), p. 116.

256   "When I asked him why . . ." Ibid., pp. 116–117.

257   " . . . the entire corpus of mythology . . ." Ibid., p. 117.

257   "He cut it into little pieces . . ." Quoted in Verin (1986), p. 81.

259   "the district administrator demolished . . ." Ibid., p. 263.

262   "They will forgive much . . ." Ibid.

262   "many speakers are tempted . . ." Ibid.

262   "There are [proverbs] applicable to virtually every emotion . . ." Fox (1990), p. 37.

262   "Much money and many cattle . . ." Ibid., p. 36.

263   "piquancy and relish" AA78, p. 428.

263   "A people is known by its proverbs . . ." AA81, p. 59.

263   "one old book of human nature" AA94, p. 190.

263   "The cover may be white, brown, or black . . ." Ibid.

263   "don't decide before God . . ." Ibid., p. 193.

263   "swim without disturbing the water" Koenig (1984), p. 41.

263   "friendship cannot be bought" AA95, p. 282.

263   "cross in a crowd, the crocodile won't eat you" Ibid., p. 287.

263   "stops not till the bottom is reached" AA94, p. 202.

263   "The thing done is like a loincloth . . ." Ibid., p. 200.

263   "plump out of sight . . ." AA78, p. 430.

263   "shame changes your face" Koenig (1984), p. 29.

263   "obstinacy makes you a dwarf " AA96, p. 442.

263   "man is his conduct" Fox (1990), p. 38.

263   "long feet will find food" AA99, p. 282.

263   "if you do not want to go to the forest . . ." Koenig (1984), p. 41.

263   "clapping won't skin the hands" AA99, p. 285.

263   "a butterfly in the sun . . ." Ibid., p. 277.

264   "How is it that only the rich are niggardly?" Ibid., p. 279.

264   "Delight is not for the poor" Koenig (1984), p. 44.

264   "only in sleep do we resemble the rich" Ibid., p. 34.

264   "like an ox's stomach in the winter . . ." AA81, p. 64.

264   "men are like the rim of the pan . . ." Koenig (1984), p. 30.

264   "many cocks within the fence . . ."AA81, p. 74.

264   "One thousand words . . ." Koenig (1984), p. 36.

264   "it is better to die tomorrow than today" Ibid., p. 40.

264   "If a rooster does not crow . . ." Ibid., p. 33.

264   "A dead man cannot see your tears" Ibid., p. 48.

264   "Do not kick away the canoe . . ." Ibid., p. 43.

264   "a hare-lipped person who whistles . . ." Ibid., p. 49.

264   "if you are just a dung beetle . . ." Ibid., p. 42.

264   "When men fight by the light of the moon . . ." Ibid., p. 46.

264   "a pig fight: a grunt or two settles it" AA81, p. 70.

264   "rats about to pass each other . . ." Koenig (1984), p. 36.

264   "when old women meet . . ." AA92, p. 415.

264   "the wife is liked, but the mother-in-law . . ." Sibree (1969), p. 250.

264   "Nothing is more precious than your baby . . ." Koenig (1984), p. 38.

264   "other people's children . . ." Ibid.

264   "wisdom literature" Fox (1990), p. 36.

265   "poetic techniques of such great refinement . . ." Ibid., p. 51.

265   "Whish-whish, shaking of the winnowing-pan . . ." Ibid., p. 52.

265   "The fringe of my lamba is damp . . ." Ibid., p. 100.

266   "That road, there in the north . . ." Ibid., p. 315.

266   "Broad fingers, large fingers . . ." Ibid., p. 254.

266   "Three ears of corn with roots . . ." Ibid., p. 286.

266   "figurative saying" Ibid., p. 51.

266   "Red ants within the bamboo . . ." Ibid., p. 56.

267   "the artistic refinement of hainteny . . ." Ibid., p. 69.

267   "'I laid my eggs,' said the bird . . ." Ibid., p. 296.

268   "My song is saturated with thy light . . ." Rabéarivelo (1975), p. 7.

268   "[T]his is what I am . . ." Ibid., p. xvii.

269   "responds to the need of the race . . ." Quoted in Fox (1990), p. 71.

269   "does not attempt to reproduce the style . . ." Ibid.

269   "Slow as a limping cow . . ." Rabéarivelo (1975), p. 55.

270   "There in the north stand two stones . . ." Ibid., p. 71.

273   "When you plant one rice seed . . ." Koenig (1984), p. 30.

273   "Without the forest, there will be . . ." Lanting (1990), p. 110.

275   "if the golden bamboo lemur had been discovered . . ." Quoted in Peters (1998a), p. 38.

275   "It is clear that [Wright] has been the predominant influence . . ." Ibid., p. 32.

276   "flawed from the start . . ." Peters (1999), p. 71.

276   "Is biodiversity preservation still good . . ." Peters (1998a), p. 24.

280   "Probably more than one half . . ." AA92, p. 434.

281   "rational forest resource management" Quoted in Peters (1999), p. 68.

281   "elevated the practice of tavy . . ." Ibid.

281   "the way of the ancestors" Ibid., p. 67.

281   "We Malagasy are free . . ." Ibid., p. 72.

286   "Your nation is committing environmental suicide" Quoted in Kull (1996), p. 61.

287   "has become a model for increasing support . . ." Lanting (1990), p. 120.

288   "According to this critique . . ." Richard and O'Connor (1997), p. 411.

288   "all too often those who benefit most . . ." Ibid., p. 415.

288   "a strongly marked and pervasive feature . . ." Ibid.

288   "The underlying assumption . . ." Peters (1998a), p. 27.

288   "[T]he sheer demographic and geographic scope . . ." Ibid., p. 27.

288   "[W]e learned that long-term success . . ." Wright (1997), p. 387.

294   "The assumption has always been . . ." Quoted in Peters (1998a), pp. 33–34.

297   "From the very beginning . . ." Ibid., p. 40.

297   "preservation of biological diversity . . ." Ibid., pp. 23–24.

297   "Only with experience have biologists . . ." Ibid., p. 40.

297   "improve the flow of information . . ." Ibid., p. 43.

303   "Madagascar is rapidly heading . . ." Quoted in Richard and O'Connor (1997), p. 411.

304   "This team has continued to bring news . . ." Wright (1997), p. 390.

304n   "The project has been criticized strongly . . ." Kull (1996), p. 77.

304n   "After nearly ten percent of the village had died . . ." Quoted in Peters

(1998a), p. 29.

305 "an implicit code of reciprocity" Peters (1999), p. 67.

305 "Outsiders have been shocked . . ." Jolly et al. (1984), p. 44.

305 "This is the primary reason why farmers . . ." Quoted in Peters (1999), p. 67.

306 "In Madagascar, we can confidently predict . . ." Richard and O'Connor (1997), p. 409.

306 "Given the population pressures . . ." Quoted in Peters (1998a), p. 26.

306 "The residents of the Ranomafana region . . ." Ibid., p. 31.

306*n* "the world champion of erosion" Kull (1996), p. 66.

315 "The sharing of entrance fees . . ." Peters (1998b), p. 525.

316 "The challenge for the next few decades . . ." From a draft outline for a book to be edited by Pat Wright.

317 "of all the rude and primitive contrivances . . ." AA99, p. 315.

319 "They are most unwilling . . ." AA82, p. 80.

322 "trees, trees, nothing but trees!" Ibid., p. 81.

322 "In these woods, the growth of the trees . . ." Sibree (1969), p. 71.

322 "of an extraordinary bulk and height . . ." Copland (1822), p. 4.

323 "Oh ye prosperous people . . ." AA89, p. 35.

325 "extremely dependent on forest resources . . ." Quoted in Peters (1998a), p. 19.

325 "In truth, rural communities end up paying . . ." Ibid., p. 19.

325 "LEMURIAN DOLLAROS" Ibid., p. 17.

326 "Why is it that there are stacks of money . . ." Ibid., pp. 17–18.

326 "Wait until you're on the path to extinction!" Ibid., p. 18.

326 "preservation of biological diversity . . ." Ibid., pp. 23–24.

326 "It appears as if the peasant cultivator . . ." Ibid., p. 24.

326 "village projects" Ibid., p. 37.

326 "As long as lemurs are given precedence . . ." Ibid., p. 30.

326*n* "I know because I was one of them" Ibid., p. 37.

327 "In the United States, there are no parks . . ." Wright (1997), p. 396.

330 "There is no country on Earth more deserving . . ." From a Conservation International press release dated March 14, 1997.

331 "arguably the most ethical . . ." Peters (1998a), p. 35.

332 "Perhaps resident peoples would choose . . ." Ibid., p. 36.

332 "How does that translate into conservation? . . ." Ibid.

333 "above all experiments, complex experiments . . ." Richard and O'Connor (1997), p. 416.

334 "If these promising young scientists . . ." Ibid., p. 406.

334 "The country will be either lost or saved . . ." From a letter to the editor published in *The Sciences*, July-August 1996, p. 7.

334 "This is not a time for despair . . ." Richard and O'Connor (1997), p. 417.

Bibliography

Attenborough, David. 1981. *Journeys to the Past: Travels in New Guinea, Madagascar, and the Northern Territory of Australia*. Guildford, Surrey: Lutterworth Press.

Barnard, Frederick L. 1848. *A Three Years' Cruize in the Mozambique Channel, for the Suppression of the Slave Trade*. London: R. Bentley.

Battistini, R., and G. Richard-Vindard, editors. 1972. *Biogeography and Ecology in Madagascar*. The Hague: W. Junk.

Bloch, Maurice. 1971. *Placing the Dead: Tombs, Ancestral Villages, and Kinship Organization in Madagascar*. New York: Seminar Press.

Boothby, Richard. 1647. *A Briefe Discovery or Description of the Most Famous Island of Madagascar or St. Laurence in Asia Neare Unto East-India*. London: Printed for John Hardesty by R. B. and Francis Lloyd, Merchants.

Bradt, Hilary. 1992. *Guide to Madagascar*. 3rd ed. Edison, N.J.: Hunter.

Brody, Jane E. 1998. "Dr. Patricia Wright: Saving Madagascar's Bounty for Its Lemurs and Its People." *New York Times*, August 11, 1998.

Brooks, Daniel R., Deborah A. McLennan, James M. Carpenter, Stephen G. Weller, and Jonathan A. Coddington. 1995. "Systematics, Ecology, and Behavior." *BioScience*, Vol. 45, No. 10.

Brown, Mervyn. 1978. *Madagascar Rediscovered: A History from Early Times to Independence*. London: Damien Tunnacliffe.

Burney, David A. 1987a. "Pre-settlement Vegetation Changes at Lake Tritrivakely, Madagascar." In *Palaeoecology of Africa and the Surrounding Islands*, Vol. 18. Rotterdam: A. A. Balkema.

———. 1987b. "Late Holocene Vegetational Change in Central Madagascar." *Quaternary Research*, Vol. 28.

———. 1987c. "Late Quaternary Stratigraphic Charcoal Records from Madagascar." *Quaternary Research*, Vol. 28.

——. 1989. "The Piece of Africa That Slipped Away." *SWARA: East African Wildlife Society*, Vol. 12, No. 4, July/August 1989.

——. 1993a. "Recent Animal Extinctions: Recipes for Disaster." *American Scientist*, Vol. 81, November/December 1993.

——. 1993b. "Late Holocene Environmental Changes in Arid Southwestern Madagascar." *Quaternary Research*, Vol. 40.

——. 1996a. "Paleoecology of Humans and Their Ancestors." In *East African Ecosystems and Their Conservation*, edited by T. R. McClanahan and T. P. Young. New York: Oxford University Press.

——. 1996b. "Climate Change and Fire Ecology as Factors in the Quaternary Biogeography of Madagascar." *Biogeographie de Madagascar*.

——. 1996c. "Historical Perspectives on Human-Assisted Biological Invasions." *Evolutionary Anthropology*, Vol. 4, No. 6.

——. 1997. "Theories and Facts Regarding Holocene Environmental Change Before and After Human Colonization." In *Natural Change and Human Impact in Madagascar*, edited by Steven M. Goodman and Bruce D. Patterson. Washington, D.C.: Smithsonian Institution Press.

——. 1999. "Rates, Patterns, and Processes of Landscape Transformation and Extinction in Madagascar." In *Extinction in Near Time*, edited by R. D. E. MacPhee. New York: Kluwer Academic/Plenum Publishers.

Burney, David A., Helen F. James, Frederick V. Grady, Jean-Gervais Rafamantanantsoa, Ramilisonina, Henry T. Wright, and James B. Cowart. 1997. "Environmental Change, Extinction, and Human Activity: Evidence from Caves in NW Madagascar." *Journal of Biogeography*, Vol. 24.

Burney, David A., and Ross D. E. MacPhee. 1988. "Mysterious Island: What Killed Madagascar's Large Native Animals?" *Natural History*, July 1988.

Burney, David A., and Ramilisonina. 1998. "The Kilopilopitsofy, Kidoky, and Bokyboky: Accounts of Strange Animals From Belo-sur-Mer, Madagascar, and the Megafaunal 'Extinction Window.' " *American Anthropologist*, Vol. 100, No. 4.

Chapman, Olive M. 1943. *Across Madagascar*. London: E. J. Burrow & Co.

Churchill, Awnsham, editor. 1732. *A Collection of Voyages and Travels*. London: J. Walthoe.

Copland, Samuel. 1822. *A History of the Island of Madagascar*. London: Burton & Smith.

Cousins, William E. 1895. *Madagascar of Today: A Sketch of the Island, with Chapters on Its Past History and Present Prospects*. London: The Religious Tract Society.

Culotta, Elizabeth. 1995. "Many Suspects to Blame in Madagascar Extinctions." *Science*, Vol. 268, June 16, 1995.

Dahl, Otto C. 1991. *Migration from Kalimantan to Madagascar*. Oslo: Norwegian University Press.

Daley, Suzanne. 1998. "Vanilla Farming? Not as Bland as You Might Think." *New York Times*, January 19, 1998.

Defoe, Daniel. 1974. *The King of Pirates: Being an Account of the Famous Enterprises of Captain Avery, with the Lives of Other Pirates and Robbers*. New York: AMS Press.

Dewar, Robert E. 1984. "Extinctions in Madagascar: The Loss of the Subfossil Fauna." In *Quaternary Extinctions: A Prehistoric Revolution*, edited by Paul S. Martin and Richard G. Klein. Tucson: University of Arizona Press.

———. 1985. "Of Nets and Trees: Untangling the Reticulate and Dendritic in Madagascar's Prehistory." *World Archaeology*, Vol. 26, No. 3.

———. 1996. "The Archaeology of the Early Settlement of Madagascar." In *The Indian Ocean in Antiquity*, edited by Julian Reade. London: Keegan Paul International.

———. 1997a. "Does It Matter That Madagascar Is an Island?" *Human Ecology*, Vol. 25.

———. 1997b. "Were People Responsible for the Extinction of Madagascar's Subfossils, and How Will We Ever Know?" In *Natural Change and Human Impact in Madagascar*, edited by Steven M. Goodman and Bruce D. Patterson. Washington, D.C.: Smithsonian Institution Press.

Diamond, Jared. 1997. *Guns, Germs, and Steel: The Fates of Human Societies*. New York: W. W. Norton.

Drury, Robert. 1831. *The Pleasant and Surprising Adventures of Robert Drury, During his Fifteen Years' Captivity on the Island of Madagascar.* London: Whittaker, Treacher, and Arnot.

Du Bois, Sieur. 1897. *The Voyages Made by the Sieur D. B. to the Islands Dauphine or Madagascar & Bourbon or Mascarenne in the Years 1669, 70, 71, & 72.* London: Printed for D. Nutt by T. and A. Constable.

Durrell, Gerald. 1992. *The Aye-Aye and I: A Rescue Mission in Madagascar*. New York: Arcade Publishing.

Ellis, Rev. William. 1859. *Three Visits to Madagascar.* Philadelphia: John Potter & Co.

Fox, Leonard. 1990. *Hainteny: The Traditional Poetry of Madagascar.* Lewisburg, Pa.: Bucknell University Press.

Gibbons, Ann. 1996. "On the Many Origins of Species." *Science*, Vol. 273, No. 5281, September 13, 1996.

Godfrey, Laurie R. 1986. "The Tale of the Tsy-Aomby-Aomby: In Which a Legendary Creature Is Revealed to Be Real." *The Sciences*, January/February 1986.

Goodman, Steven M., and Bruce D. Patterson, editors. 1997. *Natural Change and Human Impact in Madagascar*. Washington, D.C.: Smithsonian Institution Press.

Hallam, Anthony. 1967. "The Bearing of Certain Palaeozoogeographic Data on Continental Drift." *Palaeogeography, Palaeoclimatology, Palaeoecology*, Vol. 3.

Hamond, Walter. 1643. *Madagascar, the Richest and most fruitfull island in the World.* London: Printed for Nicolas Bourne.

Haring, Lee. 1992. *Verbal Arts in Madagascar: Performance in Historical Perspective*. Philadelphia: University of Pennsylvania Press.

Hevesi, Eugene. 1941. "Hitler's Plan for Madagascar." *Contemporary Jewish Record*, August 1941.

Houlder, John A. 1877. *North-east Madagascar: A Narrative of a Missionary Tour, from the Capital to Andranovelona via Andovoranto and the North-east Coast, and Back to*

Antananarivo by Way of Mandritsara and Ambatondrazaka. Antananarivo: London Missionary Society Press.

Hurvitz, David. 1986. "The 'Anjoaty' and Embouchures in Madagascar." *In Madagascar: Society and History,* edited by Conrad Phillip Kottak. Durham, N.C.: Carolina Academic Press.

Jolly, Alison. 1980. *A World like Our Own: Man and Nature in Madagascar.* New Haven: Yale University Press.

Jolly, Alison, et al., editors. 1984. *Key Environments: Madagascar.* Oxford: Pergamon Press.

Koenig, Jean-Paul. 1984. *Malagasy Customs and Proverbs.* Sherbrooke, Quebec, Canada: Naaman.

Krause, David W., Joseph H. Hartman, and Neil A. Wells. 1997. "Late Cretaceous Vertebrates from Madagascar: Implications for Biotic Change in Deep Time." In *Natural Change and Human Impact in Madagascar,* edited by Steven M. Goodman and Bruce D. Patterson. Washington, D.C.: Smithsonian Institution Press.

Kull, Christian A. 1996. "The Evolution of Conservation Efforts in Madagascar." *International Environmental Affairs,* Vol. 8, No. 1, Winter 1996.

Kus, Susan, and Victor Raharijaona. 1990. "Domestic Space and the Tenacity of Tradition Among Some Betsileo of Madagascar." In *Domestic Architecture and the Use of Space,* edited by Susan Kent. New York: Cambridge University Press.

Langrand, Olivier. 1990. *Guide to the Birds of Madagascar.* New Haven: Yale University Press.

Lanting, Frans. 1990. *Madagascar: A World out of Time.* New York: Aperture Foundation, Inc.

Mack, John. 1986. *Madagascar: Island of the Ancestors.* London: British Museum Publications Ltd.

MacPhee, R. D. E. 1994. "Morphology, Adaptations, and Relationships of *Plesiorycteropus,* and a Diagnosis of a New Order of Eutherian Mammals." *Bulletin of the American Museum of Natural History,* No. 220.

MacPhee, R. D. E., D. A. Burney, and N. A. Wells. 1985. "Early Holocene Chronology and Environment of Ampasambasimba, a Malagasy Subfossil Lemur Site." *International Journal of Primatology,* Vol. 6, No. 5.

Marcuse, Walter D. 1914. *Through Western Madagascar in Quest of the Golden Bean.* London: Hurst & Blackett.

Martin, Paul. 1984. "Prehistoric Overkill: The Global Model." In *Quaternary Extinctions: A Prehistoric Revolution,* edited by Paul S. Martin and Richard G. Klein. Tucson: University of Arizona Press.

McNeil, Donald G., Jr. 1996a. "Antananarivo Journal: A Palace Inferno Sears Madagascar's Very Soul." *New York Times,* June 22, 1996.

———. 1996b. "Madagascar Reptile Theft Hits Rarest of Tortoises." *New York Times,* July 2, 1996.

———. 1996c. "Madagascar in Crisis: The Land of Endangered Species in Danger." *New York Times,* July 21, 1996.

Mittermeier, Russell A., Ian Tattersall, William R. Konstant, David M. Meyers, and Roderic B. Mast. 1994. *Lemurs of Madagascar.* Washington, D.C.: Conservation International.

Mullens, Joseph. 1875. *Twelve Months in Madagascar.* 2nd ed. London: J. Nisbet & Co.

Murphy, Dervla. 1985. *Muddling Through in Madagascar.* Woodstock, N.Y.: The Overlook Press.

Osborn, Chase S. 1924. *Madagascar: Land of the Man-Eating Tree.* New York: Republic Publishing Co.

Peters, Joe. 1997. "Local Participation in Conservation of the Ranomafana National Park, Madagascar." *Journal of World Forest Resource Management,* Vol. 8.

———. 1998a. "Transforming the Integrated Conservation and Development Project (ICDP) Approach: Observations from the Ranomafana National Park Project, Madagascar." *Journal of Agricultural and Environmental Ethics,* Vol. 11.

———. 1998b. "Sharing National Park Entrance Fees: Forging New Partnerships in Madagascar." *Society and Natural Resources,* Vol. 11.

———. 1999. "Understanding Conflicts Between People and Parks at Ranomafana, Madagascar." *Agriculture and Human Values,* Vol. 16.

Powell, Edward A. 1925. *Beyond the Utmost Purple Rim: Abyssinia, Somaliland, Kenya Colony, Zanzibar, the Comoros, Madagascar.* New York: The Century Co.

Preston-Mafham, Ken. 1991. *Madagascar: A Natural History.* New York: Facts on File.

Quammen, David. 1996. *The Song of the Dodo.* New York: Scribner.

Rabéarivelo, Jean-Joseph. 1975. *Translations from the Night: Selected Poems of Jean-Joseph Rabéarivelo.* African Writers Series, 167. London: Heinemann Educational.

Rabinowitz, Philip D. 1983. "The Separation of Madagascar and Africa." *Science,* Vol. 220, April 1, 1983.

Rage, Jean-Claude. 1988. "Gondwana, Tethys, and Terrestrial Vertebrates During the Mesozoic and Cainozoic." In *Gondwana and Tethys,* edited by M. G. Audley-Charles and A. Hallam. New York: Oxford University Press.

Raxworthy, Christopher J., and Ronald A. Nussbaum. 1994. "A Rainforest Survey of Amphibians, Reptiles and Small Mammals at Montagne D'Ambre, Madagascar." *Biological Conservation,* Vol. 69.

———. 1995. "Systematics, Speciation, and Biogeography of the Dwarf Chameleons (*Brookesia;* Reptilia, Squamata, Chamaeleontidae) of Northern Madagascar." *Journal of Zoology, London,* Vol. 235.

———. 1996a. "Patterns of Endemism for Terrestrial Vertebrates in Eastern Madagascar." In *Biogéographie de Madagascar,* edited by W.R. Lourenço. Paris: Editions de l' ORSTOM.

———. 1996b. "Montane Amphibian and Reptile Communities in Madagascar." *Conservation Biology,* Vol. 10, No. 3.

———. 1997. "Biogeographic Patterns of Reptiles in Eastern Madagascar." In *Natural Change and Human Impact in Madagascar,* edited by Steven M. Goodman and Bruce D. Patterson. Washington, D.C.: Smithsonian Institution Press.

Richard, Alison F., and Sheila O'Connor. 1997. "Degradation, Transformation, and Conservation: The Past, Present, and Possible Future of Madagascar's Environment." In *Natural Change and Human Impact in Madagascar,* edited by Steven M. Goodman and Bruce D. Patterson. Washington, D.C.: Smithsonian Institution Press.

Richardson, James. 1877. *Lights and Shadows; or, Chequered Experiences Among Some of the Heathen Tribes of Madagascar.* Antananarivo: London Missionary Society Press.

Ruud, Jørgen. 1960. *Taboo: A Study of Malagasy Customs and Beliefs.* Oslo: Oslo University Press.

Safford, Roger. 1995. "Photospot: Ground-Rollers of Madagascar." *Bulletin of the African Bird Club,* Vol. 2.2, August 1995.

Sharp, Lesley. 1993. *The Possessed and the Dispossessed: Spirits, Identity, and Power in a Madagascar Migrant Town.* Berkeley: University of California Press.

Shoumatoff, Alex. 1988. "Look at That." *The New Yorker,* March 7, 1988.

Sibree, James. 1915. *A Naturalist in Madagascar.* Philadelphia: J. B. Lippincott Co.

———. 1969. *The Great African Island: Chapters on Madagascar.* Westport, Conn.: Negro Universities Press.

Sibree, James, et al., editors. 1875–1900. *The Antananarivo Annual and Madagascar Magazine.* Antananarivo: London Missionary Society Press.

Simons, Elwyn. 1993. "Lost Lemurs of the Crocodile Caves: Deep Inside a Mountain on Madagascar, the Bones of Giant Simian Ancestors Give Dramatic Testimony to the Evolution of Primates." *The Sciences,* November/December 1993.

Southall, Aidan. 1986. "Common Themes in Malagasy Culture." In *Madagascar: Society and History,* edited by Conrad Phillip Kottak. Durham, N.C.: Carolina Academic Press.

Standing, Herbert. 1887. *The Children of Madagascar.* London: Religious Tract Society.

Storey, Michael, John J. Mahoney, Andrew D. Saunders, Robert A. Duncan, Simon P. Kelley, and Millard F. Coffin. 1995. "Timing of Hot-Spot-Related Volcanism and the Breakup of Madagascar and India." *Science,* Vol. 267, No. 5199, February 10, 1995.

Stuenes, Solweig. 1989. "Taxonomy, Habits, and Relationships of the Subfossil Madagascan Hippopotami *Hippopotamus Lemerlei* and *H. Madagascariensis.*" *Journal of Vertebrate Paleontology,* Vol. 9, No. 3, September 1989.

Swaney, Deanna, and Robert Willcox. 1994. *Madagascar & Comoros: A Travel Survival Kit.* Oakland, Calif.: Lonely Planet Publications.

Verin, Pierre. 1986. The *History of Civilization in North Madagascar.* Rotterdam: A. A. Balkema.

Wallace, Alfred R. 1876. *The Geographical Distribution of Animals.* New York: Harper & Brothers.

———. 1881. *Island Life.* New York: Harper & Brothers.

Webster, Donovan. 1997. "The Looting and Smuggling and Fencing and Hoarding of Impossibly Precious, Feathered, and Scaly Wild Things: Inside the $10 Billion Black Market in Endangered Animals." *New York Times Magazine*, February 16, 1997.

Wright, Patricia C. 1988. "Lemurs Lost and Found: Three Species of These Primates Are Discovered Alive and Well and Living on Bamboo." *Natural History*, July 1988.

———. 1995. "Demography and Life History of Free-Ranging *Propithecus Diadema Edwardsi* in Ranomafana National Park, Madagascar." *International Journal of Primatology*, Vol. 16, No. 5.

———. 1997. "The Future of Biodiversity in Madagascar: A View from Ranomafana National Park." In *Natural Change and Human Impact in Madagascar*, edited by Steven M. Goodman and Bruce D. Patterson. Washington, D.C.: Smithsonian Institution Press.

Index